'This introductory textbook goes beyond the descriptive level and the usual engineering approximations to a deeper, yet accessible, level of fundamental composite mechanics that provides valuable insights into composite performance.'

William Curtin,
École polytechnique fédérale de Lausanne

'Composite materials combine and comprise an understanding of materials science, mechanics and general engineering.

The 3rd edition of *An Introduction to Composite Materials* by T. W. Clyne and D. Hull is a carefully revised version of the previous very successful textbook. It is a comprehensive summary of the current knowledge in composites science and technology – specially prepared as a textbook for young scientists and graduate students to get a substantial insight into this still young area.

I personally recommend this book to my graduate students.'

Karl Schulte,
Technische Universität Hamburg

'This extensively revised and expanded edition includes the latest developments in composites research and applications. The new chapters on surface coatings, highly porous materials and bio- and nano-composites are uniquely valuable.'

Tsu-Wei Chou,
University of Delaware

An Introduction to Composite Materials

Third Edition

This fully expanded and updated edition provides both scientists and engineers with all the information they need to understand composite materials, covering their underlying science and technological usage. It includes four completely new chapters on surface coatings, highly porous materials, bio-composites and nano-composites, as well as thoroughly revised chapters on fibres and matrices, the design, fabrication and production of composites, mechanical and thermal properties and industry applications. Extensively expanded referencing engages readers with the latest research and industrial developments in the field, and increased coverage of essential background science makes this a valuable self-contained text. A comprehensive set of homework questions, with model answers available online, explains how calculations associated with the properties of composite materials should be tackled, and educational software accompanying the book is available at doitpoms.ac.uk.

This is an invaluable text for final-year undergraduates in materials science and engineering, and graduate students and researchers in academia and industry.

T. W. Clyne is Professor of Mechanics of Materials in the Department of Materials Science and Metallurgy at the University of Cambridge, and the Director of the Gordon Laboratory. He is also the Director of DoITPoMS, an educational materials science website, a Fellow of the Royal Academy of Engineering and a Helmholtz International Fellow.

D. Hull is an Emeritus Professor at the University of Liverpool. He is also a Fellow of the Royal Society and a Fellow of the Royal Academy of Engineering.

An Introduction to Composite Materials

Third Edition

T. W. CLYNE
University of Cambridge

D. HULL
University of Liverpool

CAMBRIDGE
UNIVERSITY PRESS

University Printing House, Cambridge CB2 8BS, United Kingdom

One Liberty Plaza, 20th Floor, New York, NY 10006, USA

477 Williamstown Road, Port Melbourne, VIC 3207, Australia

314-321, 3rd Floor, Plot 3, Splendor Forum, Jasola District Centre, New Delhi - 110025, India

79 Anson Road, #06-04/06, Singapore 079906

Cambridge University Press is part of the University of Cambridge.

It furthers the University's mission by disseminating knowledge in the pursuit of education, learning and research at the highest international levels of excellence.

www.cambridge.org
Information on this title: www.cambridge.org/9780521860956
DOI: 10.1017/9781139050586

First and Second editions © Cambridge University Press 1981, 1996.
Third edition © T. W. Clyne and D. Hull 2019

First published 1981
Second edition 1996
Third edition 2019

A catalogue record for this publication is available from the British Library

Library of Congress Cataloging in Publication data
Names: Clyne, T. W., author. | Hull, Derek, author.
Title: An introduction to composite materials / T.W. Clyne (University of Cambridge),
    D. Hull (University of Liverpool).
Description: Cambridge ; New York, NY : Cambridge University Press, [2019] |
    Includes bibliographical references and index.
Identifiers: LCCN 2018061490 | ISBN 9780521860956 (hardback : alk. paper)
Subjects: LCSH: Composite materials.
Classification: LCC TA418.9.C6 H85 2019 | DDC 620.1/18–dc23
LC record available at https://lccn.loc.gov/2018061490

ISBN 978-0-521-86095-6 Hardback

Additional resources for this publication at www.cambridge.org/compositematerials

Contents

Preface to the Third Edition

The topic of composite materials continues to evolve in terms of range, research activity and technological importance. This was the case between publication of the first edition in 1981 and the second in 1996. The coverage of the book was expanded and broadened to reflect this. In fact, the rate of development of composites has accelerated in the period since then and hence a further substantial enlargement and evolution in coverage has been implemented. Composites certainly now constitute one of the most important and diverse classes of material. All materials scientists and engineers need to be familiar with at least the main principles and issues involved in their usage.

While this edition retains much of the structure and conceptual framework of the previous two editions, it now includes four completely new chapters. Moreover, all of the other chapters, which progressively cover the various types of fibre and matrix, the structure of composites, their elastic deformation, strength and toughness, the role of the interface and the thermal characteristics of composite systems, have all been rewritten, to a greater or lesser extent. There has, of course, been extensive updating to reflect the prodigious levels of research and industrial development in the area over the past couple of decades. The citation of references has been expanded and restructured. While previously there was a short list of sources for further information at the end of each chapter (with limited specific citation in the text), this edition provides much more comprehensive referencing, both in quantity and in terms of detail. This change is designed to improve the potential value to researchers, as well as undergraduates.

Nevertheless, much of the material remains pitched at around the level of a final-year undergraduate or a Masters course. In fact, another innovation in this edition is the provision of a large number of questions (with model answers available on the website). Many of these are derived from a third-year undergraduate course on composite materials that has been running (and evolving!) for over 30 years in the Materials Science Department at Cambridge University. A further pedagogical development concerns educational software packages that can be used in conjunction with the book. These form part of a major initiative called DoITPoMS (Dissemination of IT for the Promotion of Materials Science), hosted on the Cambridge University site (www.doitpoms.ac.uk), which comprises a large number of interactive modules covering a wide range of topics. Many are relevant to the general area of composite materials, but several are specific to topics in the book and reference to them is included in the enhanced coverage.

In addition to this expansion in terms of the range of teaching resources, attempts have been made to encompass more of the necessary background science, so as to reduce the need to consult other texts. Examples of this include more comprehensive coverage of the manipulation of stresses and strains as (second-rank) tensors, which is particularly important when treating highly anisotropic materials such as composites, and a considerably expanded chapter on fracture mechanics. These are both areas in which the background knowledge needed for a full understanding of the behaviour of composites is relatively demanding. Bringing this material within the remit of the book is aimed at creating a more coherent overall picture, within a consistent framework of nomenclature and symbolism.

In addition, the new chapters are aimed at expansion of the range of situations that can usefully be treated using the approaches of composite theory. The first of these (Chapter 11) concerns the mechanics of substrate/coating systems, a topic of considerable scientific and technological interest. It is shown how tools developed within the framework of the book can be used to obtain insights into the development of curvature in such systems, and also into the driving forces for spallation (debonding) of such coatings. The following chapter, on highly porous materials, is based on a similar philosophy – in this case showing how such materials, which are also of technological importance, can usefully be treated as special types of composite material.

Chapter 13 is also a new addition. This concerns bio-composites, such as wood and bone. Of course, these are widely recognised as (complex) composite materials, and the treatment presented here is fairly superficial. Nevertheless, information is presented on how they relate to manufactured composites, and there is some coverage of the important topics of recycling, degradation and sustainability. The final new chapter relates to scale effects in composites and to the class of materials sometimes referred to as nano-composites. Despite the enormous levels of interest and research in such materials over recent decades, levels of industrial exploitation have remained minimal – at least as far as load-bearing components are concerned. The reasons for this are outlined.

The final two chapters, as in the previous edition, concern fabrication of composites and their application. These are largely in the form of case histories of various types. There has again been considerable expansion and updating of these, reflecting the huge range of current industrial usage and the ways in which composites have penetrated numerous markets – and in many cases facilitated their expansion and raised their significance. There is extensive cross-reference in these chapters to locations in the book where details are provided about characteristics of the composites concerned that have favoured their usage.

Finally, we would again like to thank our wives, Gail and Pauline, for their invaluable support during the preparation of this book.

T. W. Clyne and D. Hull
November 2018

Nomenclature

Parameters

A	(m^2)	cross-sectional area
A	$(s^{-1} Pa^{-n})$	constant in creep equation (10.17)
a	$(-)$	direction cosine
a	(m)	radius of sphere
a	$(m^2 s^{-1})$	thermal diffusivity
b	(m)	width
Bi	$(-)$	Biot number
C	(Pa)	stiffness (tensor of fourth rank)
C	$(Pa^{-n} s^{-m-1})$	constant in creep equation (10.18)
c	$(J K^{-1} m^{-3})$	volume specific heat
c	(m)	crack length or flaw size
c	$(-)$	$\cos(\phi)$
D	(m)	fibre diameter
d	(m)	fibre/particle diameter
E	(Pa)	Young's modulus
E'	(Pa)	biaxial modulus
e	$(-)$	relative displacement
f	$(-)$	reinforcement (fibre) volume fraction
F	(N)	force
G	$(J m^{-2})$	strain energy release rate
G	(Pa)	shear modulus
g	$(-)$	fraction (undergoing pull-out)
H	(m)	thickness (of substrate)
h	(m)	thickness (of coating)
h	(m)	spacing between fibres
h	(m)	height
h	$(W m^{-2} K^{-1})$	heat transfer coefficient
I	$(-)$	unit tensor (identity matrix)
I	$(-)$	invariant (in the secular equation)
I	(m^4)	second moment of area
K	(Pa)	bulk modulus
K	$(Pa\ m^{1/2})$	stress intensity

K	$(W\ m^{-1}\ K^{-1})$	thermal conductivity
k	$(J\ K^{-1})$	Boltzmann's constant
L	(m)	sample length
L	(m)	fibre half-length
M	$(m\ N)$	bending moment
m	$(-)$	Weibull modulus
N	$(mole^{-1})$	Avogadro's number
N	$(-)$	number of loading cycles
N	(m^{-2})	number of fibres per unit area
n	$(-)$	dimensionless constant
n	$(-)$	stress exponent
P	(N)	force
P	(Pa)	pressure
P	$(-)$	probability
P	$(-)$	porosity
Q	$(m^3\ m^{-2}\ s^{-1})$	fluid flux
q	$(W\ m^{-2})$	heat flux
R	$(J\ K^{-1}\ mole^{-1})$	universal gas constant
R	(m)	far-field radial distance from fibre axis
r	(m)	radius of fibre, tube or crack tip
S	(Pa^{-1})	compliance tensor
S	$(-)$	Eshelby tensor
S	(Pa)	stress amplitude during fatigue
S	$(m^2\ m^{-3})$	specific surface area
s	$(-)$	fibre aspect ratio ($2L/d = L/r$)
s	$(-)$	$\sin(\phi)$
T	(K)	absolute temperature
T	$(N\ m)$	torque
T'	$(K\ m^{-1})$	thermal gradient
t	(m)	ply or wall thickness
t	(s)	time
U	(J)	work done during fracture
u	(m)	displacement in x direction (fibre axis)
V	(m^3)	volume
v	$(m\ s^{-1})$	velocity
W	$(J\ m^{-3})$	work of fracture
x	(m)	distance (Cartesian coordinate)
y	(m)	distance (Cartesian coordinate)
z	(m)	distance (Cartesian coordinate)
α	(K^{-1})	thermal expansion coefficient
β	$(-)$	reinforcement/matrix ratio conductivity ratio
β	$(-)$	dimensionless constant
Δ	$(-)$	relative change in volume

δ	(m)	crack opening displacement
δ	(m)	pull-out length
δ	(m)	distance from neutral axis to interface
ε	(–)	strain
ϕ	(°)	loading angle (between fibre axis and loading direction)
Φ	(°)	global loading angle (between laminate reference axis and loading direction)
γ	(–)	shear strain
γ	(J m^{-2})	surface energy
η	(–)	interaction ratio
η	(–)	dimensionless constant
η	(Pa s)	viscosity
κ	(m^{-1})	curvature
κ	(m^2)	(specific) permeability
λ	(m)	mean free path
λ	(–)	dimensionless constant
θ	(û)	wetting angle
ν	(–)	Poisson ratio
ρ	(kg m^{-3})	density
ρ	(m)	distance from fibre axis
Σ	(N m^2)	beam stiffness
σ	(Pa)	stress
τ	(Pa)	shear stress
ψ	(°)	phase angle (mode mix)
ξ	(–)	dimensionless constant

Subscripts

0	initial
1	x direction (along fibre axis)
2	y direction
3	z direction
A	applied
a	air
b	background
b	buckling
c	coated
c	composite
c	critical
d	debonding
e	fibre end
f	failure
f	fibre (reinforcement)

fr   frictional sliding
g   global
H   hoop
H   hydrostatic
i   interfacial
k   kink band
L   liquid
m   matrix
n   network
p   pull-out
p   particle
r   radial
s   survival
RoM   rule of mixtures
t   stress transfer
th   threshold
trans transverse
u   failure (ultimate tensile)
u   uncoated
v   volume
Y   yield (0.2% proof stress often taken)
θ   hoop
*   critical (e.g. debonding or fracture)

Superscripts

ax   axial
C   constrained
T   transformation
T*   misfit
tr   transverse

1 General Introduction

The usage of composite materials continues to expand rapidly. The current world-wide market value is not easy to estimate, but is certainly more than US$100 billion. Composites now constitute one of the broadest and most important classes of engineering materials – second only to steels in industrial significance and range of applications. There are several reasons for this. One is that they often offer highly attractive combinations of stiffness, strength, toughness, lightness and corrosion resistance. Another is that there is considerable scope for tailoring their structure to suit service conditions. This concept is well illustrated by biological materials such as wood, bone, teeth and hide, which are all composites with complex internal structures that have been designed (via evolutionary processes) to give mechanical properties well suited to the performance requirements. This versatility is, of course, attractive for many industrial purposes, although it also leads to complexity that needs to be well understood if they are to be used effectively. In fact, adaptation of manufactured composite structures for different engineering purposes requires input from several branches of science. In this introductory chapter, an overview is given of the types of composites that have been developed.

1.1 Types of Composite Material

Many materials are effectively composites. This is particularly true of natural biological materials, which are often made up of at least two constituents. In many cases, a strong and stiff component is present, often in elongated form, embedded in a softer and more compliant constituent forming the *matrix*. For example, wood is made up of fibrous chains of cellulose molecules in a matrix of lignin, while bone and teeth are both essentially composed of hard inorganic crystals (hydroxy-apatite or osteones) in a matrix of a tough organic constituent called collagen. Many of the complexities of the structure and properties of bone are well illustrated in the extensive work of Currey [1,2] and a brief survey of biological composites is presented later in this book (Chapter 13).

Commonly, such composite materials show marked *anisotropy* – that is to say, their properties vary significantly when measured in different directions. This usually arises because the harder (and stiffer) constituent is in fibrous form, with the fibre axes preferentially aligned in particular directions. In addition, one or more of the constituents may exhibit inherent anisotropy as a result of their crystal structure. In natural materials, such anisotropy of mechanical properties is often exploited within the structure. For example, wood is much stronger in the direction of the fibre tracheids,

1

which are usually aligned parallel to the axis of the trunk or branch, than it is in the transverse directions. High strength is required in the axial direction, since a branch becomes loaded like a cantilevered beam by its own weight and the trunk is stressed in a similar way by the action of the wind. Such beam bending causes high stresses along the length, but not through the thickness.

In making artificial composite materials, this potential for controlled anisotropy offers considerable scope for integration between the processes of material specification and component design. This is an important point about use of composites, since it represents a departure from conventional engineering practice. An engineer designing a component commonly takes material properties to be isotropic. In fact, this is often inaccurate even for conventional materials. For example, metal sheet usually has different properties in the plane of the sheet than those in the through-thickness direction, as a result of *crystallographic texture* (preferred orientation) produced during rolling, although such variations are in many cases relatively small. In a composite material, on the other hand, large anisotropies in stiffness and strength are possible and must be taken into account during design. Not only must variations in strength with direction be considered, but the effect of any anisotropy in stiffness on the stresses created in the component under external load should also be taken into account. The material can thus be produced bearing in mind the way it will be loaded when it is made into a component, with the processes of material production and component manufacture being integrated into a single operation. This happens when biological materials are produced. In fact, the fine-scale structure of a natural material such as wood is often influenced during its creation by stresses acting on it at the time.

There are several different types of composite. Examples of possible configurations with different types of reinforcement are shown in Fig. 1.1. The matrix material may be polymeric, metallic or ceramic, although by far the largest proportion of composites in

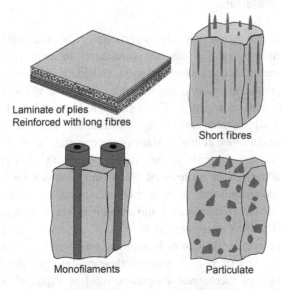

Laminate of plies
Reinforced with long fibres

Short fibres

Monofilaments

Particulate

Fig. 1.1 Schematic depiction of different types of reinforcement configuration.

industrial use are based on polymers – predominantly thermosets (resins), although in some cases thermoplastics are used. Reinforcements are usually ceramics of some sort, most commonly long fibres of carbon or glass. It should, however, be appreciated that other types of fibre, including polymeric and metallic forms, are used in commercial composite materials and that there are also materials containing short fibres or particulate reinforcement. Composites with metallic or ceramic matrices (MMCs and CMCs) are also of industrial significance. Furthermore, it should be noted that there is extensive ongoing research into novel types of composite. As might be expected, property enhancements sought by the introduction of reinforcement into metals or ceramics are often less pronounced that those for polymers, with improvements in high-temperature performance or tribological properties often of interest for MMCs. With CMCs, on the other hand, the objective is usually to enhance the toughness of the matrix. With all three types of matrix, there is enormous potential for achieving property combinations that are unobtainable in monolithic materials.

In considering the formulation of a composite material for a particular type of application, a starting point is clearly to consider the properties exhibited by the potential constituents. Properties of particular interest include the stiffness (Young's modulus), strength and toughness. Density is also of great significance in many situations, since the mass of the component may be of critical importance. Thermal properties, such as expansivity and conductivity, must also be taken into account. In particular, because composite materials are subject to temperature changes (during manufacture and/or in service), a mismatch between the thermal expansivities of the constituents leads to internal residual stresses. These can have a strong effect on the mechanical behaviour.

Some indicative property data are shown in Table 1.1 for a few engineering materials, including some composites. These values are very approximate, but they immediately confirm that some attractive property combinations (for example, high stiffness/strength/toughness, in combination with low density) can be obtained with

Table 1.1 Overview of properties of some engineering materials, including composites.

Material	Density ρ (kg m^{-3})	Young's modulus E (GPa)	Tensile strength σ_* (MPa)	Fracture energy G_c (kJ m^{-2})	Thermal conductivity K (W m^{-1} K^{-1})	Thermal expansivity α ($\mu\varepsilon$ K^{-1})
Mild steel	7800	208	400	100	40	17
Concrete	2400	40	20	0.01	2	12
Spruce (// to grain)	600	16	80	4	0.5	3
Spruce ($\perp$ to grain)	600	1	2	0.2	0.3	10
Chopped strand mat (in-plane)	1800	20	300	30	8	20
Carbon fibre composite (// to fibres)	1600	200	1500	10	10	0
Carbon fibre composite ($\perp$ to fibres	1600	10	40	0.2	2	30
Al-20% SiC$_p$ (MMC)	2800	90	300	2	140	18

composites. They also, of course, highlight the potential significance of anisotropy in the properties of certain types of composite. An outline of how such properties can be predicted from those of the individual constituents forms a key part of the contents of this book.

1.2   Property Maps and Merit Indices for Composite Systems

Selecting the constituents and structure of a composite material for a particular application is not a simple matter. The introduction of reinforcement into a matrix alters all of its properties (assuming that the volume fraction, f, can be regarded as significant, which usually means more than a few per cent). It may also be necessary to take account of possible changes in the microstructure of the matrix resulting from the presence of the reinforcement. The generation of residual stresses (for example, from differential thermal contraction during manufacture) may also be significant. Before considering such secondary effects, it is useful to take a broad view of the property combinations obtainable with different composite systems. This can be visualised using *property maps*. Two examples are presented in Fig. 1.2. These shows plots of: (a) Young's modulus, E; and (b) hardness,[1] H, against density, ρ. A particular material (or type of material) is associated with a point or a region in such maps. This is a convenient method of comparing the property combinations offered by potential matrices and reinforcements with those of alternative conventional materials.

Of course, in general, attractive combinations of these two pairs of properties will lie towards the top-left of these diagrams, although in the case of hardness it should be appreciated that this is a relatively complex 'property' that depends to some extent on microstructure (whereas both stiffness and density are more or less independent of microstructure for a particular type of material). Once the properties of a particular type of composite have been established, then they can, of course, be included in maps of this type. An example is shown in Fig. 1.3, which compares approximate combinations of E and ρ expected for some composite materials with those for materials such as steel, titanium and alumina.

This concept can often be taken a little further by identifying a *merit index* for the performance required, in the form of a specified combination of properties. Appropriate models can then be used to place upper and lower bounds on the composite properties involved in the merit index, for a given volume fraction of reinforcement. The framework for such manipulations was set out in Ashby's seminal work [3], which he also oriented specifically towards composites [4]. An example is shown in Fig. 1.4 for three different fibres and a polymer matrix. The shaded areas joining the points corresponding

---

[1] Hardness is not really a 'genuine' property, although it is a measure of the resistance that the material offers to plastic deformation and is related to the yield stress, σ_Y, and the work-hardening characteristics. It is obtained from the size of an indent produced by an applied load. This is a simple and quick procedure, but hardness values vary with indenter shape and load, since these affect the plastic strains induced. If work-hardening is neglected, then the (Vickers) hardness is expected to have a value of around $3\sigma_Y$.

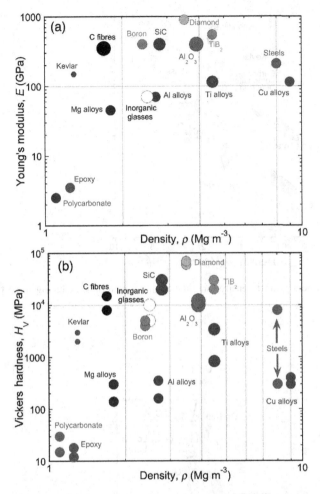

Fig. 1.2 Maps of (a) stiffness and (b) hardness against density, for a selection of metals, ceramics and polymers.

to a fibre to that of the matrix (epoxy resin) represent the possible combinations of E and ρ obtainable from a composite of the two constituents concerned. (The density of a composite is given simply by the weighted mean of the constituents; the stiffness, however, can only be identified as lying between upper and lower bounds – see Chapter 4 – unless more information is given about fibre orientation.) As can be seen in the figure, it is also possible to carry out this operation with the 'reinforcement' being holes – i.e. to consider the creation of foams. As with fibres, the architecture of the porosity is important, and could be such as to cause anisotropy.

Also shown in Fig. 1.4 are lines corresponding to constant values of the ratios E/ρ, E/ρ^2 and E/ρ^3. These ratios represent the merit indices to be maximised to obtain minimum component weight consistent with a maximum permissible deflection for different component shapes and loading configurations. For example, the lightest square

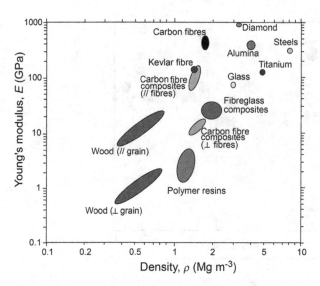

Fig. 1.3 Map of stiffness against density for some common materials, including some fibres and composites.

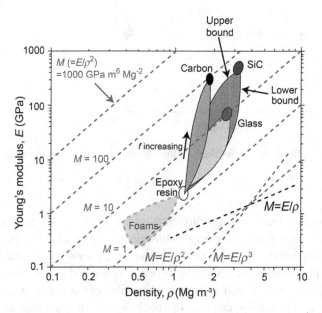

Fig. 1.4 Predicted map of stiffness against density for composites of glass, carbon or SiC fibres in a matrix of epoxy resin. Also shown is the effect of introducing porosity (to create foams). The shaded areas are bounded by the axial and transverse values of E predicted for these composite systems. The diagonal dotted lines represent constant values of three merit indices (E/ρ, E/ρ^2 and E/ρ^3). For E/ρ^2, several lines are shown corresponding to different values of the ratio.

section beam able to support a given load without exceeding a specified deflection is the one made of the material with the largest value of E/ρ^2. It can be seen from the figure that, while the introduction of carbon and silicon carbide fibres would improve the E/ρ ratio in similar fashions, carbon fibres would be much the more effective of the two if the ratio E/ρ^3 were the appropriate merit index. Also notable is that a foam could perform better (i.e. give a lighter component capable of bearing a certain type of load) than the monolithic matrix, particularly if the loading configuration is such that the merit index is E/ρ^3.

1.3   The Concept of Load Transfer

Central to an understanding of the mechanical behaviour of a composite is the concept of *load-sharing* between the matrix and the reinforcing phase. The stress may vary sharply from point to point (particularly with short fibres or particles as reinforcement), but the proportion of the external load borne by each of the individual constituents can be gauged by volume-averaging the load within them. Of course, at equilibrium, the external load must equal the sum of the volume-averaged loads borne by the constituents (e.g. the matrix and the fibre).[2] This gives rise to the condition

$$f\bar{\sigma}_f + (1-f)\bar{\sigma}_m = \sigma_A \tag{1.1}$$

governing the volume-averaged matrix and fibre stresses $(\bar{\sigma}_m, \bar{\sigma}_f)$ in a composite under an external applied stress σ_A, containing a volume fraction f of reinforcement. Thus, for a simple two-constituent composite under a given applied load, a certain proportion of that load will be carried by the fibres and the remainder will be carried by the matrix. Provided the response of the composite remains elastic, this proportion will be independent of the applied load and it represents an important characteristic of the material. It depends on the volume fraction, shape and orientation of the reinforcement and on the elastic properties of both constituents. The reinforcement may be regarded as acting efficiently if it carries a relatively high proportion of the externally applied load. This can result in higher strength, as well as greater stiffness, because the reinforcement is usually stronger, as well as stiffer, than the matrix. Analysis of the load-sharing that occurs in a composite is central to an understanding of the mechanical behaviour of composite materials.

The above concept constitutes an important criterion for distinguishing between a genuine composite and a material in which there is an additional constituent – for example, there might be a fine dispersion of a precipitate – that is affecting the properties (such as the yield stress and hardness), but is present at too low a volume fraction to carry a significant proportion of an applied load.

---

[2] In the absence of an externally applied load, the individual constituents may still be stressed (due to the presence of residual stresses), but these must balance each another according to Eqn (1.1).

References

1. Currey, JD, The many adaptations of bone. *Journal of Biomechanics* 2003; **36**: 1487–1495.
2. Currey, JD, The structure and mechanics of bone. *Journal of Materials Science* 2012; **47**: 41–54.
3. Ashby, MF, *Materials Selection in Mechanical Design*. Pergamon Press, 1992.
4. Ashby, MF, Criteria for selecting the components of composites: overview No. 106. *Acta Metallurgica et Materialia* 1993; **41**: 1313–1335.

2 Fibres, Matrices and Their Architecture in Composites

In this chapter, an overview is provided of the types of fibre and matrix in common use and of how they are assembled into composites. Many types of reinforcement, mostly fibres, are available commercially. Their properties are related to atomic structure and the presence of defects, which must be controlled during manufacture. Matrices may be based on polymers, metals or ceramics. Choice of matrix is usually related to required properties, component geometry and method of manufacture. Certain composite properties may be sensitive to the nature of the reinforcement/matrix interface; this topic is covered in Chapter 7. Properties are also dependent on the arrangement and distribution of fibres, i.e. the fibre architecture, an expression that encompasses intrinsic features of the fibres, such as their diameter and length, as well as their volume fraction, alignment and spatial distribution. Fibre arrangements include laminae (sheets containing aligned long fibres) and laminates that are built up from these. Other continuous fibre systems, such as woven configurations, are also covered. Short fibre systems can be more complex and methods of characterising them are also briefly described.

2.1   Reinforcements

Many reinforcements are available, some designed for particular matrix systems. The reader is referred below to more specialised publications for details about their production, structure and properties. Nevertheless, an overview of certain characteristics is useful here and this is provided in Table 2.1 for a range of fibres in common use. Most of these have relatively high stiffness and low density. Carbon, glass and, to some extent, aramid fibres (such as *Kevlar*) are used extensively in polymer matrix composites. Ceramic fibres (and particles) can be used to reinforce metals, while both metallic and ceramic fibres are commonly used in ceramic-based composites. The latter include *carbon–carbon* composites, which sound a little unlikely, but are in fact of considerable importance for applications such as aircraft brakes – see Section 16.5.1.

Among the points to note in this table is that some fibres are highly anisotropic in certain properties (such as stiffness), although for most purposes the axial properties are much more important than those in other directions. It may also be noted that the tensile strength data are very approximate. This is unavoidable, since the fracture strength of ceramic (brittle) materials is sensitive to the presence of flaws in the sample concerned.

9

Table 2.1 Overview of diameters and properties of several different types of fibre.

Fibre	Density ρ (kg m^{-3})	Axial modulus E_1 (GPa)	Transverse modulus E_2 (GPa)	Shear modulus G_{12} (GPa)	Poisson ratio ν_{12}	Axial strength σ_* (GPa)	Axial CTE α_1 ($\mu\varepsilon$ K^{-1})	Transverse CTE α_2 ($\mu\varepsilon$ K^{-1})
E-glass ($d\sim$10 μm)	2600	76	76	31	0.22	3–4	5	5
Kevlar ($d\sim$12 μm)	1470	150	4	3	0.35	2–3	−4	54
HS (PAN) carbon ($d\sim$8 μm)	1750	250–300	14	14	0.20	3–6	−1	10
HM (PAN) carbon ($d\sim$8 μm)	1940	400–800	6	78	0.20	2–4	−1	10
SiC monofilament ($d\sim$150 μm)	3200	400	400	170	0.20	3	5	5
SiC whisker ($d\sim$0.5 μm)	3200	550	350	170	0.17	6	4	4
α Al$_2$O$_3$ long ($d\sim$10 μm)	3900	385	385	150	0.26	2	8	8
δ Al$_2$O$_3$ staple ($d\sim$3 μm)	3400	300	300	120	0.26	2	8	8
Stainless steel (304) ($d\sim$50–500 μm)	7800	200	200	80	0.27	1	17	17
Tungsten ($d\sim$50–500 μm)	19 300	413	413	155	0.28	3	5	5
Flax (~65% cellulose) ($d\sim$50 μm)	1500	80	10	10	0.3	2	–	–

In fact, the relatively high values for many of these strengths (relative to what might be expected for corresponding bulk material) is largely due to the fact that (fine) fibres tend to contain very few large flaws.

2.1.1   Carbon Fibres

Carbon fibres are largely composed of graphene planes oriented so that they lie parallel to the axis of the fibre. Apart from this condition, and a tendency for a number of adjacent planes to lie parallel to each other, the arrangement is rather disordered. It is illustrated [1] in Fig. 2.1. Such an arrangement is often referred to as 'turbostratic', a term used to describe a structure in which basal planes have slipped out of alignment. The details of the structure do depend on the way the fibres are produced, but the key point is that graphene has very strong bonding within the plane, so the alignment in carbon fibres ensures that the axial stiffness and strength are high.[1] (The in-plane Young's modulus of a perfect graphene crystal, normal to the c-axis, is ~1000 GPa,

---

[1] There has been considerable interest in making 'micro-fibrils' from carbon nanotubes (since it is very difficult to produce composites with significant volume fractions of well-dispersed nanotubes); however, the structure of large assemblies of nanotubes is likely to resemble that of conventional turbostratic carbon fibre and indeed stiffness and strength values of such micro-fibrils have tended to be no higher than those of standard carbon fibre – see Chapter 14.

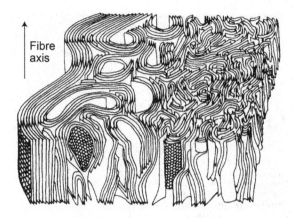

Fibre axis

Fig. 2.1 Schematic representation of the (turbostratic) structure of a carbon fibre [1].

while that along the c-axis is only ~35 GPa.) Carbon fibres are thus highly anisotropic, not only in stiffness and strength, but also in properties such as thermal expansivity (which is higher transversely) and thermal conductivity (which is higher in the axial direction). Their thermal properties are described in Sections 10.1.1 and 10.3.1. For many purposes, this anisotropy is not very significant. For example, the design of fibre composites is often such that applied loads are borne primarily along fibre axes. (In fact, under transverse loading of a uniaxial composite, its stiffness and strength are in any event low, since the matrix, which is usually weak and compliant, bears much of the load, so that fibre stiffness and strength in that direction is unimportant.)

The detailed structure of carbon fibres can vary significantly, depending on exactly how they are made. There are two main processing routes, which are briefly outlined below. An overview [2] of the key properties (stiffness and tensile strength) for these two types of commercial product can be seen in Fig. 2.2, although it should be noted that these data are from the manufacturers (and might hence be slightly optimistic in some cases). It is clear from this plot that there is, in general, a choice between *high modulus* (HM) or *high strength* (HS) products. Use of lower processing temperatures leads to somewhat less dense products that are strong, but not so stiff, while higher temperatures give higher density and stiffness, but accompanied by greater brittleness (lower strains to failure and lower strength).

From PolyAcryloNitrile (PAN)

PAN-based carbon fibres date from around 1960 and are now in extensive use, with annual production of over 60 000 tonnes. PAN resembles polyethylene, but with one of the two hydrogen atoms on every other carbon backbone atom replaced by a nitrile ($-C\equiv N$) group. Bulk PAN is drawn down to a fibre and stretched to produce alignment of the molecular chains. When the stretched fibre is heated, the nitrile groups react to produce a *ladder polymer*, consisting of rows of six-membered rings. While the fibre is still under tension, it is heated in an oxygen-containing environment. This causes further chemical reaction and the formation of cross-links between the ladder molecules.

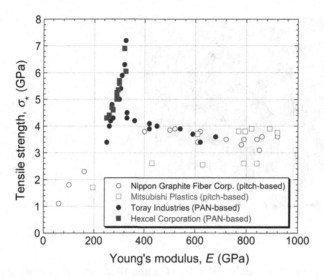

Fig. 2.2 Fibre strength as a function of modulus for commercially available carbon fibres (taken from manufacturers' pamphlets [2]).

The oxidised PAN is then reduced to give the carbon ring structure, which is converted to turbostratic graphite by heating at higher temperatures (in the absence of air – i.e. *pyrolysing*). The fibres usually have a thin skin of circumferential basal planes and a core with randomly oriented crystallites. Detailed information about PAN fibres is available in the literature [2,3].

From Mesophase Pitch

This process originated in the mid-1960s and, while these fibres are less extensively used than PAN-derived variants, they are still of considerable commercial significance. Pitch, which occurs naturally, is a complex mixture of thousands of different species of hydrocarbon and heterocyclic molecules. When heated above 350°C, condensation reactions occur, creating large, flat molecules that tend to align parallel to one another. This is often termed '*mesophase* pitch', a viscous liquid exhibiting local molecular alignment (i.e. a *liquid crystal*). This liquid is extruded through a multi-hole spinneret to produce 'green' yarn, which is made infusible by oxidation at temperatures below its softening point. These fibres can be converted thermally, without applied tension, into a graphitic fibre with a high degree of axial preferred orientation. The basal planes are often oriented radially, as well as being aligned along the fibre axis. This conversion is carried out at ~2000°C. The resulting structures are highly graphitic – more so than for PAN fibres. This affects some thermo-physical properties. For example, the Young's modulus is often high (~700–800 GPa), as is the thermal conductivity (~1000 W m^{-1} K^{-1}). This latter feature is advantageous for certain applications, such as the carbon–carbon composites used for aircraft brakes (Section 16.5.1). The tensile strength, on the other hand, is usually only moderate (~2–4 GPa). Several publications provide detailed coverage of pitch-based carbon fibres [4,5]

2.1.2 Glass Fibres

Glass fibres are used for many different purposes. Most of them are based on silica (SiO_2), with additions of oxides such as those of calcium, boron, sodium, iron and aluminium. The atomic-scale structure is amorphous. The main building block is SiO_4 tetrahedra. A key issue is the extent to which what would, in the case of quartz (pure silica) glass, be a rigid network of these tetrahedra gets disrupted by the presence of 'network-modifying' (lower valence) oxides, such as Na_2O. Such additions reduce the glass transition temperature, so that the material is easier to deform (draw) at relatively low temperatures, although of course this has the effect of reducing the maximum use temperature. Other properties, such as stiffness, are also affected by the composition, although only to a limited degree. Many different compositions are in use, but most fibres used in making composites are of a type designated *E-glass*, which has a combination of properties that is well-suited to this application.

Glass fibres are produced by melting the raw materials in a reservoir and feeding into a series of platinum bushings, each of which has several hundred holes in its base. The glass flows under gravity and fine filaments are drawn mechanically downwards as the glass extrudes from the holes. The fibres are wound onto a drum at speeds of several thousand metres per minute. Control of the fibre diameter is achieved by adjusting the head of the glass in the tank, the viscosity of the glass (dependent on composition and temperature), the diameter of the holes and the winding speed. The diameter of E–glass is usually between 8 and 15 μm.

The structure, and hence the properties, of glass are isotropic. A key property is, of course, the tensile strength. As for all brittle materials, this depends on the presence of flaws, which are predominantly located at the surface. All brittle fibres tend to get stronger as their diameter is reduced, since this is naturally associated with having finer flaws. On the other hand, various practical difficulties arise if the diameter becomes very small (sub-micron) – see Chapter 14 – and a value in the vicinity of 10 μm is popular. This tends to offer a good combination of strength and flexibility – i.e. such fibres can be handled (usually as bundles of some sort) easily and safely. Freshly drawn E-glass fibres have very few surface flaws and their tensile strength is high (~3–4 GPa). However, they are rather susceptible to the accidental introduction of surface damage, and hence to a sharp reduction in strength. In fact, very high fibre strength is not necessarily required for good composite properties: in addition to stiffness, the toughness of a composite is a key property in terms of its engineering usage, and this is more sensitive to issues related to the interface and to the promotion of energy-absorbing mechanisms such as fibre pull-out – see Chapter 9. Nevertheless, a fibre strength that is at least fairly high is likely to be useful and it is certainly helpful to ensure that its surface is protected from mechanical damage and chemical attack, both during fibre handling and *in situ* in the composite material.

To minimise such damage, fibres for use in composites are usually treated with a *size* at an early stage in manufacture. This is a thin coating applied by spraying the fibres with water containing an emulsified polymer. The size actually serves several purposes: not only does it protect the surface from damage, but it also helps hold the fibres together during handling and promotes bonding with the matrix in the composite – i.e. it acts as a *coupling agent*. This is considered further in Section 7.3.1, although the reader is referred to other publications [6–8] for the details of glass fibre production, including treatment with a size.

2.1.3   Polymeric Fibres

Fibres containing long-chain (polymeric) molecules are of considerable significance for use in composite materials, particularly if *cellulose* (in wood and other natural materials) is included in this category. The most important polymeric fibre for use in manufactured composites is the aromatic polyamide (*aramid*) that is produced under the trade name *Kevlar*. In fact, the structures of cellulose and Kevlar, shown in Fig. 2.3, are rather similar. Both have phenyl rings in the backbone and hydrogen bonding between adjacent chains. Chain alignment is produced in Kevlar during drawing and stretching operations, giving it an axial modulus of ~150 GPa. As for carbon fibres, the transverse stiffness is much lower. One characteristic of polymeric fibres, in contrast to glass and carbon, is that they tend to be relatively tough. This is because their failure often involves at least some plastic deformation, associated with a degree of sliding between chains. While this sounds like an important attribute, in practice composite materials often have a high toughness, which is not dependent on the inherent toughness of the fibre (or of the matrix) – see Chapter 9. Nevertheless, there are some applications for which relatively tough fibres offer a distinct advantage. A good example of this is bullet-proof vests, which commonly contain Kevlar fibres. On the other hand, polymeric fibres do have limitations, including a relatively low stiffness and also very limited tolerance of heat. Details about the production and structure of aramid fibres are available in the literature [9,10].

Certain other polymeric fibres are of commercial importance (although not used very widely in composite materials). These include polyethylene and nylon. These are not usually very stiff, but they do have fairly high tensile strengths and they are cheap

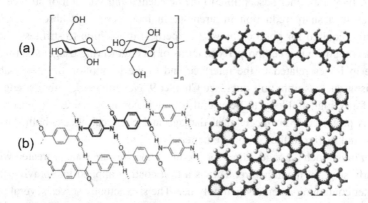

Fig. 2.3 Structures of (a) cellulose and (b) Kevlar (polyparaphenylene terephthalamide) molecules.

and easy to produce. There is also a wide range of natural composites – mostly reinforced by cellulose fibres, which are formed by *in situ* polymerisation of glucose molecules. Cellulose fibres (micro-fibrils) can be extracted from a range of plants, such as cotton, flax and jute, as well as timber [11]. It can be seen from the data for flax in Table 2.1 that the properties of cellulose fibres compare fairly well with those of many artificial fibres, particularly if density is important. Some manufactured composites are based on such fibres. Bio-composites, and also certain recycling issues, are covered in Chapter 13.

2.1.4   Silicon Carbide

Other reinforcements are used in composites, although at much lower levels than carbon and glass fibres. These include silicon carbide, which has a similar structure to diamond and offers a similarly attractive combination of low density and high stiffness, combined with good thermal stability and thermal conductivity. It is much easier to synthesise than diamond and can readily be produced in large quantities in a crude form such as powder. Bulk production of fibres is more problematic. Several different forms have been developed, although it should be appreciated that none of these are currently in very extensive commercial use.

Large-diameter (~100–150 μm) fibres, often termed *monofilaments*, are made by chemical vapour deposition (CVD) onto fine core fibre substrates, usually carbon (~30 μm diameter) or tungsten (~10 μm diameter). The core fibre is fed continuously through a reaction chamber. A gaseous carbon-containing silane, such as methyl-trichloro-silane (CH_3SiCl_3), is passed through the chamber. The core fibre is heated, usually by passing an electrical current through it, and the gas dissociates thermally at the fibre surface to deposit the SiC. Surface layers, designed to improve the resistance to handling damage and the compatibility with the matrix (usually a metal, intermetallic or ceramic), are often deposited in a second reactor. For example, graphitic layers are commonly applied. The process is also employed commercially for production of boron fibres. Details are available elsewhere concerning both fibre production [12] and use of Ti-based composites containing this type of reinforcement [13].

Fibres that are primarily SiC are made by a polymer precursor route analogous to the PAN-based method for carbon fibres (Section 2.1.1). Fibres about 15 μm in diameter are produced in this way, using *polycarbosilane* (PCS) as a precursor. The best-known fibre produced by this route, which has the trade name *Nicalon*, was first developed in the late 1970s, although it has been further developed since then. Polycarbosilane is produced in a series of chemical steps involving the reaction of dichlorodimethylsilane with sodium to produce polydimethylsilane, which yields polycarbosilane on heating in an auto-clave. This is spun into fibres, which are then pyrolysed at temperatures up to 1300°C. The final product contains significant levels of SiO_2 and free carbon, as well as SiC. These fibres thus tend to have inferior properties to the (much purer) monofilaments, although their smaller diameter means they are much more flexible and easier to handle (as bundles), as well as being somewhat cheaper. Details of their production, structure, properties and usage (in ceramic matrix composites) are available elsewhere [14,15].

2.1.5   Oxide Fibres

Oxide fibres offer potential for good resistance to oxidative degradation, which is often a problem for other fibres when used at high temperature (in metal- or ceramic matrix composites). Those in widest use are predominantly combinations of alumina and silica. Fibres containing approximately 50% of each of these (i.e. aluminosilicates) comprise by far the greatest tonnage of refractory fibres and are in extensive use for high-temperature insulation purposes. Such fibres are usually glassy. Alumina fibres with much lower silica contents, which are crystalline, are more expensive to manufacture, but have greater resistance to high temperature and higher stiffness and strength [16].

Both continuous and short alumina fibres are available. An example of the former type is the 'FP' fibre (20 μm diameter) produced by Du Pont [17], while the latter group includes the 'Saffil' fibre (3 μm diameter), which was originally marketed as a replacement for asbestos but has since been investigated quite extensively as a reinforcement in metal matrix composites (MMCs). However, commercial usage of both types remains relatively limited.

2.1.6   Ceramic Whiskers and Nanotubes

There has been extensive interest in fine-scale (sub-micron) reinforcement for composites, mostly in the form of single crystals. This started some considerable time ago (in the 1950s), when it became clear that various ceramics (and metals) can be grown (often from the vapour phase) in the form of elongated single crystals, usually termed *whiskers*. Since these are so fine (perhaps around 100 nm in diameter), and have no grain boundaries, they often have very high strengths. Of course, these are not easy to measure, but they certainly can be of the order of 6–8 GPa, which is probably starting to approach the theoretical limit (corresponding to atomic planes shearing over one another, without dislocations being present; this can only be estimated, but it is around 3–5% of the shear modulus). Such strength values have caused considerable excitement over the years, recently reanimated with the intense interest in carbon nanotubes, which are similar in concept, but even finer (approximately a few nanometres in diameter). There has certainly been a lot of work on production of whiskers (particularly SiC and Si_3N_4 [18–20]) and carbon nanotubes [21,22].

However, this has not, so far, led to commercially viable composites. This topic is covered in more detail in Chapter 14, but the main difficulties are clear. One of the problems is that such fine fibres are expensive to produce, and also difficult to handle – partly because they readily become airborne and are then a potential health hazard. It is also very difficult to disperse significant levels of ultra-fine fibres within a matrix. Moreover, and more fundamentally, the key properties of composites containing such ultra-fine reinforcement are unlikely to be attractive, despite their high tensile strength values. In particular, the toughness of the composite is likely to be relatively low with very fine reinforcement. This issue is covered in Chapter 9.

2.1.7   Particulate

Particles are attractive in some respects as reinforcement, and, while they are quite commonly single crystals and can be relatively fine, they constitute a distinctly different category from the previous one. Powder particles offer advantages in terms of cost and ease of handling and processing. Such material is in fact used in a wide variety of composite materials, often simply as a cheap filler. For example, many engineering polymers and rubbers contain additions of powders such as talc, clays, mica, silica and silicates. In some cases, the mechanical properties of individual particles are of little or no concern. In other situations, density reduction might be a primary aim, as with the additions of hollow glass or ceramic microspheres. However, there are types of composite, notably for certain MMCs, in which particles of high stiffness and/or good resistance to fracture are needed. Chemical compatibility with the matrix is also relevant in many cases and there are likely to be optimal ranges of particle size, commonly of the order of a few microns or tens of microns. An example is provided by the incorporation of SiC grit, which has been in widespread commercial use for decades as an abrasive, into Al-based MMCs at levels around 20 vol%, with these materials often designed primarily for good wear resistance [23,24].

2.2   Statistics of Fibre Tensile Strength

2.2.1   Fracture of Brittle Materials

Most fibres are brittle. That is to say, they sustain little or no plastic deformation or damage up to the point when they fracture. Put another way, cracks can propagate with very little energy absorption. The *fracture energy* (or *critical strain energy release rate*, G_c) of glass is very low – not much more than the minimum for any material (i.e. 2γ, where γ is the surface energy), and with a typical value of around 10 J m^{-2}. It differs little for the various forms in which glass is manufactured. The tensile strength of a particular component made of such material is wholly dependent on the size (and orientation) of the largest flaw that is present, which is commonly somewhere on the free surface. The relationship between the size, c, of a surface flaw (oriented normal to the stress axis) and the tensile strength, σ_*, for a brittle material, is provided by the well-known *Griffith equation* – see Section 9.1.1, which can be expressed as

$$\sigma_* = \sqrt{\left(\frac{G_c E}{\pi c}\right)} \qquad (2.1)$$

where E is the Young's modulus (~75 GPa for glass). It follows that the strength is around 500 MPa when a 1 μm flaw is present and 1.5 GPa for a 0.1 μm flaw. On the other hand, a piece of glass with a scratch in it, which could easily have a depth of 100 μm or more, will break under a moderate stress of a few tens of megapascal. In fact, freshly drawn glass fibres, with a diameter of the order of 10 μm, commonly have no flaw above a few tens of nanometres in size and so can have strengths of several GPa.

However, flaws above this size are readily introduced and it is clear that quoting a well-defined tensile strength is potentially misleading. In fact, a population of flaws is expected along the length of a fibre, so significant variations in strength are expected.

2.2.2   Weibull Analysis

This situation can be treated on a statistical basis. The approach, pioneered by Weibull in 1939, involves conceptually dividing a length L of fibre into a number of incremental lengths, ΔL_1, ΔL_2 etc. When a stress σ is applied, the parameter n_σ defines the number of flaws per unit length sufficient to cause failure under this stress. The fibre fractures when it has at least one incremental element with such a flaw and for this reason the analysis is often known as a *weakest link theory* (WLT). The probability of any given element failing depends on n_σ and on the length of the element. For the first element

$$P_{f1} = n_\sigma \Delta L_1 \tag{2.2}$$

The probability of the entire fibre surviving under this stress is the product of the probabilities of survival of each of the N individual elements that make up the fibre

$$P_S = (1 - P_{f1})(1 - P_{f2})\ldots\ldots(1 - P_{fN}) \tag{2.3}$$

Since the length of the elements can be taken as vanishingly small, the corresponding P_f values must be small. Using the approximation $(1-x) \approx \exp(-x)$, applicable when $x \ll 1$, leads to

$$P_S = \exp\left[-(P_{f1} + P_{f2}\ldots\ldots + P_{fN})\right] \tag{2.4}$$

Substituting from Eqn (2.2), and the corresponding equations for the other elements

$$P_S = \exp\left[-Ln_\sigma\right] \tag{2.5}$$

An expression for n_σ is required if this treatment is to be of any use. Weibull proposed that most experimental data for failure of brittle materials conforms to an equation of the form

$$n_\sigma L_0 = \left(\frac{\sigma}{\sigma_0}\right)^m \tag{2.6}$$

in which m is usually termed the *Weibull modulus* and σ_0 is a normalising strength, which may for our purposes be taken as the most probable strength expected from a fibre of length L_0. Making this assumption, the probability of failure of a fibre of length L, for an applied stress σ, is

$$P_f = 1 - \exp\left[-\left(\frac{L}{L_0}\right)\left(\frac{\sigma}{\sigma_0}\right)^m\right] \tag{2.7}$$

The Weibull modulus is an important parameter for characterising the strength distribution exhibited by the fibre (or any other brittle material). If the value of m is large (say >20), then it can be seen from Eqn (2.6) that stresses even slightly below the normalising value σ_0 would lead to a low probability of failure, while if they were slightly above then a high probability would be predicted. Conversely, a low Weibull modulus (say, <5) would introduce much more uncertainty about the strength of a fibre. In practice, many ceramic materials exhibit Weibull moduli in the range 2–10, representing considerable uncertainty about the stress level at which any given specimen is likely to fail.

To check whether a set of strength values conforms to Eqn (2.7), it is convenient to rearrange the equation into a form in which a linear relationship is predicted. This is usually obtained via the logarithm of the probability of survival ($P_S = 1 - P_f$)

$$\ln (P_S) = - \left(\frac{L}{L_0} \right) \left(\frac{\sigma}{\sigma_0} \right)^m \tag{2.8}$$

so that

$$\ln \left(\frac{1}{P_S} \right) = \left(\frac{L}{L_0} \right) \left(\frac{\sigma}{\sigma_0} \right)^m \tag{2.9}$$

Taking logs again then gives

$$\ln \left(\ln \left(\frac{1}{P_S} \right) \right) = \ln (L) - \ln (L_0) + m \ln (\sigma) - m \ln (\sigma_0) \tag{2.10}$$

A plot, in this form, of data for P_S as a function of σ, should give a straight line with a gradient of m. An example is shown in Fig. 2.4, which gives data [25] for the strength distributions of three types of SiC monofilament. It can be seen that these data do

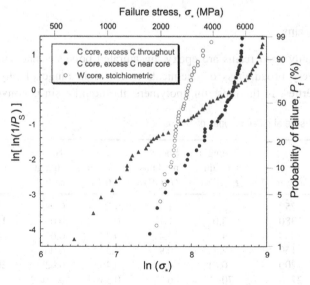

Fig. 2.4 Weibull plot [25] of failure strength data from three types of SiC monofilament, each manufactured under different conditions. These data were obtained by testing a large number of individual fibres of each type. The gradients (Weibull moduli, m) of the three plots are about 2, 4 and 8.

conform approximately to Eqn (2.10), in that the plots are more or less linear. The two carbon-cored fibres have about the same average strength, but rather different variabilities (m values of 2 and 4). The tungsten-cored fibre, on the other hand, has a lower average strength, but much lower variability ($m = 8$). These differences can be attributed to the nature and distribution of the flaws that are present.

The variability of strength exhibited by most ceramic fibres has important consequences for the mechanical behaviour of composite materials. It means, for example, that points of fibre fracture are often fairly randomly distributed and do not necessarily become concentrated in a single crack plane that propagates through the material. This leads to wide distributions of damage and promotes fibre pull-out (see Section 8.2), enhancing the toughness.

2.3   Matrices

The properties exhibited by various types of matrix are presented in Table 2.2. Information of this type, when considered together with data for reinforcements, allows potential systems to be appraised. For example, glass is evidently of no use for reinforcement of metals if enhancement of stiffness is a primary aim. Slightly more subtle points, such as whether fibre and matrix have widely differing thermal expansion coefficients (and would hence be prone to differential thermal contraction stresses), may also be explored. In practice, however, ease of manufacture (see Chapter 15) often assumes considerable importance. In the following sections some points are made concerning the factors that affect the choice of matrix.

2.3.1   Thermosetting Resins

The most commonly used resins are epoxy, unsaturated polyester and vinyl esters, which cover a very broad class of chemicals and a wide range of physical and mechanical properties. In thermosetting polymers, the liquid resin is converted into

Table 2.2 Overview of properties of several different types of matrix.

Matrix	Density ρ (kg m^{-3})	Young's modulus E (GPa)	Shear modulus G (GPa)	Poisson ratio ν	Tensile strength σ_* (GPa)	Thermal expansivity α ($\mu\varepsilon$ K^{-1})
Epoxy	1250	3.5	1.27	0.38	0.04	58
Polyester	1380	3.0	1.1	0.37	0.04	150
Polyether ether ketone (PEEK)	1300	4	1.4	0.37	0.07	45
Polycarbonate	1150	2.4	0.9	0.33	0.06	70
Polyurethane rubber	1200	0.01	0.003	0.46	0.02	200
Aluminium	2710	70	26	0.33	0.1–0.3	24
Magnesium	1740	45	7.5	0.33	0.1–0.2	26
Titanium	4510	115	44	0.33	0.4–1.0	10
Borosilicate glass	2230	64	28	0.21	0.05	3.2

a hard, rigid solid by chemical cross-linking, leading to formation of a tightly bound 3D network. This is usually done while the composite is being formed. The mechanical properties depend on the molecular units making up the network and on the length and density of the cross-links. The former depends on the chemicals used and the latter on the cross-links formed during curing. This can be done at room temperature, but it is usual to use a cure schedule that involves heating at one or more temperatures for predetermined times, to achieve optimum cross-linking and hence optimum properties. A relatively high-temperature final post-cure treatment is often given to minimise any changes in properties during service. Shrinkage during curing, and thermal contraction on cooling after curing, can lead to residual stresses in the composite – see Chapter 10.

It can be seen from the data in Table 2.2 that thermosets have slightly different properties from thermoplastics. Notable among these are much lower strains to failure. Thermosets are brittle, while thermoplastics can undergo appreciable plastic deformation. However, there are also significant differences between different types of thermoset. For example, epoxies are in general tougher than unsaturated polyesters or vinyl esters. In fact, epoxies are superior in most respects to alternative thermosetting systems, which are sometimes preferred simply on the grounds of lower cost.

2.3.2   Thermoplastics

Unlike thermosetting resins, thermoplastics are not cross-linked. They derive their strength and stiffness from inherent properties of the monomer units and high molecular weight. In glassy thermoplastics, such as poly(methyl methacrylate) (PMMA), there is a high concentration of molecular entanglements, which act like cross-links. Heating of amorphous materials above the *glass transition temperature*, T_g, creates molecular mobility, allowing these entanglements to unravel, so the material changes from a rigid solid to a viscous liquid. In crystalline polymers, heating leads to melting of the crystals (at T_m). Many thermoplastics are semi-crystalline, with $T_g < T_m$, and, for some of these, such as polyethylene, room temperature is above T_g, but below T_m. Such materials tend to be soft, and less brittle than fully glassy materials (which must operate below T_g). All of these polymers may have anisotropic structure and properties, depending on processing conditions. In amorphous regions, molecular alignment can be created by shear stresses arising during moulding or subsequent plastic deformation, while crystallographic texture (non-random crystallite orientation distributions) can arise from features of crystal nucleation and growth, or again from imposed plastic deformation, causing crystals to become reoriented.

Thermoplastics tend to exhibit good resistance to attack by chemicals and generally good thermal stability, particularly for certain advanced thermoplastics used in composites. Polyether ether ketone (PEEK), a semi-crystalline polymer, is a good example. The stiffness and strength of PEEK is little affected by heating up to 150°C, a temperature at which most polymers have become substantially degraded. Composites such as PEEK–60% carbon fibre are widely used in the aerospace industry. Other high-performance thermoplastics include polysulphones, polysulphides and polyimides.

Most of these are amorphous polymers. Many thermoplastics also show good resistance to absorption of water. All thermoplastics yield and undergo large deformations before final fracture and their mechanical properties are strongly dependent on temperature and strain rate. Another important feature of all thermoplastics is that under constant load conditions the strain tends to increase with time – i.e. creep occurs. This means that there may be a redistribution of the load between matrix and fibres during deformation and under in-service loading conditions.

One of the most significant features of thermoplastic composites is that processing tends to be more difficult than with thermosets. This is essentially because they are already polymeric, and hence highly viscous even when liquid, before the composite is fabricated. Although T_g and T_m are in many cases quite low, the melts they produce have high viscosities and cannot easily be impregnated into fine arrays of fibres. Usually it is necessary to ensure that flow distances are short, for example by inter-leaving thin polymer sheets with fibre preforms, and to apply substantial pressures for appreciable times (see Section 15.1). Once fibre and matrix have been brought together in some way, then various shaping operations, such as injection moulding, can be carried out.

2.3.3   Metal Matrices

The development of metal matrix composites has largely been concentrated on three metals: aluminium, magnesium and titanium. Metals are normally alloyed with other elements to improve their physical and mechanical properties, and a wide range of alloy compositions is available. Final properties are strongly influenced by thermal and mechanical treatments, which determine the microstructure. Some typical properties of common metal matrices are given in Table 2.2. The metals used for composites are usually ductile and essentially isotropic. Unlike polymers, the increases in stiffness achieved by incorporation of the reinforcement are often relatively small. However, important improvements are often achieved in properties such as wear characteristics, creep performance and resistance to thermal distortion. All three metals are very reactive, with a strong affinity for oxygen. This has implications for the production of composites, particularly in regard to chemical reactions at the interface between the matrix and the reinforcement, which has proved especially troublesome for titanium.

2.3.4   Ceramic Matrices

Four main classes of ceramic have been used for ceramic matrix composites. *Glass ceramics* are complex glass-forming oxides, such as boro-silicates and alumino-silicates, which have been heat treated so that a crystalline phase precipitates to form a fine dispersion in the glassy phase. Glass ceramics have lower softening temperatures than crystalline ceramics and are easier to fabricate – this is an especially important consideration for ceramic composites. *Engineering ceramics*, such as SiC, Si_3N_4, Al_2O_3 and ZrO_2, are fully crystalline and have the normal structure of crystalline grains randomly oriented relative to each other. There has been considerable interest in reinforcement of *cement* and *concrete*, usually adding short fibres in such a way that moulding capabilities are not severely impaired. Finally, carbon–carbon composites, produced by vapour infiltration of an array of carbon fibres, form a specialised, but commercially important, subclass of composite material. They find use not only in the long-standing application of aircraft brakes, but also in various other components in the aerospace industry with demanding requirements [26].

The objective of adding reinforcement to ceramics is usually to improve their toughness. Ceramics are very brittle and even a small increase in toughness may be worthwhile. When ceramics are added, the toughness increase often comes from repeated crack deflection at interfaces, in which case the nature of the interface assumes overriding importance [27]. However, it should be noted that substantial toughness enhancement can often be achieved by introducing metallic reinforcement, particularly in the form of fibres. This may reduce the high-temperature performance of the material, although some metal fibres, such as certain stainless steels, have very good stability at elevated temperatures. Systems such as alumina reinforced with ~15 vol% of stainless steel fibres can offer attractive combinations of toughness and potential for usage at high temperatures [28,29], and their usage is likely to increase in the future.

2.4   Long Fibre Composite Architectures

2.4.1   Laminates

In many situations, a relatively high value of the fibre volume fraction, f, is desirable. The easiest way to achieve this is to align (long) fibres in a given direction. In principle, an assembly of close-packed cylinders can occupy over 90% of the volume. In practice, this is not achievable, partly because it would require many of the fibres to be in contact with their neighbours, creating several problems. However, values of around 60–70% are realistic, consistent with the need to keep fibres apart and with manufacturing constraints. Furthermore, for many applications, while good properties (primarily stiffness and strength) are commonly required in various directions within a sheet or plate, properties normal to this plane (i.e. in the through-thickness direction) are usually much less important (because the component is unlikely to be loaded in that direction). The basic building block for composite structures is thus often a thin sheet containing fibres aligned in one direction, termed a *lamina* or *ply*.

In order to achieve suitable properties in various directions within the plane (often tailored to the expected loading of the component during service) *laminae* are stacked in a predetermined sequence to create a *laminate*. For the prediction of elastic properties of the component as a whole, each lamina may be regarded as homogeneous in the sense that the fibre distribution and volume fraction are uniform throughout. The properties of the laminate can thus be predicted for any given *stacking sequence* – see Chapter 4. A sequence is shown in Fig. 2.5, with an indication of the nomenclature used to describe it.

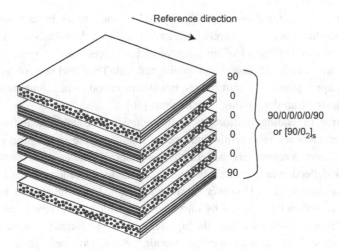

Fig. 2.5 Schematic depiction of a fibre laminate (stack of plies), illustrating the nomenclature system.

Fig. 2.6 SEM micrograph of a woven roving assembly of long glass fibres.

2.4.2   Woven and Planar Random Fibre Assemblies

Continuous fibres can be assembled in ways other than stacking of unidirectional plies. Much of this is done using technology originally developed for textile processes – i.e. *weaving*, *braiding* and *knitting*. The arrangement of fibres in a typical woven assembly is shown in Fig. 2.6. In this case, the angle between the warp and weft directions is 90°. The flexibility of this type of fibre assembly allows draping and shaping to occur, facilitating use in non-planar structures. The angle between the warp and weft directions will affect these characteristics. Of course, this type of structure leads to a rather inhomogeneous distribution of the fibres. Furthermore, a high fibre content is not possible – the maximum volume fraction is usually not much more than about 20–25%. Therefore, while use of this type of starting material is convenient in terms

Fig. 2.7 SEM micrograph of a chopped strand mat glass fibre preform.

of handling and processing, it is not normally used for very demanding applications. Nevertheless, usage is quite extensive.

A commonly used form of fibre distribution, particularly for low-cost applications, is *chopped strand mat*. Bundles of relatively long fibres are assembled together with random in-plane orientations, as shown in Fig. 2.7. These are created by sedimentation of fibre bundles from suspension in a fluid. The material is easy to handle as a preform and the resultant composite material has isotropic in-plane properties. However, the fibre volume fraction is limited to relatively low values (<~20%). Moreover, the scope for tailoring of the properties in different in-plane directions, which is considerable for laminates and exists to a limited degree with woven assemblies, is not available for this type of material.

2.5 Short Fibre Configurations

While most high-performance composites tend to be based on long (continuous) fibres, there are many applications in which, for various reasons, short fibres, or even particles, are preferred. This tends to allow more versatility in the processing, so that, for example, the composite material itself can be subjected to forming processes such as extrusion, drawing, rolling, etc. These processes are, of course, widely used for metals, but cannot normally be applied to long fibre composites. Characterisation of the fibre architecture is more complex for short fibre composites, since they may not be aligned within a plane and there is more likelihood of them displaying large local variations in orientation. Moreover, while clustering (local variations in position) may be an issue for long fibres, such effects are commonly more pronounced with short fibres (and particles).

2.5.1 Fibre Orientation Distributions

There are several ways of both measuring and characterising fibre orientation distributions. Some involve examining polished sections, although these are rather time-consuming and outdated. The most powerful technique, now in widespread use, is that of computed tomography, which involves analysis of absorption images obtained by passing an X-ray beam through a sample in different directions. Associated software commonly allows the resultant fibre architecture, which can be visualised in 3D, to be transferred directly into numerical (FEM) models, where it can be meshed and used to predict various characteristics.

Orientation distributions in 3D are commonly represented on a *stereographic projection* (stereogram). Thus, *texture* information for polycrystals can be presented as *pole figures*, which depict the relative frequencies of the orientation of specified crystallographic directions, relative to the external frame of reference. There are in fact several issues and possible options when obtaining and presenting such information [30]. Representation of fibre distributions is simpler than for the texture of polycrystals, since only the orientation of the fibre axis is required and each fibre is represented as a point on the stereogram. In fact, a random (isotropic) 3D distribution of orientations does not plot as a uniform density of points; the points are clustered near the centre and are sparse towards the edges. This is illustrated in Fig. 2.8, which shows (a) how two directions, 1 and 2, plot as P_1 and P_2 on the projection and (b) how an isotropic

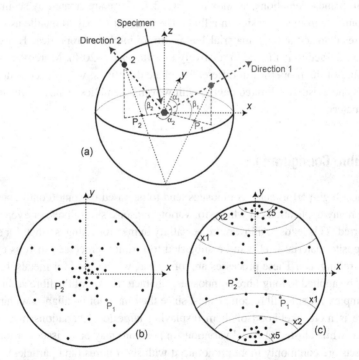

Fig. 2.8 Representation of orientation distributions in 3D, using the stereographic projection. (a) Construction of the stereogram, showing how two directions, 1 and 2, are plotted as points P_1 and P_2 where the lines from 1 and 2 to the 'south pole' intersect the 'equatorial plane'. (b) Stereogram of a set of randomly oriented directions. (c) Stereogram of a set of directions, with a systematic bias towards the reference direction y, with superimposed contours separating regions in which the population densities are different multiples of the random case.

distribution of directions plots as a non-uniform density of points. The most effective way to present fibre distribution information is in the form of a series of contours, representing the ratio of the local density of points to that expected for an isotropic distribution. Such contours, which are normally constructed using a computer program, are shown in Fig. 2.8(c). This allows the strength of any preferred orientation to be characterised by a single figure, since the value of the highest contour present can readily be established.

An example [31] of the application of such procedures to a fibre orientation distribution is presented in Fig. 12.5 for the case of a metallic fibre network material – i.e. a 'composite' for which the 'matrix' is air. (Such highly porous materials are covered in Chapter 12.) Of course, this case has the advantage of very high 'contrast' (difference between the X-ray absorption rates of fibre and matrix), although in practice it is usually adequate for most composite systems. Visualisations are shown in Fig. 12.5 for two different fibre network materials, both made by sintering assemblies of (relatively coarse) stainless steel fibres, (a) without and (b) with a compression procedure that created a non-random distribution of the orientation of individual fibre segments (between joints). Also shown are corresponding extracted fibre orientation distribution data for the two materials, taken from the two stereograms. It can be seen that compression has created a marked tendency for the fibres to lie at relatively large angles to the unique (pressing) direction, while the uncompressed sample was approximately isotropic. This is clearly as expected, although the strength of the effect will depend on fibre segment length, and perhaps also on fibre yielding and work hardening characteristics, as well as on the compression pressure.

It may also be noted that, under certain circumstances, variations in local fibre orientation can have a significant effect on certain properties for long fibre composites, despite the fact that these variations tend to be relatively small. An example is provided by the axial compressive strength of uniaxial composites ('struts'), which is sensitive to fibre 'waviness' along the length of the component – see Section 8.1.3.

2.5.2 Clustering of Fibres and Particles

Examination of laminae and laminate cross-sections usually reveals that the positional distribution of the fibres is not entirely uniform and this tendency is often considerably more marked with short fibres and particles. This is unlikely to have much influence on most 'macroscopic' properties, such as stiffness, thermal expansivity and thermal conductivity. However, it could be relevant to the onset of damage and failure, since local inhomogeneities can influence both the initiation and propagation of cracks and might also be relevant to localised plasticity. There are therefore certain circumstances in which it can be helpful to characterise the severity of clustering (which is likely to be affected by the details of the processing route).

An example of this is provided by the plot [32] shown in Fig. 2.9, which confirms that a clear correlation could be established between the ductility of a particular type of MMC (an Al-9Si-0.5Mg alloy containing 20 vol% of SiC particulate, with average size ~12 μm) and the severity of the clustering of these particles. Different variants of this

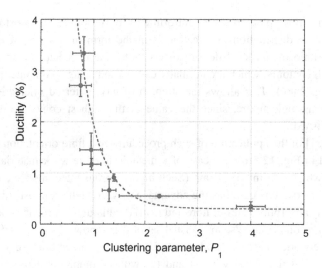

Fig. 2.9 Measured ductility of various composite materials, all based on an Al alloy matrix containing 20 vol% of SiC particulate, as a function of a parameter characterising the severity of clustering of the particles in the microstructure. This varied as a result of changes in the casting and thermo-mechanical processing procedures employed during production [32].

material were produced by altering the processing procedures and conditions, ensuring in each case that there was no porosity present – all samples were subjected before testing to a hot isostatic pressing operation – and that the matrix microstructure was essentially the same in all cases. The clustering parameter was obtained via analysis of a series of metallographic sections, with the locations of particle centres being established and a Dirichlet tessellation procedure then being used to characterise their distribution. The clustering severity parameter used in the plot was the ratio of the variance of the distribution of cell areas to that for a random distribution with the same average areal density of particles. Of course, this is a limited analysis, being confined, for example, to 2D examination, but it does at least confirm that clustering can be an issue (in tending to promote cracking, at least in this type of composite).

References

1. Bennett, SC and DJ Johnson, Structural heterogeneity in carbon fibres, in *Proc. 5th London Carbon and Graphite Conference*. Society for the Chemical Industry, 1978, pp. 377–386.
2. Tanaka, F and T Okabe, Historical review of processing, microstructures, and mechanical properties of PAN-based carbon fibers, in *Comprehensive Composite Materials II*, Gdoutos, EE, editor. Elsevier, 2018, pp. 66–85.
3. Yusof, N and AF Ismail, Post spinning and pyrolysis processes of polyacrylonitrile (PAN)-based carbon fiber and activated carbon fiber: a review. *Journal of Analytical and Applied Pyrolysis* 2012; **93**: 1–13.
4. Huang, XS, Fabrication and properties of carbon fibers. *Materials* 2009; **2**(4): 2369–2403.

5. Bermudez, V, S Lukubira and AA Ogale, Pitch precursor-based carbon fibers, in *Comprehensive Composite Materials II*, Gdoutos, EE, editor. Elsevier, 2018, pp. 41–65.

6. DiBenedetto, AT, Tailoring of interfaces in glass fiber reinforced polymer composites: a review. *Materials Science and Engineering A: Structural Materials, Properties, Microstructure and Processing* 2001; **302**(1): 74–82.

7. Li, H, T Charpentier, JC Du and S Vennam, Composite reinforcement: recent development of continuous glass fibers. *International Journal of Applied Glass Science* 2017; **8**(1): 23–36.

8. Dwight, DW and S Begum, Glass fiber reinforcements, in *Comprehensive Composite Materials II*, Gdoutos, EE, editor. Elsevier, 2018, pp. 243–268.

9. Singh, TJ and S Samanta, Characterization of Kevlar fiber and its composites: a review. *Materials Today: Proceedings* 2015; **2**(4–5): 1381–1387.

10. Yang, HM, Glass fiber reinforcements, in *Comprehensive Composite Materials II*, Gdoutos, EE, editor. Elsevier, 2018, pp. 187–217.

11. Medina, LA and J Dzalto, Natural fibers, in *Comprehensive Composite Materials II*, Gdoutos, EE, editor. Elsevier, 2018, pp. 269–294.

12. Wawner, FE, Boron and silicon carbide fibers (CVD), in *Comprehensive Composite Materials II*, Gdoutos, EE, editor. Elsevier, 2018. pp. 167–186.

13. Moongkhamklang, P, VS Deshpande and HNG Wadley, The compressive and shear response of titanium matrix composite lattice structures. *Acta Materialia* 2010; **58**(8): 2822–2835.

14. Zhao, DF, HZ Wang and XD Li, Development of polymer-derived SiC fiber. *Journal of Inorganic Materials* 2009; **24**(6): 1097–1104.

15. Ichikawa, H and T Ishikawa, Silicon carbide fibers (organometallic pyrolysis), in *Comprehensive Composite Materials II*, Gdoutos, EE, editor. Elsevier, 2018, pp. 127–166.

16. Berger, MH and AR Bunsell, Oxide fibers, in *Comprehensive Composite Materials II*, Gdoutos, EE, editor. Elsevier, 2018, pp. 218–242.

17. Dhingra, AK, Alumina Fiber FP. *Philosophical Transactions of the Royal Society A: Mathematical Physical and Engineering Sciences* 1980; **294**(1411): 411–417.

18. Seeger, T, P Kohler-Redlich and M Ruhle, Synthesis of nanometer-sized SiC whiskers in the arc-discharge. *Advanced Materials* 2000; **12**(4): 279–282.

19. Silva, PC and JL Figueiredo, Production of SiC and Si3N4 whiskers in C+SiO2 solid mixtures. *Materials Chemistry and Physics* 2001; **72**(3): 326–331.

20. Li, J, T Shirai and M Fuji, Rapid carbothermal synthesis of nanostructured silicon carbide particles and whiskers from rice husk by microwave heating method. *Advanced Powder Technology* 2013; **24**(5): 838–843.

21. Zhang, Q, JQ Huang, WZ Qian, YY Zhang and F Wei, The road for nanomaterials industry: a review of carbon nanotube production, post-treatment, and bulk applications for composites and energy storage. *Small* 2013; **9**(8): 1237–1265.

22. Lu, W, Q Li and TW Chou, Carbon nanotube based fibers, in *Comprehensive Composite Materials II*, Gdoutos, EE, editor. Elsevier, 2018, pp. 13–40.

23. Deuis, RL, C Subramanian and JM Yellup, Abrasive wear of aluminium composites: a review. *Wear* 1996; **201**(1–2): 132–144.

24. Alpas, AT, S Bhattacharya and IM Hutchings, Wear of particulate metal matrix composites, in *Comprehensive Composite Materials II*, Clyne, TW, editor. Elsevier, 2018, pp. 137–172.

25. Martineau, P, M Lahaye, R Pailler, R Naslain, IM Couzi and F Cruege, SiC filament/titanium matrix composites regarded as model composites. Part 1: filament microanalysis and strength characterization. *Journal of Materials Science* 1984; **19**: 2731–2748.

26. Taylor, RL, SBV Siva and PSR Sreekanth, Carbon matrix composites, in *Comprehensive Composite Materials II*, Ruggles-Wrenn, MB, editor. Elsevier, 2018, pp. 339–378.
27. Krenkel, W and F Reichert, Design objectives and design philosophies, interphases and interfaces in fiber-reinforced CMCs, in *Comprehensive Composite Materials II*, Ruggles-Wrenn, MB, editor. Elsevier, 2018, p. 1–18.
28. Pemberton, SR, EK Oberg, J Dean, D Tsarouchas, AE Markaki, L Marston and TW Clyne, The fracture energy of metal fibre reinforced ceramic composites (MFCs). *Composites Science and Technology* 2011; **71**(3): 266–275.
29. Clyne, TW and LW Marston, Metal fibre-reinforced ceramic composites and their industrial usage, in *Comprehensive Composite Materials II*, Clyne, T W, editor. Elsevier, 2018, pp. 464–481.
30. Callahan, PG, M Echlin, TM Pollock, S Singh and M De Graef, Three-dimensional texture visualization approaches: theoretical analysis and examples. *Journal of Applied Crystallography* 2017; **50**: 430–440.
31. Tan, JC, JA Elliott and TW Clyne, Analysis of tomography images of bonded fibre networks to measure distributions of fibre segment length and fibre orientation. *Advanced Engineering Materials* 2006; **8**(6): 495–500.
32. Murphy, AM, SJ Howard and TW Clyne, Characterisation of the severity of particle clustering and its effect on the fracture of particulate MMCs. *Materials Science and Technology* 1998; **14**: 959–968.

3 Elastic Deformation of Long Fibre Composites

In the previous chapter, some background was provided about types of reinforcement and their distribution within different matrices. Attention is now turned to predicting the behaviour of the resulting composites. The prime concern is with mechanical properties. The reinforcement is usually designed to enhance the stiffness and strength of the matrix. The details of this enhancement can be rather complex. The simplest starting point is the elastic behaviour of a composite with aligned long (continuous) fibres. This arrangement creates high stiffness (and strength) in the fibre direction. However, it is also important to understand the behaviour when loaded in other directions, so the treatment also covers transverse loading. In this chapter, and in the following one, perfect bonding is assumed at the fibre/matrix interface. Details concerning this region, and consequences of imperfect bonding, are considered in Chapter 6.

3.1   Axial Young's Modulus

The simplest treatment of the elastic behaviour of aligned long fibre composites is based on the premise that the material can be treated as if it were composed of parallel slabs of the two constituents bonded together, with relative thicknesses in proportion to the volume fractions of matrix and fibre. This is illustrated in Fig. 3.1, which compares the assumptions imposed by use of the 'slab model' with the situation in an actual long fibre composite under different types of applied load. The two slabs are constrained to have the same lengths parallel to the bonded interface. Thus, if a stress is applied in the direction of fibre alignment (termed the 1 direction throughout this book), then both constituents experience the same strain in this direction, ε_1. This 'equal strain' condition is valid for loading along the fibre axis, provided there is no interfacial sliding.

It is now a simple matter to derive an expression for the Young's modulus of the composite, E_1. The axial strain in the fibre and the matrix must correspond to the ratio between the stress and the Young's modulus for each of the two components, so that

$$\varepsilon_1 = \varepsilon_{1f} = \frac{\sigma_{1f}}{E_f} = \varepsilon_{1m} = \frac{\sigma_{1m}}{E_m} \tag{3.1}$$

Hence, for a composite in which the fibres are much stiffer than the matrix ($E_f \gg E_m$), the reinforcement bears a much higher stress than the matrix ($\sigma_{1f} \gg \sigma_{1m}$) and there is a

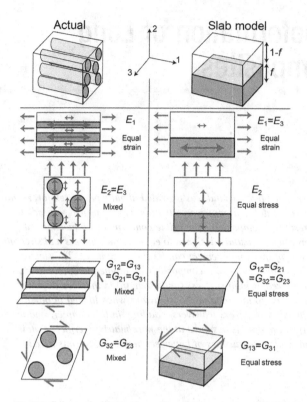

Fig. 3.1 Schematic illustration of loading geometry and distributions of stress and strain, and effects on the Young's moduli and shear moduli, for a uniaxial fibre composite and for the slab model representation.

redistribution of the load. The overall composite stress σ_1 can be expressed in terms of the two contributions being made to the load

$$\sigma_1 = (1 - f)\sigma_{1m} + f\sigma_{1f} \tag{3.2}$$

The Young's modulus of the composite can now be written as

$$E_1 = \frac{\sigma_1}{\varepsilon_1} = \frac{(1-f)\sigma_{1m} + f\sigma_{1f}}{\sigma_{1f}/E_f} = E_f\left[\frac{(1-f)\sigma_{1m}}{\sigma_{1f}} + f\right]$$

Using the ratio between the stress in the two constituents given by Eqn (3.1), this simplifies to

$$E_1 = (1 - f)E_m + fE_f \tag{3.3}$$

This well-known *rule of mixtures* indicates that the composite stiffness is simply a weighted mean between the moduli of the two components, depending only on the volume fraction of fibres. It is expected to be valid to a high degree of precision, providing the fibres are long enough for the equal strain assumption to apply. (The details of this condition are examined in Chapter 5.) Very minor deviations from the equation are expected as a result of effects that arise when the Poisson ratios of the two constituents are not equal. More advanced treatments (such as the Eshelby model described in Chapter 5) show that the predicted discrepancies are extremely small in all cases. The rule of mixtures can readily be confirmed experimentally for uniaxial long fibre composites. The equal strain treatment is often described as a *Voigt model*.

3.2   Transverse Young's Modulus

Accurate prediction of the transverse stiffness is far more difficult than for the axial value. In addition, experimental measurement is more prone to error, partly as a result of higher stresses in the matrix – which can, for example, cause polymeric matrices to creep under modest applied loads. The simplest approach is to assume that the system can again be represented using the 'slab model' depicted in Fig. 3.1. In the fibre composite shown in the left side of Fig. 3.1, both 2 and 3 directions are transverse to the fibres. An obvious problem with the slab model is that the two transverse directions are not identical; direction 3 is equivalent to the axial direction (direction 1). In reality the matrix is subjected to an effective stress intermediate between the full applied stress operating on the matrix when it is normal to the plane of the slab interface and the reduced value calculated in Section 3.1 for loading parallel to this interface – i.e. a 'mixed' condition applies, as indicated in Fig. 3.1. Before considering this any further, the limiting case of the *'equal stress'* model will be examined. When a stress is applied in the 2 direction

$$\sigma_2 = \sigma_{2f} = \varepsilon_{2f}E_f = \sigma_{2m} = \varepsilon_{2m}E_m \tag{3.4}$$

The overall net strain can be written in terms of the two contributions to it

$$\varepsilon_2 = f\varepsilon_{2f} + (1 - f)\varepsilon_{2m} \tag{3.5}$$

from which the composite modulus can be expressed as

$$E_2 = \frac{\sigma_2}{\varepsilon_2} = \frac{\sigma_{2f}}{f\varepsilon_{2f} + (1 - f)\varepsilon_{2m}}$$

Substituting expressions for ε_{2f} and ε_{2m} derived from Eqn (3.4) leads to

$$E_2 = \left[\frac{(1 - f)}{E_m} + \frac{f}{E_f} \right]^{-1} \tag{3.6}$$

This equal stress treatment, giving an 'inverse rule of mixtures', is often described as a *'Reuss model'*.

Although this treatment is simple and convenient, it gives a relatively poor approximation for E_2. It is instructive to consider the true nature of the stress and strain distributions in a fibre composite during this type of loading, which is depicted schematically in Fig. 3.1 (next to the '$E_2 = E_3$, Mixed' label). Regions of the matrix 'in series' with the fibres (close to them and in line along the loading direction) are subjected to a high stress similar to that carried by the reinforcement, whereas regions

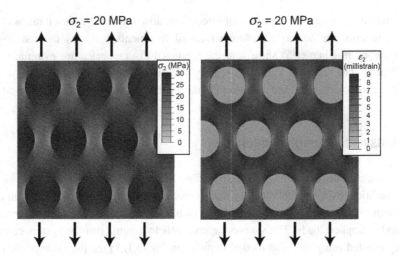

Fig. 3.2 Stress and strain fields in a transverse section of a polyester–40 vol% glass long fibre composite, with the fibres in a hexagonal array, subject to a transverse (vertical) stress of 20 MPa. The elastic properties used for fibre and matrix are those in Tables 2.1 and 2.2. (Courtesy of Dr M. Burley.)

'in parallel' with the fibres (adjacent laterally) are constrained to have the same (low) strain as the reinforcement and hence carry a low stress, as shown in the figure.

A degree of quantification of this picture is provided by Fig. 3.2, which shows stress and strain fields obtained using the finite element method (FEM). These relate to a long fibre composite under a transverse stress of 20 MPa. It may be noted that, while this is a moderate stress, the levels thus created at certain locations in the matrix (near to the interface) might be sufficient to cause cracking or debonding. It can also be seen that the distribution of matrix strain is very inhomogeneous, with the peak levels (approaching 1%) quite possibly being high enough to cause some kind of failure (at or near the interface). In practice, composite components are usually designed to ensure that transverse loads on unidirectionally reinforced material remain relatively low.

The non-uniform distribution of stress and strain during transverse loading means that the simple equal stress model is inadequate for many purposes. The slab model gives an underestimate of the Young's modulus, which constitutes a lower bound. Various empirical or semi-empirical expressions designed to give more accurate estimates have been proposed. The most successful of these is that due to Halpin and Tsai [1]. This is not based on rigorous elasticity theory, but broadly takes account of enhanced fibre load-bearing, relative to the equal stress assumption. Their expression for the transverse stiffness is

$$E_2 = \frac{E_\mathrm{m}(1 + \xi \eta f)}{(1 - \eta f)}, \quad \text{in which } \eta = \frac{\left(\dfrac{E_\mathrm{f}}{E_\mathrm{m}} - 1\right)}{\left(\dfrac{E_\mathrm{f}}{E_\mathrm{m}} + \xi\right)} \tag{3.7}$$

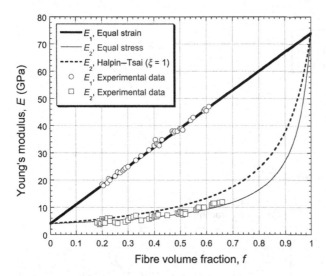

Fig. 3.3 Comparison between experimental data [2] for the axial and transverse Young's moduli, E_1 and E_2, for polyester–glass fibre composites and corresponding predictions from the equal strain model (Eqn (3.3)) for E_1 and the equal stress (Eqn (3.6)) and Halpin–Tsai (Eqn (3.7)), with $\xi = 1$) models for E_2. The experimental E_2 values have been affected by inelastic deformation of the matrix.

The value of ξ (xi) may be taken to be adjustable, but its magnitude is of the order of unity. The expression gives the correct values in the limits of $f = 0$ and $f = 1$ and in general gives good agreement with experiment over the complete range of fibre content.

A comparison is presented in Fig. 3.3 between the predictions of Eqns (3.3), (3.6) and (3.7) and experimental data for a glass fibre–polyester system. It is clear that the equal strain treatment (Eqn (3.3)) is in close agreement with data for the axial modulus. For the transverse modulus, the situation is less clear. Firstly, the experimental data show considerable scatter; some of the values actually lie below the equal stress prediction (Eqn (3.6)), which should constitute a lower bound. Secondly, many of the values appear to lie closer to the equal stress curve than to the Halpin–Tsai prediction, although this is less obvious for the high fibre contents. This behaviour is almost certainly the result of inelastic deformation of the matrix. These values were obtained by mechanical loading experiments in which relatively large stresses were present for appreciable times. (This is much less significant during axial loading, since the matrix stresses are so low.) Plastic deformation and creep may occur during a transverse test of this type, particularly with thermoplastic polymers, and this will lead to an underestimate of the true stiffness. In general, tests with stronger and more creep-resistant matrices, or under conditions where all the stresses are kept low and are of short duration (as with dynamic methods of stiffness measurement [3]), have confirmed that the transverse moduli of long fibre composites agree quite well with the Halpin–Tsai prediction (Eqn (3.7)).

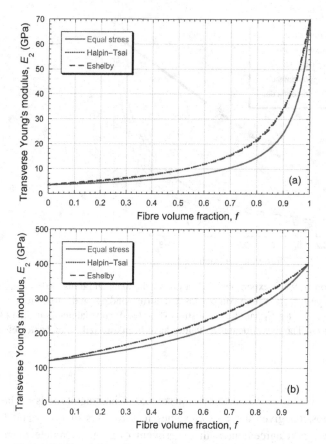

Fig. 3.4 Predicted dependence on fibre volume fraction of the transverse Young's moduli of continuous fibre composites, according to the equal stress (Eqn (3.6)), Halpin–Tsai, (Eqn (3.7)) and Eshelby models for (a) glass fibres in epoxy and (b) silicon carbide fibres in titanium.

Beyond these simple models for predicting the transverse modulus, there are power-ful, but complex, analytical tools such as the Eshelby equivalent homogeneous inclusion approach (see Chapter 5) and of course numerical techniques such as finite element modelling (FEM), which is now ubiquitous in many areas of science and engineering. The plots shown in Fig. 3.4 give an idea of the errors likely to be introduced in real cases by use of the equal stress expression, as compared with the Eshelby method, which is expected to be much more reliable. It can be seen that Eqn (3.6) gives a significant underestimate for both polymer matrix composites (PMCs) and metallic matrix com-posites (MMCs), which respectively have large and small fibre/matrix modulus ratios. The Halpin–Tsai expression (Eqn (3.7)), on the other hand, is fairly accurate.

In practice, the behaviour may be influenced by other factors, which are difficult to incorporate into simple models. These include the effects of fibre misalignment, elastic anisotropy of the fibre (or of the matrix – e.g. for a textured polycrystalline metal) or the early onset of an inelastic response. Nevertheless, it should be noted that, even in the absence of any such complications, use of the equal stress model introduces significant errors: this should be borne in mind, for example, if it is being used in laminate elasticity analysis (see Chapter 4).

3.3   Other Elastic Constants

3.3.1   Shear Moduli

The shear moduli of composites can be predicted in a similar way to the axial and transverse stiffnesses, using the slab model. This is done by evaluating the net shear strain induced when a shear stress is applied to the composite, in terms of the individual displacement contributions from the two constituents. It is important to understand the nomenclature convention that is used. A shear stress designated τ_{ij} $(i \neq j)$ refers to a stress acting in the i direction on the plane with a normal in the j direction. Similarly, a shear strain γ_{ij} is a rotation towards the i direction of the j axis. The shear modulus G_{ij} is the ratio of τ_{ij} over γ_{ij}. Since the composite body is not rotating, the condition $\tau_{ij} = \tau_{ji}$ must hold. In addition, $G_{ij} = G_{ji}$, so that $\gamma_{ij} = \gamma_{ji}$. Also, the 2 and 3 directions are equivalent in an aligned fibre composite, so it follows that there are two shear moduli, with $G_{12} = G_{21} = G_{13} = G_{31} \neq G_{23} = G_{32}$.

There are also two shear moduli for the slab model (Fig. 3.1), but these are unlikely to correspond closely with the values for the fibre composite. The stresses τ_{12} and τ_{21} are assumed to operate equally within both of the constituents. The derivation is similar to the equal stress treatment leading to Eqn (3.6) for transverse stiffness

$$\tau_{12} = \tau_{12f} = \gamma_{12f}G_f = \tau_{12m} = \gamma_{12m}G_m$$

where γ_{12f} and γ_{12m} are the individual shear strains in the two constituents. The total shear strain is found by summing the two contributions to the total shear displacement in the 1 direction

$$\gamma_{12} = \frac{(u_{1f} + u_{1m})}{f + (1-f)} = f\gamma_{12f} + (1-f)\gamma_{12m}$$

$$\therefore G_{12} = \frac{\tau_{12}}{\gamma_{12}} = \frac{\tau_{12f}}{f\gamma_{12f} + (1-f)\gamma_{12m}} = \left[\frac{f}{G_f} + \frac{(1-f)\gamma_{12m}}{\tau_{12f}}\right]^{-1}$$

i.e.
$$G_{12} = \left[\frac{f}{G_f} + \frac{(1-f)}{G_m}\right]^{-1} \tag{3.8}$$

The other shear modulus shown by the slab model, $G_{13} = G_{31}$ in Fig. 3.1, corresponds to an equal shear strain condition and is analogous to the axial tensile modulus case. It is readily shown that

$$G_{13} = fG_f + (1-f)G_m \tag{3.9}$$

which is similar to Eqn (3.3). It may be noted that neither the equal stress condition nor the equal strain condition are close to the situation during shearing of the fibre composite, in which the strain partitions unevenly within the matrix. Therefore neither of the above equations is expected to be very reliable, particularly the equal strain expression.

It is not obvious just how poor the approximation represented by Eqn (3.8) is likely to be, nor even which of the two actual shear moduli it will approach more closely. In fact, more rigorous methods predict that the values of G_{12} and G_{23} are rather close to each other, with G_{12} slightly larger in magnitude. Eqn (3.8) gives a significant underestimate relative to both of them, while Eqn (3.9) is a gross overestimate. In view of this, the semi-empirical expressions of Halpin and Tsai [1], mentioned in the previous section, are frequently employed. In this case, the appropriate equation is:

$$G_{12} = \frac{G_m(1 + \xi \eta f)}{(1 - \eta f)}, \quad \text{in which } \eta = \frac{\left(\dfrac{G_f}{G_m} - 1\right)}{\left(\dfrac{G_f}{G_m} + \xi\right)} \tag{3.10}$$

and the parameter ξ is again often taken to have a value of around unity. This has been done for the curves in Fig. 3.5, which shows comparisons between the predictions of Eqn (3.10) and those of the equal stress (Eqn (3.8)) and Eshelby models for both PMCs and MMCs. It can be seen that the Halpin–Tsai expression represents a fairly good approximation to the axial shear modulus (G_{12}). A striking feature of both the transverse and the shear moduli for polymer matrix composites (Figs 3.4(a) and 3.5(a)) is that they are close to the matrix values up to relatively high fibre volume fractions, although in both cases the true modulus is not as low as the prediction of the equal stress model.

3.3.2   Poisson Ratios

The Poisson ratio ν_{ij} refers to loading in the i direction and is defined as

$$\nu_{ij} = -\frac{\varepsilon_j}{\varepsilon_i} \tag{3.11}$$

For a uniaxially aligned fibre composite, there are three different Poisson ratios, as illustrated in Fig. 3.6.

This brings the total number of elastic constants identified so far for this type of material to seven. However, as outlined in Chapter 4, only five independent elastic constants are needed to describe the behaviour of such a transversely isotropic material. It follows that there must be relationships between these seven values. One of these is the so-called 'reciprocal relationship', which is derived in Chapter 4 and may be written

$$\frac{\nu_{12}}{E_1} = \frac{\nu_{21}}{E_2} \tag{3.12}$$

Estimation of the ν_{ij} values using the slab model presents difficulties because of the greater degree to which the Poisson strains of the two constituents must match when compared with the real composite. The effect of this is that, although three Poisson

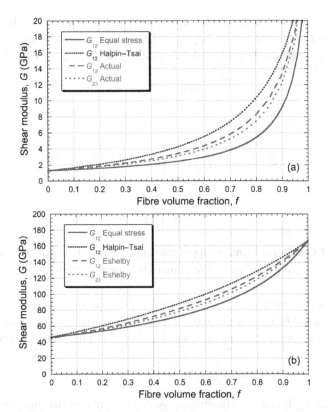

Fig. 3.5 Predicted dependence on fibre volume fraction of the shear moduli of continuous fibre composites, according to the equal stress equation for G_{12} (Eqn (3.8)), the Halpin–Tsai expression for G_{12} (Eqn (3.10)) and the Eshelby model (G_{12} and G_{23}). Data are shown for (a) glass fibres in epoxy and (b) silicon carbide fibres in titanium.

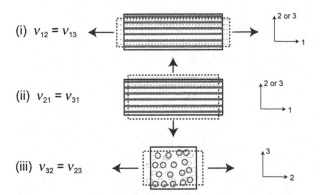

Fig. 3.6 Schematic illustration of how the three Poisson ratios are defined for a uniaxially aligned fibre composite.

ratios can be identified for the slab model, a meaningful calculation can only be done for the equal imposed strain case, giving v_{12} – see Fig. 3.6. In this case, the Poisson strains for the two constituents can be evaluated independently and summed. Thus

$$\varepsilon_{2f} = -v_f \varepsilon_{1f} = -v_f \frac{\sigma_{1f}}{E_f} \quad \text{and} \quad \varepsilon_{2m} = -v_m \varepsilon_{1m} = -v_m \frac{\sigma_{1m}}{E_m}$$

so that

$$\varepsilon_2 = -\left[\frac{f v_f \sigma_{1f}}{E_f} + \frac{(1-f) v_m \sigma_{1m}}{E_m} \right] = -\varepsilon_1 [f v_f + (1-f) v_m]$$

and

$$v_{12} = -\frac{\varepsilon_2}{\varepsilon_1} = f v_f + (1-f) v_m \tag{3.13}$$

A simple rule of mixtures is therefore applicable and, because the equal strain assumption is accurate for axial loading of the composite, this is expected to be a reliable prediction.

In fact, simple expressions can also be derived to give fairly realistic predictions for the other two ratios. The ratio of the axial strain to the transverse strain, when loading transversely, v_{21}, is readily obtained from the above reciprocal relationship (Eqn (3.12))

$$v_{21} = \frac{E_2}{E_1} [f v_f + (1-f) v_m] \tag{3.14}$$

This is lower than v_{12} because, on loading transversely, the fibres offer strong resistance to axial straining. This leads to pronounced straining in the other transverse direction, so that v_{23} is expected to be high. An expression for v_{23} can be obtained by considering the overall volume change experienced by the material [4]

$$\Delta = \varepsilon_1 + \varepsilon_2 + \varepsilon_3 = \frac{\sigma_H}{K} \tag{3.15}$$

in which σ_H is the applied hydrostatic stress and K is the bulk modulus of the composite. In this case, a single stress, σ_2, is being applied, so that

$$\sigma_H \left(= \frac{\sigma_1 + \sigma_2 + \sigma_3}{3} \right) = \frac{\sigma_2}{3}$$

It follows that

$$\varepsilon_3 = \frac{\sigma_2}{3K} - \varepsilon_1 - \varepsilon_2$$

leading to

$$v_{23} = -\frac{\varepsilon_3}{\varepsilon_2} = -\frac{\sigma_2}{3K\varepsilon_2} + \frac{\varepsilon_1}{\varepsilon_2} + 1$$

so that

$$v_{23} = 1 - v_{21} - \frac{E_2}{3K} \tag{3.16}$$

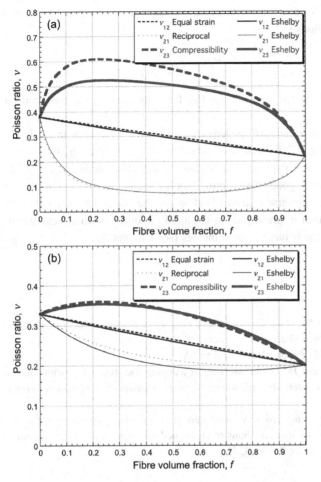

Fig. 3.7 Predicted dependence on fibre volume fraction of the three Poisson ratios of long fibre composites, according to the equal strain condition (Eqn (3.13)) for v_{12}, the reciprocal relation (Eqn (3.14)) for v_{21}, the compressibility expression (Eqn (3.16)) for v_{23} and the Eshelby model (Chapter 5) for all three. Predictions are shown for (a) glass fibres in epoxy and (b) silicon carbide fibres in titanium.

The bulk modulus of the composite can be estimated using an equal stress assumption, which should be quite accurate in this case, so that

$$\sigma_H = \Delta_f K_f = \Delta_m K_m$$

and

$$\Delta = f\Delta_f + (1-f)\Delta_m$$

giving

$$K = \frac{\sigma_H}{\Delta} = \left[\frac{f}{K_f} + \frac{(1-f)}{K_m}\right]^{-1} \tag{3.17}$$

The bulk moduli of the constituents are related to their other elastic constants by standard expressions such as

$$K_f = \frac{E_f}{3(1 - 2\nu_f)}$$

so it is straightforward to evaluate ν_{23} from the standard elastic constants of the constituents.

The accuracy of Eqn (3.16) is determined largely by the error in E_2. In the comparisons shown in Fig. 3.7, the Halpin–Tsai values of E_2, predicted by Eqn (3.7), were used to obtain values for ν_{21} and ν_{23}. It can be seen that agreement with the Eshelby predictions is fairly good. These plots convey a clear impression of the pronounced tendency under transverse (tensile) loading for the composite to contract in the other transverse direction in preference to the axial direction. Such effects are of particular significance for the behaviour of laminates (see Chapter 4).

References

1. Halpin, JC and SW Tsai, *Environmental Factors in Composite Design*, Report AFML-TR-67–423, Air Force Materials Laboratory, 1967.
2. Brintrup, A, PhD thesis, RWTH Aachen, 1975.
3. Wolfenden, A and JM Wolla, Mechanical damping and dynamic modulus measurements in alumina and tungsten fibre reinforced aluminium. *Journal of Materials Science* 1989; **24**: 3205–3212.
4. Clyne, TW, A compressibility-based derivation of simple expressions for the transverse Poisson's ratio and shear modulus of an aligned long fibre composite. *Journal of Materials Science Letters* 1990; **9**: 336–339.

4 Tensor Analysis of Anisotropic Materials and the Elastic Deformation of Laminae

In the previous chapter, it was shown that an aligned composite is usually stiff along the fibre axis, but much more compliant in the transverse directions. Sometimes, this is all that is required. For example, in a slender beam, such as a fishing rod, the loading is often predominantly axial and transverse or shear stiffness are not important. However, there are many applications in which loading is distributed within a plane: these range from panels of various types to cylindrical pressure vessels. Equal stiffness in all directions within a plane can be produced using a planar random assembly of fibres. This is the basis of chopped strand mat. However, demanding applications require material with higher fibre volume fractions than can readily be achieved in a planar random (or woven) array. The approach adopted is to stack and bond together a sequence of thin 'plies' or 'laminae', each composed of long fibres aligned in a single direction, into a laminate. It is important to be able to predict how such a construction responds to an applied load. In this chapter, attention is concentrated on the stress distributions that are created and the elastic deformations that result. This involves consideration of how a single lamina deforms on loading at an arbitrary angle to the fibre direction. A summary is given first of some matrix algebra and analysis tools used in elasticity theory.

4.1 Tensor Representation of Elastic Deformation

4.1.1 Stress and Strain as Second-Rank Tensors

While stress and strain can sometimes be handled as if they were simple scalars (i.e. numbers with no directions associated with them), more rigorous treatment is often required, particularly when studying composites. Stress and strain are in fact second-rank tensors. The word 'tensor' comes from the Latin for 'extension',[1] and tensors are strongly associated with mechanics, although they are also widely used in other branches of science and engineering. The utility of tensors is mainly concerned with treating differences in the response or characteristics of a material in different directions within it. Their usage is thus particularly required when treating *anisotropic* materials – and composites are often highly anisotropic. However, even for isotropic materials,

---

[1] The term is also used in Robert Hooke's famous expression '*Ut tensio, sic vis* – As the extension, so the force'. This is the origin of the simple form of Hooke's law, relating the normal stress on a material to the resultant normal strain.

tensor analysis is necessary, or at least very helpful, when treating many types of mechanical loading.

A tensor is an n-dimensional array of values, where n is the 'rank' of the tensor. The simplest type is thus a tensor of zeroth rank, which is a *scalar* – i.e. a single numerical value. Properties like temperature and density are scalars. They are not associated with any particular direction in the material concerned, and the variable does not require any associated index (suffix[2]). A first-rank tensor is a *vector*. This is a 1D array of values. There are normally three values in the array, each corresponding to one of three (orthogonal) directions. Each value has a single suffix, specifying the direction concerned. These suffixes are commonly numerical (1, 2 and 3), although sometimes other nomenclature, such as x, y and z or r, θ and z, may be used. Force and velocity are examples of vectors. The components of a vector can thus be written in a form such as

$$\mathbf{F} = F_i = [F_1 \ F_2 \ F_3] \qquad (4.1)$$

with each of the suffixes (1, 2 and 3) referring to a specific direction, such as x, y and z.

There are other variables, including *stress*, for which each component requires the specification of two directions, rather than one, so that two suffixes are needed. In the case of stress, these two suffixes specify, first, the direction in which a force is being applied and, second, the normal of the plane on which the force is acting. Stress is thus a *second-rank tensor* and the components form a 2D array of values.

$$\sigma_{ij} = \begin{bmatrix} \sigma_{11} & \sigma_{12} & \sigma_{13} \\ \sigma_{21} & \sigma_{22} & \sigma_{23} \\ \sigma_{31} & \sigma_{32} & \sigma_{33} \end{bmatrix} \qquad (4.2)$$

When the suffixes i and j are the same, the force acts parallel to the plane normal, and so the component concerned is a *normal stress*. When they are different, it is a *shear stress* (and sometimes the symbol τ is used instead of σ for such components).

Some of the stresses that could act on a body are depicted in Fig. 4.1. Provided that the body is in static equilibrium, which is commonly assumed, then the normal forces acting on opposite faces so as to generate a normal stress (e.g. σ_{33} in Fig. 4.1) must be equal in magnitude and anti-parallel in direction. (If this were not the case, then the body would *translate*.) For shear stresses, a further condition applies. Not only must the two forces generating the σ_{23} stress (see Fig. 4.1) be equal and opposite, but the magnitude of the σ_{32} stress must be equal to that of the σ_{23} stress. (If this were not the case, then the body would *rotate*.) Shear stresses thus act in pairs. This applies to all shear stresses, so that $\sigma_{ij} = \sigma_{ji}$, and the tensor represented in Eqn (4.2) must be *symmetrical*.

4.1.2 Transformation of Axes

It follows that there are just six independent components in a general stress state – three normal stresses and three shear stresses. Their magnitude will, of course, depend on the

---

[2] The term 'suffix' is in common use for these indices, although they are employed as subscripts.

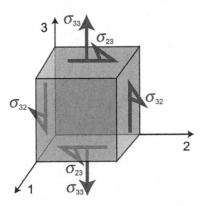

Fig. 4.1 Illustration of the nomenclature of stresses acting on a body.

directions of the axes chosen to provide the frame of reference. However, the state of stress itself will clearly be unaffected if we choose an alternative frame of reference. Any tensor can be *transformed* so as to be referred to a new set of axes, provided the orientation of these with respect to the original set is specified. Furthermore, any stress state can be expressed solely in terms of normal stresses (i.e. all shear stresses are zero), provided that a certain set of axes is chosen. Partly because it is often helpful to express a stress state in terms of this unique set of normal stresses, the procedures for transforming tensors are important. They are illustrated first for a vector (force) and then for a stress.

Consider a vector (a force, for example), F (= $[0, F_2, F_3]$), with components that are referred to the axis set (1, 2, 3). A specific reorientation of this set of axes is now introduced, namely a rotation by an angle ϕ about the 1-axis, to create a new axis set (1′, 2′, 3′) – see Fig. 4.2. In this case, the new 1′-axis coincides with the old 1-axis, but the 2′- and 3′-axes have been rotated with respect to the 2- and 3-axes. The values of $F_{2'}$ and $F_{3'}$ are found by resolving the components F_2 and F_3 onto the 2′- and 3′-axes and adding these resolved components together

$$F_{2'} = F_2 \cos(2' \wedge 2) + F_3 \cos(2' \wedge 3)$$
$$F_{3'} = F_2 \cos(3' \wedge 2) + F_3 \cos(3' \wedge 3)$$

(4.3)

where the symbolism $(x \wedge y)$ represents the angle between the x and y axes. In terms of the angle ϕ, these two equations can be written as

$$F_{2'} = F_2 \cos(-\phi) + F_3 \cos(-90 - \phi) = cF_2 - sF_3$$
$$F_{3'} = F_2 \cos(90 - \phi) + F_3 \cos(-\phi) = sF_2 + cF_3$$

(4.4)

in which c and s represent $\cos\phi$ and $\sin\phi$ respectively.

Clearly, these cosines (of angles between new and old axes) are central to such transformations. They are commonly termed *direction cosines* and represented by a_{ij}, which is conventionally the cosine of the angle between the new i direction (= i') and the

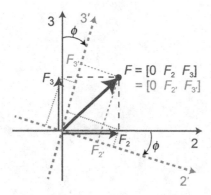

Fig. 4.2 Rotation, in the 2–3 plane of the axes forming the reference frame for a vector F.

old j direction. Of course, the rationale can be extended to cases in which all three axes have been reoriented, leading to the following set of equations

$$F_{1'} = a_{11}F_1 + a_{12}F_2 + a_{13}F_3$$
$$F_{2'} = a_{21}F_1 + a_{22}F_2 + a_{23}F_3 \tag{4.5}$$
$$F_{3'} = a_{31}F_1 + a_{32}F_2 + a_{33}F_3$$

It can be seen that the direction cosines form a matrix and this set of equations can be written more compactly in matrix form

$$\begin{vmatrix} F_{1'} \\ F_{2'} \\ F_{3'} \end{vmatrix} = |T| \begin{vmatrix} F_1 \\ F_2 \\ F_3 \end{vmatrix} \tag{4.6}$$

in which the *transformation matrix* is given by

$$|T| = \begin{vmatrix} a_{11} & a_{12} & a_{13} \\ a_{21} & a_{22} & a_{23} \\ a_{31} & a_{32} & a_{33} \end{vmatrix} \tag{4.7}$$

Sets of equations such as Eqn (4.6) can be written even more concisely by using the *Einstein summation convention*. This states that, when a suffix occurs twice in the same term, then this indicates that summation should be carried out with respect to that term. For example, in the equation

$$F_{i'} = a_{ij}F_j \tag{4.8}$$

j is a *dummy suffix*, which is to be summed (from 1 to 3). The i suffix, on the other hand, is a *free suffix*, which can be given any chosen value. For example, Eqn (4.8) could be used to create the equation

$$F_{1'} = a_{11}F_1 + a_{12}F_2 + a_{13}F_3 \tag{4.9}$$

and also the two other equations, corresponding to i being equal to 2 or 3.

4.1.3 Transformation of Second-Rank Tensors

Extension of the above treatment to second-rank tensors, such as stress, follows quite logically. However, for a stress we are concerned, not just with resolving three single components into new directions (e.g. F_1, F_2 and F_3 in the case of a force), but rather we need to take into account the fact that both the force and the area on which it is acting will change when we refer them to different axes. In other words, the operation we carried out with respect to a single suffix in the case of force, F_i, has to be implemented with respect to two suffixes for a stress, σ_{ij}. Each stress component thus needs to be multiplied by two direction cosines, rather than one. The set of equations corresponding to Eqn (4.8) can thus be written

$$\sigma'_{ij} = a_{ik} a_{jl} \sigma_{kl} \tag{4.10}$$

in which the prime notation is now applied to the symbol, rather than the individual suffixes, to denote the transformed version. Eqn (4.10) can be expanded into nine equations, each giving one of the nine components of the stress tensor when referred to the new set of axes (although only six are required, since the stress tensor is always symmetrical). Both k and l are dummy suffixes in this equation (since they are repeated), while i and j are free suffixes. The first equation of the set, corresponding to $i = 1$ and $j = 1$, can thus be expanded to

$$\sigma'_{11} = a_{11} a_{11} \sigma_{11} + a_{11} a_{12} \sigma_{12} + a_{11} a_{13} \sigma_{13}$$
$$+ a_{12} a_{11} \sigma_{21} + a_{12} a_{12} \sigma_{22} + a_{12} a_{13} \sigma_{23}$$
$$+ a_{13} a_{11} \sigma_{31} + a_{13} a_{12} \sigma_{32} + a_{13} a_{13} \sigma_{33}$$

while that corresponding to $i = 3$ and $j = 2$ is given by

$$\sigma'_{32} = a_{31} a_{21} \sigma_{11} + a_{31} a_{22} \sigma_{12} + a_{31} a_{23} \sigma_{13}$$
$$+ a_{32} a_{21} \sigma_{21} + a_{32} a_{22} \sigma_{22} + a_{32} a_{23} \sigma_{23}$$
$$+ a_{33} a_{21} \sigma_{31} + a_{33} a_{22} \sigma_{32} + a_{33} a_{23} \sigma_{33}$$

One motivation for treating stresses within this mathematical framework is that it facilitates identification of the *principal stresses*. These are the normal stresses acting on the *principal planes*, which are the planes on which there are no shear stresses. (These principal stresses are the *eigenvalues* of the stress tensor.) Any state of stress can thus be transformed such that it can be expressed in the form

$$\sigma_{ij} = \begin{bmatrix} \sigma_1 & 0 & 0 \\ 0 & \sigma_2 & 0 \\ 0 & 0 & \sigma_3 \end{bmatrix} \tag{4.11}$$

with the single suffix being commonly used to denote a principal stress. (It is important to recognise, however, that these are still second-rank tensors, and should thus, strictly speaking, always have two suffixes.) Obtaining these principle stresses, for a general 3D stress state, requires *diagonalising* of the stress tensor. This is done by setting the determinant of the coefficients equal to zero. The solutions for the principal stresses are thus found from

$$\begin{vmatrix} \sigma_{11} - \lambda & \sigma_{12} & \sigma_{13} \\ \sigma_{21} & \sigma_{22} - \lambda & \sigma_{23} \\ \sigma_{31} & \sigma_{32} & \sigma_{33} - \lambda \end{vmatrix} = 0 \tag{4.12}$$

This is a cubic equation in λ, the roots of which are the principal stresses. It is often termed the *secular equation*. Expanding the terms in Eqn (4.12) leads to

$$\lambda^3 - I_1\lambda^2 + I_2\lambda - I_3 = 0 \tag{4.13}$$

in which the coefficients (termed the *invariants*, since they do not vary as the axes are changed) are given by

$$I_1 = \sigma_{11} + \sigma_{22} + \sigma_{33}$$
$$I_2 = \sigma_{11}\sigma_{22} + \sigma_{22}\sigma_{33} + \sigma_{33}\sigma_{11} - \sigma_{12}^2 - \sigma_{23}^2 - \sigma_{31}^2$$
$$I_3 = \sigma_{11}\sigma_{22}\sigma_{33} + 2\sigma_{12}\sigma_{23}\sigma_{31} - \sigma_{11}\sigma_{23}^2 - \sigma_{22}\sigma_{13}^2 - \sigma_{33}\sigma_{12}^2 \tag{4.14}$$

4.1.4   Use of Mohr's Circle

It is therefore fairly straightforward, if a little cumbersome, to find the principal stresses, and the orientation of the (normals of the) principal planes, for a 3D stress state specified with respect to an arbitrary set of axes. It just requires a cubic equation to be solved. However, in practice it is common to know one principle plane (direction), but to be interested in finding the principal directions within that plane, or in establishing how the normal and shear stresses vary with direction in that plane. Provided the plane concerned is a principal one, this problem reduces to solving a quadratic equation, rather than a cubic. It can therefore be tackled via a geometrical construction, equivalent to using some simple trigonometry. This is the basis of Mohr's circle, which was proposed in 1892 by Christian Otto Mohr, a German civil engineer who became a professor of mechanics in Dresden.

Provided the 1–2 plane is a principal plane, the appropriate version of Eqn (4.12) is

$$\begin{vmatrix} \sigma_{11} - \lambda & \sigma_{12} & 0 \\ \sigma_{21} & \sigma_{22} - \lambda & 0 \\ 0 & 0 & \sigma_{33} - \lambda \end{vmatrix} = 0$$

and one principal stress is clearly σ_{33} (= σ_3). Since $\sigma_{12} = \sigma_{21}$; the secular equation reduces to

$$(\sigma_{11} - \lambda)(\sigma_{22} - \lambda) - \sigma_{12}^2 = 0$$
$$\therefore \lambda^2 - (\sigma_{11} + \sigma_{22})\lambda + (\sigma_{11}\sigma_{22} - \sigma_{12}^2) = 0$$

This has the solution

$$\sigma_1, \sigma_2 = \left(\frac{\sigma_{11} + \sigma_{22}}{2}\right) \pm \sqrt{\left(\frac{\sigma_{11} - \sigma_{22}}{2}\right)^2 + \sigma_{12}^2} \tag{4.15}$$

It is also possible to start with the tensor in diagonalised form and then transform it to give σ_{11}', σ_{22}' and σ_{12}', for a specified orientation

$$\begin{vmatrix} \sigma_{11}' & \sigma_{12}' & 0 \\ \sigma_{12}' & \sigma_{22}' & 0 \\ 0 & 0 & \sigma_3 \end{vmatrix} = |T| \begin{vmatrix} \sigma_1 & 0 & 0 \\ 0 & \sigma_2 & 0 \\ 0 & 0 & \sigma_3 \end{vmatrix}$$

Since we are rotating about a principal axis, the form of Eqn (4.10) applicable in this case reduces to

$$\sigma_{11}' = a_{11}a_{11}\sigma_1 + a_{12}a_{12}\sigma_2 + a_{13}a_{13}\sigma_3$$
$$\sigma_{22}' = a_{21}a_{21}\sigma_1 + a_{22}a_{22}\sigma_2 + a_{23}a_{23}\sigma_3$$
$$\sigma_{12}' = a_{11}a_{21}\sigma_1 + a_{12}a_{22}\sigma_2 + a_{13}a_{23}\sigma_3$$

The set of direction cosines applicable here can be written as

$$\begin{vmatrix} a_{11} & a_{12} & a_{13} \\ a_{21} & a_{22} & a_{23} \\ a_{31} & a_{32} & a_{33} \end{vmatrix} = \begin{vmatrix} \cos\phi & -\sin\phi & 0 \\ \sin\phi & \cos\phi & 0 \\ 0 & 0 & 1 \end{vmatrix}$$

so it follows that these stresses are given by

$$\sigma_{11}' = \cos^2\phi\,\sigma_1 + \sin^2\phi\,\sigma_2$$
$$\sigma_{22}' = \sin^2\phi\,\sigma_1 + \cos^2\phi\,\sigma_2$$
$$\sigma_{12}' = \sin\phi\cos\phi\,\sigma_1 - \sin\phi\cos\phi\,\sigma_2$$

These equations can also be written in a form involving 2ϕ, rather than ϕ

$$\sigma_{11}' = \left(\frac{\sigma_1 + \sigma_2}{2}\right) + \left(\frac{\sigma_1 - \sigma_2}{2}\right)\cos 2\phi$$
$$\sigma_{22}' = \left(\frac{\sigma_1 + \sigma_2}{2}\right) - \left(\frac{\sigma_1 - \sigma_2}{2}\right)\cos 2\phi \tag{4.16}$$
$$\sigma_{12}' = \left(\frac{\sigma_1 - \sigma_2}{2}\right)\sin 2\phi$$

It can now be seen how these equations can be solved using a geometrical construction. These expressions (i.e. the normal and shear stresses on a plane rotated by an angle ϕ from that on which σ_1 acts) are given by the coordinates of a point rotated by 2ϕ around the circumference of a circle centred at the mean of the two principal stresses, and with a radius equal to half their difference. This is illustrated in Fig. 4.3. This construction provides a convenient method of calculating the stresses acting on

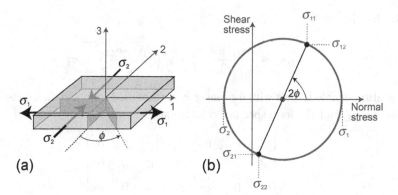

Fig. 4.3 Mohr's circle construction for a principal plane (1–2 plane), showing (a) principal stresses in the plane (σ_1 and σ_2, for a case in which σ_1 is tensile and σ_2 compressive); and (b) the corresponding Mohr's circle, giving, in addition to the principal stresses, the normal and shear stresses acting on a plane rotated about the 3-axis by an angle ϕ from that on which σ_1 acts.

particular planes, establishing principal stresses and their orientations, finding the planes on which peak shear stresses[3] operate, etc.

4.1.5   Representation of Strain

The application of a set of stresses causes all points in the body to be displaced, relative to their initial locations, so that their coordinates, in a given frame of reference, will change. Provided that the body is macroscopically homogeneous, these displacements can be related to the stress state via a description of the elastic properties of the material, which may, of course, be anisotropic. The *relative displacement tensor* (sometimes, rather confusingly, called the *deformation tensor*) indicates how any point in the body becomes displaced. It is a second-rank tensor, with the first suffix representing the direction in which the displacement has occurred and the second one the reference direction. This is illustrated in Fig. 4.4. All of the terms are (dimensionless) ratios of two distances. The shear terms (e_{ij}, with $i \neq j$) can also be considered as angles – since they are small, they are approximately equal to their tangents. Care is needed in identifying the *strain tensor*, since an applied set of stresses will in general create some *rigid body rotation*, as well as genuine deformation (shape change) of the body. The strain tensor reflects only the shape change.

   The sign of a shear component of the relative displacement tensor is taken as positive when the positive axis is rotated towards the positive direction of the other axis. For example, in Fig. 4.4(b) the e_{23} component would be positive. Separation of the strain

---

[3] Confusion occasionally arises about the sign of shear stresses (and strains), and it is sometimes implied that the Mohr circle has positive and negative sides in the vertical direction. However, there is no need for this, since the sign of a shear stress (or strain) is solely an issue of convention, and has no physical significance (unlike the sign of a normal stress or strain). Since both stress and strain tensors are symmetric, it is simplest to take all shear ($i \neq j$) terms as always positive. The only possible issue is that a consistent convention may be needed regarding the sense of rotation in Mohr space and physical space.

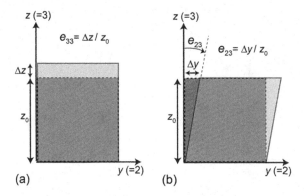

Fig. 4.4 Illustration, in the y–z (2–3) plane of how the terms of the relative displacement tensor are defined, showing (a) a normal term, e_{33}, and (b) a shear term, e_{23}.

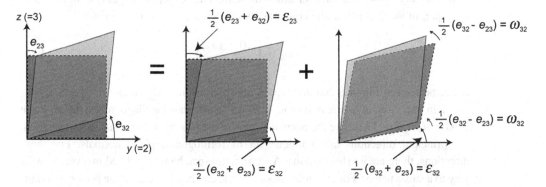

Fig. 4.5 Illustration of how, in the 2–3 plane, the relative displacements e_{23} and e_{32} can be represented as the sum of a strain, ε_{23} ($=\varepsilon_{32}$) and a rigid body rotation, ω_{23} ($=-\omega_{32}$).

from the rigid body rotation is illustrated in Fig. 4.5 for shear deformation in the y–z (2–3) plane. This separation can be expressed more generally. Any second-rank tensor can be separated into symmetrical and anti-symmetrical components. For the relative displacement tensor, this can be written as

$$e_{ij} = \varepsilon_{ij} + \omega_{ij} = \frac{1}{2}\left(e_{ij} + e_{ji}\right) + \frac{1}{2}\left(e_{ij} - e_{ji}\right) \qquad (4.17)$$

Since the *strain*, ε_{ij}, is a symmetrical tensor

$$\varepsilon_{ij} = \frac{1}{2}\left(e_{ij} + e_{ji}\right) = \varepsilon_{ji} \qquad (4.18)$$

whereas for ω_{ij}, the *rotation tensor*, which is anti-symmetrical

$$\omega_{ij} = \frac{1}{2}\left(e_{ij} - e_{ji}\right) = -\omega_{ji} \qquad (4.19)$$

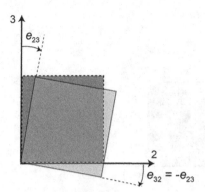

Fig. 4.6 Illustration of how an anti-symmetrical relative displacement tensor, in the 2–3 plane, represents only a rotation, with the body not being subject to any strain.

It is readily shown that such an anti-symmetrical tensor represents only a rotation. For example, in the 2–3 plane, the components of such a tensor can be written

$$\omega_{ij} = \begin{vmatrix} 0 & e_{23} \\ -e_{23} & 0 \end{vmatrix} \tag{4.20}$$

It can be seen in Fig. 4.6 that this represents solely rigid body rotation.

As in the case of a stress tensor, a strain tensor can be diagonalised to give the *principal strains*. These are the normal strains in the principal directions. As with stress, the principal directions are a unique set of (orthogonal) plane normals. For these directions, there are no shear strains. A strain tensor can be manipulated in a very similar way to a stress tensor. For example, Mohr's circle can be used to find the principal strains or to facilitate other calculations or visualisations concerning the strain.

A point worth noting at this stage is that the strain tensor can itself be divided into two components – one representing the change in volume of the body and the other the change in its shape. These are commonly termed the *hydrostatic* and *deviatoric* components. Consider a cube subjected to principal strains ε_1, ε_2 and ε_3. The change in volume, often called the *dilation*, can be written as

$$\Delta = (1 + \varepsilon_1)(1 + \varepsilon_2)(1 + \varepsilon_3) - 1 \approx \varepsilon_1 + \varepsilon_2 + \varepsilon_3 \tag{4.21}$$

In fact, we can write $\Delta = \varepsilon_{ii}$ for any strain tensor, since the sum of the diagonal terms of a second-rank tensor is an invariant, and hence independent of the reference axes – see Eqn (4.14). The hydrostatic strain tensor has one-third of the dilation for all three of the normal terms, and no shear terms. The residual part is the deviatoric component

$$\begin{vmatrix} \varepsilon_{11} & \varepsilon_{12} & \varepsilon_{13} \\ \varepsilon_{21} & \varepsilon_{22} & \varepsilon_{23} \\ \varepsilon_{31} & \varepsilon_{32} & \varepsilon_{33} \end{vmatrix} = \begin{vmatrix} \frac{1}{3}\Delta & 0 & 0 \\ 0 & \frac{1}{3}\Delta & 0 \\ 0 & 0 & \frac{1}{3}\Delta \end{vmatrix} + \begin{vmatrix} \varepsilon_{11} - \frac{1}{3}\Delta & \varepsilon_{12} & \varepsilon_{13} \\ \varepsilon_{21} & \varepsilon_{22} - \frac{1}{3}\Delta & \varepsilon_{23} \\ \varepsilon_{31} & \varepsilon_{32} & \varepsilon_{33} - \frac{1}{3}\Delta \end{vmatrix} \tag{4.22}$$

4.2   Stress–Strain Relationships and Engineering Constants

4.2.1   Stiffness and Compliance Tensors

The tensors we have been treating so far have all been *field tensors*. These are imposed on bodies, or regions of space, in some way. Relationships between field tensors, for example between stress and resultant strain, depend on properties of the material concerned. These are characterised by *matter tensors*. Matter tensors reflect the symmetry exhibited by the material. The rank of a matter tensor is equal to the sum of the ranks of the two field tensors that it links.

The relationship between a stress (tensor) and the strain (tensor) generated by it can be written as

$$\sigma_{ij} = C_{ijkl}\varepsilon_{kl} \tag{4.23}$$

C_{ijkl} is the *stiffness tensor*. This is a fourth-rank tensor, with $3^4 (= 81)$ components. (The rank of a tensor is equal to the number of its suffixes, although there are some situations in which a convention may be used such that this is not the case – an example is provided by principal stresses and strains being given a single suffix.) Eqn (4.23) is a generalised expression of *Hooke's law*. It represents nine equations, generated according to the Einstein summation convention. For example, the first of these is

$$\begin{aligned}
\sigma_{11} = \ &C_{1111}\varepsilon_{11} + C_{1112}\varepsilon_{12} + C_{1113}\varepsilon_{13} \\
&+ C_{1121}\varepsilon_{21} + C_{1122}\varepsilon_{22} + C_{1123}\varepsilon_{23} \\
&+ C_{1131}\varepsilon_{31} + C_{1132}\varepsilon_{32} + C_{1133}\varepsilon_{33}
\end{aligned} \tag{4.24}$$

The stress–strain relationship can also be expressed in the inverse sense

$$\varepsilon_{ij} = S_{ijkl}\sigma_{kl} \tag{4.25}$$

in which S_{ijkl} is the *compliance tensor*. (The symbols conventionally used for stiffness and compliance are thus the reverse of the initial letters of these words.)

4.2.2   Relationship to Elastic Constants

The above expressions look a little cumbersome and daunting, but in practice the treatment can be simplified. The symmetry of stress and strain tensors when the body is in static equilibrium means that

$$C_{ijkl} = C_{ijlk} = C_{jikl} = C_{jilk} \tag{4.26}$$

reducing the number of independent components from 81 to 36. Furthermore, the symmetry exhibited by the material commonly results in further reductions in this number. In fact, for an isotropic material, it is reduced to just two, so the elastic behaviour of isotropic materials is fully specified by the values of two constants.

In terms of engineering constants, the relationship between (normal) stress and strain can be obtained by considering the application of a single stress σ_1, generating a strain ε_1, so that

$$\sigma_1 = E\varepsilon_1 \tag{4.27}$$

in which E is the *Young's modulus*. One might be tempted to deduce that E is given by C_{1111}. However, this is incorrect, since the application of a stress σ_1 generates not only a (direct) strain ε_1, but also *Poisson strains* ε_2 and ε_3 (see below), which would appear in the full (tensorial) equation for σ_1. Expressed in terms of compliance, however, the equation we need (from Eqn (4.25)) is

$$\varepsilon_1 = S_{1111}\sigma_1 \quad \therefore E = \frac{1}{S_{1111}} = \left(\frac{1}{S_{2222}} = \frac{1}{S_{3333}} \right) \tag{4.28}$$

The second elastic constant specified for an isotropic material is commonly the *Poisson ratio*, ν. This gives the transverse contraction strain that accompanies an axial extension strain. Again considering a single (normal) stress σ_1, generating principal strains ε_1, ε_2 and ε_3, the Poisson ratio is defined by

$$\nu = -\frac{\varepsilon_2}{\varepsilon_1} = -\frac{\varepsilon_3}{\varepsilon_1} \tag{4.29}$$

The tensorial expression for ε_2, with only σ_1 applied, is

$$\varepsilon_2 = S_{2211}\sigma_1 (= S_{1122}\sigma_1)$$

so that

$$\nu = \frac{-\varepsilon_2}{\varepsilon_1} = \frac{-S_{1122}\sigma_1}{(\sigma_1/E)} = -E S_{1122} \tag{4.30}$$

Since Poisson strains simply superimpose during multi-axial loading, we can write the following expressions for the (principal) strains arising from a set of (principal) stresses

$$\varepsilon_1 = \frac{(\sigma_1 - \nu(\sigma_2 + \sigma_3))}{E}$$

$$\varepsilon_2 = \frac{(\sigma_2 - \nu(\sigma_1 + \sigma_3))}{E} \tag{4.31}$$

$$\varepsilon_3 = \frac{(\sigma_3 - \nu(\sigma_1 + \sigma_2))}{E}$$

4.2.3 Engineering Shear Strains, Shear Modulus and Bulk Modulus

A minor complication arises when considering shear strains and the shear modulus. One might imagine that the *shear modulus*, G, would be defined as shear stress divided by shear strain, e.g. $\sigma_{12}/\varepsilon_{12}$. Unfortunately, $G \neq \sigma_{12}/\varepsilon_{12}$. The problem is associated with the difference between a *tensorial shear strain*, such as ε_{12}, and the corresponding *engineering shear strain*, γ_{12}. The shear modulus is conventionally defined by

$$G = \frac{\sigma_{12}}{\gamma_{12}} \tag{4.32}$$

in which the value of γ_{12} is simply the measured strain, with no account being taken of the fact that it really represents the sum of two shear strains, one being generated

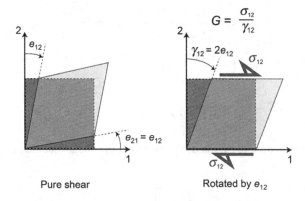

Pure shear   Rotated by e_{12}

Fig. 4.7 Illustration of how a pure shear deformation, in the 1–2 plane, when rotated about the 3-axis, can be viewed as a simple shear, and used to define the shear modulus, G, in terms of the engineering shear strain, γ_{12}. The presence of the other shear stress acting in this plane, σ_{21}, is neglected in this definition.

by each of the pair of shear stresses which is being applied. This is illustrated in Fig. 4.7.

The array of engineering strains (ε_{ii} and γ_{ij}) thus do not constitute a genuine tensor and cannot be transformed using the standard procedures for second-rank tensors. However, we can still define the shear modulus in terms of components of the compliance tensor. With only σ_{12} and σ_{21} acting, Eqn (4.25) gives

$$\varepsilon_{12} = S_{1212}\sigma_{12} + S_{1221}\sigma_{21} = 2S_{1212}\sigma_{12}$$

$$\therefore G = \frac{\sigma_{12}}{\gamma_{12}} = \frac{2\sigma_{12}}{\varepsilon_{12}} = \frac{1}{S_{1212}} \tag{4.33}$$

In fact, similar procedures can be used to establish all of the components of the compliance and stiffness tensors, in terms of (measured) engineering elastic constants. Of course, the procedure is more complex for anisotropic materials, for which there are more than two independent elastic constants, but it can still be carried out – see below.

It may finally be noted that the bulk modulus, K, is defined by

$$K = \frac{\sigma_H}{\Delta} = \frac{\frac{1}{3}(\sigma_1 + \sigma_2 + \sigma_3)}{(\varepsilon_1 + \varepsilon_2 + \varepsilon_3)} \tag{4.34}$$

where σ_H is the hydrostatic component of the stress state. The bulk modulus is a measure of the resistance of the material to volume change.

4.2.4 Relationships between Elastic Constants

Since only two elastic constants are required to fully define the behaviour of an isotropic material, it follows that there must be inter-relationships between the four that have been considered here. For example, adding up the three equations in Eqn (4.31) gives

$$\varepsilon_1 + \varepsilon_2 + \varepsilon_3 = \left(\frac{1-2v}{E}\right)(\sigma_1 + \sigma_2 + \sigma_3)$$

$$\text{i.e. } \Delta = \left(\frac{1-2v}{E}\right)3\sigma_H$$

$$\therefore K = \frac{\sigma_H}{\Delta} = \frac{E}{3(1-2v)} \tag{4.35}$$

Other relationships are commonly quoted; for example, the shear modulus is given by

$$G = \frac{E}{2(1+v)} \tag{4.36}$$

4.3   Off-Axis Elastic Constants of Laminae

4.3.1   Matrix Notation

Before specifically treating the case of a unidirectional composite, such as a single ply (lamina), it should be noted that it is common to use *matrix notation*, which involves representing the stiffness and compliance tensors as 2D (6×6) matrices, rather than 4D ($3 \times 3 \times 3 \times 3$) ones, so that

$$\sigma_p = C_{pq}\varepsilon_q \tag{4.37}$$

While this matrix does fully represent the elastic behaviour of the material, and is very convenient in terms of the ease of writing down the information, it is not in fact a tensor and it needs to be used rather carefully (particularly when transforming axes).

$$\begin{vmatrix} \sigma_1 \\ \sigma_2 \\ \sigma_3 \\ \tau_{23} \\ \tau_{31} \\ \tau_{12} \end{vmatrix} = \begin{vmatrix} C_{11} & C_{12} & C_{13} & C_{14} & C_{15} & C_{16} \\ C_{21} & C_{22} & C_{23} & C_{24} & C_{25} & C_{26} \\ C_{31} & C_{32} & C_{33} & C_{34} & C_{35} & C_{36} \\ C_{41} & C_{42} & C_{43} & C_{44} & C_{45} & C_{46} \\ C_{51} & C_{52} & C_{53} & C_{54} & C_{55} & C_{56} \\ C_{61} & C_{62} & C_{63} & C_{64} & C_{65} & C_{66} \end{vmatrix} \begin{vmatrix} \varepsilon_1 \\ \varepsilon_2 \\ \varepsilon_3 \\ \gamma_{23} \\ \gamma_{31} \\ \gamma_{12} \end{vmatrix} \tag{4.38}$$

This matrix is obtained from C_{ijkl} using the following conversion scheme

tensor	11	22	33	23, 32	13, 31	12, 21
matrix	1	2	3	4	5	6

It can be shown (using strain energy considerations) that C_{pq} is a *symmetric matrix* ($C_{pq} = C_{qp}$), so that, in the general case, it contains 21 independent constants.

The compliance tensor can also be converted to matrix notation in a similar way, so that Eqn (4.25) can be written

$$\varepsilon_p = S_{pq}\sigma_q \tag{4.39}$$

In this case, there are additional conversion factors, such that

$$S_{pq} = S_{ijkl} \qquad \text{when both } p \text{ and } q \text{ are } 1, 2 \text{ or } 3$$

$$S_{pq} = 2S_{ijkl} \qquad \text{when either } p \text{ or } q \text{ is } 4, 5 \text{ or } 6$$

$$S_{pq} = 4S_{ijkl} \qquad \text{when both } p \text{ and } q \text{ are } 4, 5 \text{ or } 6$$

These factors arise from the relation between engineering and tensorial shear strains.

4.3.2 Effect of Material Symmetry

Only in the case of a material exhibiting no symmetry at all (a triclinic single crystal) would the stiffness tensor actually contain 21 independent values. The introduction of symmetry leads to a reduction in this number. This can readily be seen by examination of the set of stress–strain equations. As outlined in Section 4.2.2, the number falls to two for isotropic materials, so that their elastic properties are fully defined by just two (engineering) elastic constants – the Young's modulus, E, and the Poisson ratio, v, are often used. The stiffness and compliance tensors also contain only two independent components. Materials exhibiting intermediate degrees of symmetry require a number between these two extremes, as illustrated in Fig. 4.8 [1] . This also shows how these constants are distributed between the elements of the stiffness (or compliance) matrix.

Taking the simple example of an isotropic material, the appropriate version of Eqn (4.39) is

$$\begin{vmatrix} \varepsilon_1 \\ \varepsilon_2 \\ \varepsilon_3 \\ \gamma_{23} \\ \gamma_{31} \\ \gamma_{12} \end{vmatrix} = \begin{vmatrix} S_{11} & S_{12} & S_{12} & 0 & 0 & 0 \\ S_{12} & S_{11} & S_{12} & 0 & 0 & 0 \\ S_{12} & S_{12} & S_{11} & 0 & 0 & 0 \\ 0 & 0 & 0 & 2(S_{11} - S_{12}) & 0 & 0 \\ 0 & 0 & 0 & 0 & 2(S_{11} - S_{12}) & 0 \\ 0 & 0 & 0 & 0 & 0 & 2(S_{11} - S_{12}) \end{vmatrix} \begin{vmatrix} \sigma_1 \\ \sigma_2 \\ \sigma_3 \\ \tau_{23} \\ \tau_{31} \\ \tau_{12} \end{vmatrix} \tag{4.40}$$

It follows that, with a single applied stress σ_1

$$\varepsilon_1 = S_{11}\sigma_1$$

$$\varepsilon_2 = \varepsilon_3 = S_{12}\sigma_1 \tag{4.41}$$

$$\gamma_{23} = \gamma_{31} = \gamma_{12} = 0$$

The elements of these matrices can always be expressed in terms of engineering elastic constants; in this case it is evident that

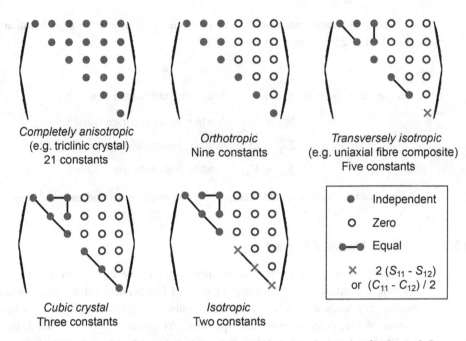

Fig. 4.8 Indication of the form of the S_{pq} and C_{pq} matrices (matrix notation for S_{ijkl} and C_{ijkl} tensors), for materials exhibiting different types of symmetry [1]. All of the matrices are symmetrical about the leading diagonal.

$$S_{11} = \frac{1}{E}, \qquad S_{12} = \frac{-\nu}{E} \tag{4.42}$$

It can also be seen in Fig. 4.8 that a transversely isotropic material, such as a uniaxial fibre composite, requires five elastic constants for full characterisation of its response. In Chapter 3, seven constants (E_1, E_2 ($=E_3$), G_{12} ($=G_{21}= G_{13}= G_{31}$), G_{23} ($=G_{32}$), ν_{12} ($=\nu_{13}$), ν_{21} ($=\nu_{31}$) and ν_{23} ($=\nu_{32}$)) were identified for such a composite. In fact, there are others, such as a bulk modulus, but defining any five of these will fix all others.

4.3.3   Off-Axis Loading of a Lamina

Certain assumptions are commonly made when treating the elastic behaviour of laminae (and laminates). The main ones are often termed the *Kirchoff assumptions*. These are: (1) a state of *plane stress* applies and (2) there are *no edge effects*. The first of these implies that stresses other than those in the plane of the lamina are negligible, so that $\sigma_3 = \tau_{23} = \tau_{31}$. This is expected to be a good approximation for a thin, isolated lamina. It may be less accurate for a laminate, in which through-thickness constraint effects can arise (see Section 5.2.3), but it is still expected to be acceptable for many purposes. A simplified form of Eqn (4.39) can now be used, focusing on in-plane stresses and strains[4] and considering initially the case of loading parallel or transverse to the fibre axis.

$$\begin{vmatrix} \varepsilon_1 \\ \varepsilon_2 \\ \gamma_{12} \end{vmatrix} = \begin{vmatrix} S_{11} & S_{12} & 0 \\ S_{12} & S_{22} & 0 \\ 0 & 0 & S_{66} \end{vmatrix} \begin{vmatrix} \sigma_1 \\ \sigma_2 \\ \tau_{12} \end{vmatrix} \tag{4.43}$$

It should be recalled here that, following the convention outlined in Chapter 3, the 1 direction is taken to be the fibre axis (and the 2 direction is normal to this, in the plane of the lamina). By inspection of the individual equations, it can be seen that

$$S_{11} = \frac{1}{E_1}, \qquad S_{12} = \frac{-\nu_{12}}{E_1} = \frac{-\nu_{21}}{E_2}$$

$$S_{22} = \frac{1}{E_2}, \qquad S_{66} = \frac{1}{G_{12}} \tag{4.44}$$

Similarly, the version of Eqn (4.37) that applies in this situation is

$$\begin{vmatrix} \sigma_1 \\ \sigma_2 \\ \tau_{12} \end{vmatrix} = \begin{vmatrix} C_{11} & C_{12} & 0 \\ C_{12} & C_{22} & 0 \\ 0 & 0 & C_{66} \end{vmatrix} \begin{vmatrix} \varepsilon_1 \\ \varepsilon_2 \\ \gamma_{12} \end{vmatrix} \tag{4.45}$$

and inspection of individual equations leads to

$$C_{11} = \frac{E_1}{1 - \nu_{12}\nu_{21}}, \qquad C_{12} = \frac{\nu_{12}E_2}{1 - \nu_{12}\nu_{21}} = \frac{\nu_{21}E_1}{1 - \nu_{12}\nu_{21}}$$

$$C_{22} = \frac{E_2}{1 - \nu_{12}\nu_{21}}, \qquad C_{66} = G_{12} \tag{4.46}$$

An immediate point to note about Eqns (4.43) and (4.45) is that there is no *interaction* between normal and shear behaviour; a normal stress gives rise only to normal strains and a shear stress produces only shear strains. However, this is not the case when the lamina is loaded in some arbitrary direction within the plane. The situation is illustrated in Fig. 4.9.

The first step in determining the strains within the lamina is to establish the induced stresses,[5] referred to the fibre axis (σ_1, σ_2 and τ_{12}), in terms of the externally applied stress system (σ_x, σ_y and τ_{xy}). This transformation of axes is carried out using Eqn (4.10), leading to equations such as

---

[4] It may be noted that |S| now contains four independent elastic constants. From Fig. 4.8, a total of three might have been expected, since both S_{13} (=S_{23}) and S_{44} (=S_{55}) have been removed by excluding the through-thickness direction. However, the relation $S_{66} = 2(S_{11} - S_{12})$, and the corresponding one for C_{66}, do not now apply, because the exclusion eliminates this symmetry from the lamina.

[5] Note that these stresses parallel and normal to the fibre axis will not now, in general, be the principal stresses in the system. There is thus potential for confusion in view of the common convention of denoting the principal stresses by σ_1, σ_2 and σ_3.

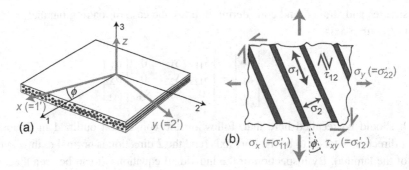

Fig. 4.9 (a) Relationship between the fibre-related axes in a lamina (1, 2 and 3) and the coordinate system (x, y and z) for an arbitrary in-plane set of applied stresses. (b) Illustration of how such an applied stress state σ'_{ij} (σ_x, σ_y and τ_{xy}) generates stresses in the fibre-related framework of σ_{ij} (σ_1, σ_2 and τ_{12}).

$$\sigma_{11} = a_{11}a_{11}\sigma'_{11} + a_{11}a_{12}\sigma'_{12}$$
$$+ \, a_{12}a_{11}\sigma'_{12} + a_{12}a_{12}\sigma'_{22}$$

which gives the normal stress parallel to the fibre direction, σ_{11} (often written as σ_1) in terms of the applied stresses σ'_{11} ($=\sigma_x$), σ'_{22} ($=\sigma_y$) and σ'_{12} ($=\tau_{xy}$). The angle ϕ is that between the fibre axis (1) and the stress axis (x). Referring to Fig. 4.9(a), these direction cosines take the values

$$a_{11} = \cos\phi \qquad\qquad a_{12} = \cos(90 - \phi) = \sin\phi$$
$$a_{21} = \cos(90 + \phi) = -\sin\phi \qquad a_{22} = \cos\phi$$

Carrying out this operation for all three stresses,

$$\begin{vmatrix} \sigma_1 \\ \sigma_2 \\ \tau_{12} \end{vmatrix} = |T| \begin{vmatrix} \sigma_x \\ \sigma_y \\ \tau_{xy} \end{vmatrix} \tag{4.47}$$

where

$$|T| = \begin{vmatrix} c^2 & s^2 & 2cs \\ s^2 & c^2 & -2cs \\ -cs & cs & c^2 - s^2 \end{vmatrix} \tag{4.48}$$

with $c = \cos\phi$ and $s = \sin\phi$. The same matrix can be used to transform tensorial strains, so that

$$\begin{vmatrix} \varepsilon_1 \\ \varepsilon_2 \\ \varepsilon_{12} \end{vmatrix} = |T| \begin{vmatrix} \varepsilon_x \\ \varepsilon_y \\ \varepsilon_{xy} \end{vmatrix} \tag{4.49}$$

However, to work in terms of engineering strains, using $\gamma_{xy} = 2\varepsilon_{xy}$, etc. (see Section 4.2.3), $|T|$ must be modified (by halving the elements t_{13} and t_{23} and doubling elements t_{31} and t_{32}) so as to give

$$
\begin{vmatrix} \varepsilon_1 \\ \varepsilon_2 \\ \gamma_{12} \end{vmatrix} = |T'| \begin{vmatrix} \varepsilon_x \\ \varepsilon_y \\ \gamma_{xy} \end{vmatrix} \tag{4.50}
$$

in which

$$
|T'| = \begin{vmatrix} c^2 & s^2 & cs \\ s^2 & c^2 & -cs \\ -2cs & 2cs & c^2 - s^2 \end{vmatrix} \tag{4.51}
$$

The procedure is now a straightforward progression from the stress–strain relationship when the lamina is loaded along its principal axes, Eqn (4.43), to a general one involving a *transformed compliance matrix* $|\bar{S}|$, which will depend on ϕ. The first step is to write the inverse of Eqn (4.50), giving the strains relative to the loading direction (i.e. the information required) in terms of the strains relative to the fibre direction. This involves using the inverse of the matrix $|T'|$, written as $|T'|^{-1}$

$$
\begin{vmatrix} \varepsilon_x \\ \varepsilon_y \\ \gamma_{xy} \end{vmatrix} = |T'|^{-1} \begin{vmatrix} \varepsilon_1 \\ \varepsilon_2 \\ \gamma_{12} \end{vmatrix} \tag{4.52}
$$

in which

$$
|T'|^{-1} = \begin{vmatrix} c^2 & s^2 & -cs \\ s^2 & c^2 & cs \\ 2cs & -2cs & c^2 - s^2 \end{vmatrix} \tag{4.53}
$$

The strains relative to the fibre direction can now be expressed in terms of the stresses in those directions, via the on-axis stress–strain relationship for the lamina, Eqn (4.43), giving

$$
\begin{vmatrix} \varepsilon_x \\ \varepsilon_y \\ \gamma_{xy} \end{vmatrix} = |T'|^{-1}|S| \begin{vmatrix} \sigma_1 \\ \sigma_2 \\ \tau_{12} \end{vmatrix} \tag{4.54}
$$

Finally, the original transform matrix of Eqn (4.47) can be used to express these stresses in terms of those being externally applied, to give the result

$$
\begin{vmatrix} \varepsilon_x \\ \varepsilon_y \\ \gamma_{xy} \end{vmatrix} = |T'|^{-1}|S||T| \begin{vmatrix} \sigma_x \\ \sigma_y \\ \tau_{xy} \end{vmatrix} = |\bar{S}| \begin{vmatrix} \sigma_x \\ \sigma_y \\ \tau_{xy} \end{vmatrix} \tag{4.55}
$$

The elements of $|\bar{S}|$ are thus obtained by multiplication of the matrices $|T'|^{-1}$, $|S|$ and $|T|$. The following expressions are obtained

$$\bar{S}_{11} = S_{11}c^4 + S_{22}s^4 + (2S_{12} + S_{66})c^2s^2$$
$$\bar{S}_{12} = S_{12}(c^4 + s^4) + (S_{11} + S_{22} - S_{66})c^2s^2$$
$$\bar{S}_{22} = S_{11}s^4 + S_{22}c^4 + (2S_{12} + S_{66})c^2s^2$$
$$\bar{S}_{16} = (2S_{11} - 2S_{12} - S_{66})c^3s - (2S_{22} - 2S_{12} - S_{66})cs^3 \qquad (4.56)$$
$$\bar{S}_{26} = (2S_{11} - 2S_{12} - S_{66})cs^3 - (2S_{22} - 2S_{12} - S_{66})c^3s$$
$$\bar{S}_{66} = (4S_{11} + 4S_{22} - 8S_{12} - 2S_{66})c^2s^2 + S_{66}(c^4 + s^4)$$

It can be seen that $|\bar{S}| \to |S|$ as $\phi \to 0$. The behaviour of the lamina is still fully described by four independent elastic constants (see footnote 4), since these six elements can all be expressed in terms of S_{11}, S_{12}, S_{22} and S_{66}. A similar procedure can be used to derive the elements of $|\bar{C}|$, the *transformed stiffness matrix*

$$\bar{C}_{11} = C_{11}c^4 + C_{22}s^4 + (2C_{12} + 4C_{66})c^2s^2$$
$$\bar{C}_{12} = C_{12}(c^4 + s^4) + (C_{11} + C_{22} - 4C_{66})c^2s^2$$
$$\bar{C}_{22} = C_{11}s^4 + C_{22}c^4 + (2C_{12} + 4C_{66})c^2s^2$$
$$\bar{C}_{16} = (C_{11} - C_{12} - 2C_{66})c^3s - (C_{22} - C_{12} - 2C_{66})cs^3 \qquad (4.57)$$
$$\bar{C}_{26} = (C_{11} - C_{12} - 2C_{66})cs^3 - (C_{22} - C_{12} - 2C_{66})c^3s$$
$$\bar{C}_{66} = (C_{11} + C_{22} - 2C_{12} - 2C_{66})c^2s^2 + C_{66}(c^4 + s^4)$$

4.3.4   Engineering Constants

Either of the above two matrices fully define the elastic response of the material. However, it is often more convenient to represent these characteristics in terms of the conventional engineering constants. These can be obtained from the stiffness or compliance matrices by inspection of the relationships presented as Eqns (4.43) and (4.45). The relationships are simpler if the compliance matrix is used. Thus

$$E_x = \frac{1}{\bar{S}_{11}}$$

$$E_y = \frac{1}{\bar{S}_{22}}$$

$$G_{xy} = \frac{1}{\bar{S}_{66}} \qquad (4.58)$$

$$v_{xy} = -E_x\bar{S}_{12}$$

$$v_{yx} = -E_y\bar{S}_{12}$$

Some examples of the behaviour predicted by Eqn (4.56), for two different composites, are shown in Fig. 4.10. The dependence of the Young's and shear moduli of a lamina on ϕ is illustrated by Fig. 4.10(a) for a polymer matrix composite (PMC), using both equal stress and Halpin–Tsai expressions for E_2 and G_{12}. The equal stress

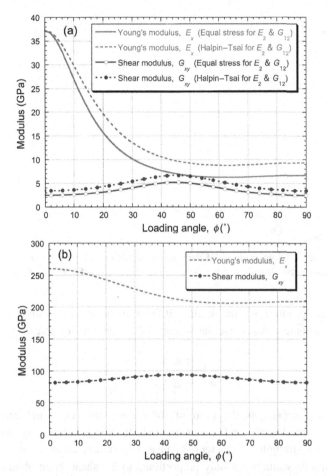

Fig. 4.10 Variation with loading angle ϕ of the Young's modulus E_x and shear modulus G_{xy} for laminae of (a) epoxy–50% glass and (b) titanium–50% SiC monofilament. For (a), the values of E_2 and G_{12} were obtained using the equal stress (Eqns (3.6) and (3.8)) and Halpin–Tsai (Eqns (3.7) and (3.10)), with $\xi = 1$ expressions, but for (b) only Halpin–Tsai expressions were used.

assumption introduces quite significant errors over a wide range of loading angles, although the predictions do not differ in qualitative terms. The Young's modulus remains close to the $\phi = 0°$ value if the stress axis is within a few degrees of the fibre axis, but if ϕ is more than about 5° then it decreases rapidly. The variations are smaller for the metal matrix composite (MMC) (Fig. 4.10(b)).

The shear stiffness is less sensitive than the Young's modulus to ϕ, but a pronounced peak is always exhibited at 45°. This efficiency of stiff diagonal (45°) members in resisting shear forces is important in many engineering situations – see, for example, the discussions by Gordon [2].

An important feature of the off-axis loading of laminae is the appearance of non-zero 'interaction' terms ($\bar{S}_{16}$ and $\bar{S}_{26}$), indicating that normal stresses produce shear strains

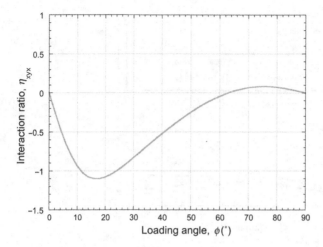

Fig. 4.11 Variation with loading angle, ϕ, of the interaction ratio η_{xyx} for a lamina of epoxy containing 50% glass fibres, obtained using Halpin–Tsai expressions for E_2 and G_{12}.

and shear stresses produce normal strains. It is convenient to introduce two other engineering constants to characterise the strength of this interaction effect

$$\eta_{xyx} = E_x \bar{S}_{16}$$
$$\eta_{xyy} = E_y \bar{S}_{26}$$

(4.59)

The parameter η_{xyx} represents the ratio of the shear strain (γ_{xy}), induced by the application of a normal stress (σ_x), to the normal strain (ε_x) induced by this stress. It characterises the strength of the tensile–shear interaction, although it should be borne in mind that its value is not only proportional to the shear strain observed for a given applied normal stress, but also becomes larger as the tensile stiffness in the loading direction increases. The parameters η_{xyx} and η_{xyy} are often termed *interaction ratios*.

As an example, Fig. 4.11 shows the dependence of η_{xyx} on ϕ for a particular lamina. It can be seen that substantial changes occur as the loading angle is changed. As expected from simple symmetry arguments, the interaction term is zero at $\phi = 0°$ and $\phi = 90°$, but at intermediate angles the effect can be pronounced. For some values of ϕ, the interaction shear strain is similar in magnitude to the direct normal strain ($\eta_{xyx} \sim 1$).

These characteristics are of considerable practical significance and are illustrated in Fig. 4.12, which includes a photograph of a demonstration macro-model. This is made up of four separate laminae, each composed of aligned aluminium rods in a polyurethane rubber matrix. These metal 'fibres' are oriented so that the stress axis forms angles of 0°, 30°, 60° and 90° to the fibre direction. It can be seen that there is a strong shear distortion for the 30° lamina, but this is negligible for the others. This is consistent with the plot in the figure, which shows the predicted dependence on ϕ of the interaction

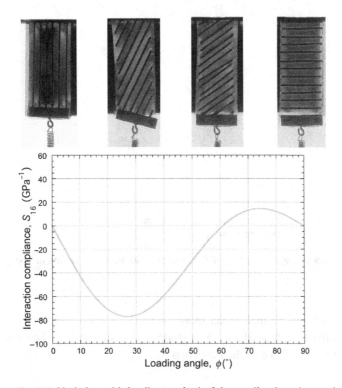

Fig. 4.12 Variation with loading angle ϕ of the tensile–shear interaction compliance, for rubber–5% Al fibre, and photos of four specimens (between crossed polars) under axial tension, lined up at the appropriate values of ϕ: these show marked tensile–shear distortion only for the $\phi = 30°$ lamina (as predicted).

compliance, for this system. In fact, there is at least approximate quantitative agreement. The applied stress in the case of the 30° lamina was about 2 MPa. Since the predicted value of is about 80 GPa^{-1}, it follows that

$$\gamma_{xy} = \bar{S}_{16}\sigma_x \approx (80)(0.002) \approx 0.16 \text{ radians } (\approx 10°) \tag{4.60}$$

This is broadly consistent with what is being observed (although there is a tendency for the grip to rotate, because the applied force is constrained from moving laterally). It may also be noted that the degree to which the laminae become extended is broadly consistent with theory, with the 60° and 90° cases being very similar, as predicted by Fig. 4.10.

An insight can be obtained about physical interpretation of these effects by recognising the role of matrix shear. The matrices in the models in Fig. 4.12 are photo-elastic and they show that significant matrix shear strains are induced in the inclined cases. These shear strains are also partly responsible for the effects illustrated in Fig. 4.13, which shows that the value of the Poisson ratio ν_{xy} peaks at an intermediate angle.

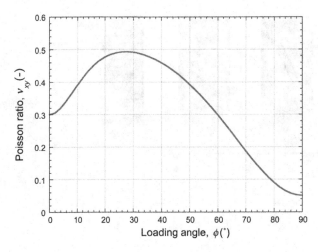

Fig. 4.13 Variation with loading angle ϕ of the Poisson ratio ν_{xy}, for a lamina of epoxy–50% glass fibre.

Large shear strains in the matrix can induce both shear distortions and large lateral contractions. Effects such as these are important in understanding the behaviour of laminates, which is explored in the next chapter.

References

1. Nye, JF, *Physical Properties of Crystals: Their Representation by Tensors and Matrices.* Clarendon, 1985.
2. Gordon, JE, *Structures, or Why Things Don't Fall Down.* Pitman Publishing Ltd, 1979.

5 Elastic Deformation of Laminates

In the previous chapter, procedures were described that allow prediction of the elastic response of a single lamina when loaded at an arbitrary angle to the fibre axis. It was shown that such uniaxial sheets tend to be highly anisotropic, with much greater stiffness when loaded parallel to the fibres than at significant angles to them. Moreover, other aspects of the elastic response are very different in different directions. For these reasons, it is common to stack laminae into bonded sets (laminates), making the elastic properties (and also the strength) more isotropic, and also opening up the possibility of tailoring the properties of a component to the loads that it will experience in service – potentially a major advantage of composites when compared with more conventional materials. In this chapter, the treatment of individual laminae is extended to the case of a laminate with an arbitrary stacking sequence, supplying an analytical tool of considerable value in the design of composite materials.

5.1 Loading of a Stack of Plies

5.1.1 Derivation of the Global Stiffness and Compliance Tensors

It is evident from Fig. 4.10 that individual laminae containing aligned fibres tend to exhibit highly anisotropic elastic properties. This anisotropy can be reduced by stacking a number of laminae (or plies) with different fibre orientations and bonding them together to form a laminate. The elastic properties of such a laminate can be predicted from those of the component plies, provided the assembly is taken to be flat and thin, with no through-thickness stresses, and edge effects are neglected (sometimes termed the *Kirchhoff assumptions*). While it is now common to use the finite element method (FEM) to investigate the behaviour of components, including laminates, having complex shapes and under various types of loading, the treatment presented here (sometimes described as *laminate theory*) is still of wide utility.

Calculation of the elastic constants of the laminate is carried out according to the scheme illustrated in Fig. 5.1. Note that the loading angle between the stress axis (x direction) and the reference direction for the orientation of the plies (i.e. the $\phi = 0°$ direction) is now expressed as Φ (upper case phi). The fibre direction for the kth ply therefore lies at an angle ($\phi - \Phi$) to the stress axis. The overall stress in the x direction can be written both as a function of the stresses in the individual plies and in terms of the overall compliances and strains

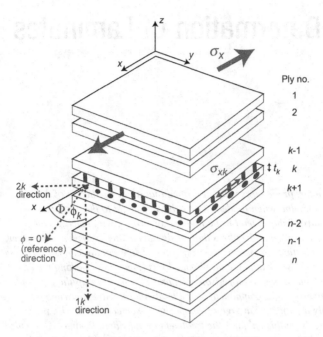

Fig. 5.1 Schematic depiction of the loading angle Φ between the x direction (stress axis) and the reference direction ($\phi = 0°$), for a laminate of n plies. Also shown is the angle ϕ_k between the reference direction and the fibre axis of the kth ply ($1k$ direction).

$$\sigma_{xg} = \frac{\sum\limits_{k=1}^{n} (\sigma_{xk} t_k)}{\sum\limits_{k=1}^{n} t_k} = \bar{C}_{11g} \varepsilon_{xg} + \bar{C}_{12g} \varepsilon_{yg} + \bar{C}_{16g} \gamma_{xyg} \qquad (5.1)$$

in which t_k is the thickness of the kth ply and the subscript g refers to a global value for the whole laminate. Since the in-plane strains are constrained to be the same for all of the laminae, the stress in any lamina can be written as

$$\sigma_{xk} = \bar{C}_{11k} \varepsilon_{xg} + \bar{C}_{12k} \varepsilon_{yg} + \bar{C}_{16k} \gamma_{xyg}$$

Substituting this into Eqn (5.1) and equating the coefficients of ε_{xg}

$$\bar{C}_{11g} = \frac{\sum\limits_{k=1}^{n} (\bar{C}_{11k} t_k)}{\sum\limits_{k=1}^{n} t_k} \qquad (5.2)$$

Similar expressions apply for the other components. If the corresponding compliances are required, they are obtained after this operation by application of the inversion relationships

$$S_{11} = \frac{\left(C_{22}C_{66} - C_{26}^2\right)}{\Delta}$$

$$S_{22} = \frac{\left(C_{11}C_{66} - C_{16}^2\right)}{\Delta}$$

$$S_{12} = \frac{\left(C_{16}C_{26} - C_{12}C_{66}\right)}{\Delta}$$

$$S_{66} = \frac{\left(C_{11}C_{22} - C_{12}^2\right)}{\Delta}$$

$$S_{16} = \frac{\left(C_{12}C_{26} - C_{22}C_{16}\right)}{\Delta}$$

$$S_{26} = \frac{\left(C_{12}C_{16} - C_{11}C_{26}\right)}{\Delta}$$

(5.3)

in which

$$\Delta = C_{11}C_{22}C_{66} + 2C_{12}C_{16}C_{26} - C_{22}C_{16}^2 - C_{66}C_{12}^2 - C_{11}C_{26}^2 \qquad (5.4)$$

5.1.2 Predicted Elastic Behaviour

Plots of (a) the Young's modulus ($E_x = 1/\bar{S}_{11g}$) and (b) the Poisson ratio ($\nu_{xy} = -E_x\bar{S}_{12g}$) against loading angle for different laminate stacking sequences are shown in Fig. 5.2. These plots were obtained by repeated application of Eqn (5.2) for each loading angle value, using the equal stress assumption for transverse modulus. (Although this is somewhat inaccurate, its use here does not affect the trends being identified.) It is important to appreciate that plots such as these are not obtainable using an explicit analytical expression, although the procedure is well-defined and tractable (and could therefore be described as being of 'closed-form'). A numerical calculation is thus required, but it is a relatively straightforward one that is readily implemented using a simple computer program. These are now readily available and are in widespread, routine use. It is not necessary to understand the details of the mathematical techniques involved in order to use such packages.

It can be seen in Fig. 5.2 that a 0/90 ('cross-ply') laminate is less anisotropic than the corresponding unidirectional lamina, although it is not isotropic. If further plies are introduced, covering the range of orientations with close, even spacings, then the anisotropy becomes smaller and may disappear, so that the stiffness is the same when loaded in any direction. For the case shown in Fig. 5.2(a), this stiffness is about 16.5 GPa. Plots such as these indicate that it is not necessary to space the ply orientations very closely in order to achieve in-plane isotropic elastic properties. On the other hand, cross-ply laminates retain pronounced anisotropy. Other lay-up sequences, such as $-\theta/+\theta$ ('angle-ply' laminates), are, of course, possible.

5.1.3 An Educational 'Laminate Theory' Package

While some kind of financial outlay is generally needed in order to access commercial software for implementation of laminate theory (using the numerical procedures

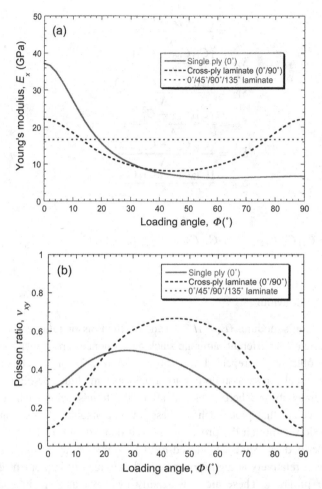

Fig. 5.2 Variation with loading angle, Φ, of (a) Young's modulus E_x and (b) Poisson ratio v_{xy} ($= - \bar{S}_{12} E_x$) for a single lamina, and for two laminates with different stacking sequences, composed of epoxy–50% glass fibres. (The equal stress expression was used for the transverse Young's modulus of a lamina.)

outlined above), certain packages with (limited) capabilities of this type are freely available for educational purposes. An example of this is the *Mechanics of Fibre Reinforced Composites* package, which is one of many within the DoITPoMS set of educational resources that is available under an open access arrangement. The URL for this particular package is www.doitpoms.ac.uk/tlplib/fibre_composites/index.php. A screenshot from the page providing access to software for prediction of the elastic constants of laminates is shown in Fig. 5.3. This package allows an arbitrary set of ply angles, and ply types, to be selected, and then displays the elastic constants as a function of loading angle.

The inverse of the stiffness tensor, the compliance tensor, is often obtained because its relationships with the elastic constants are simpler. Like before,

$$E_x = \frac{1}{\overline{S}_{11,g}}, \quad G_{xy} = \frac{1}{\overline{S}_{66,g}}, \quad \nu_{xy} = -E_x \overline{S}_{12,g}$$

Clearly the laminate will exhibit different elastic constants if the loading system were applied at an arbitrary angle, Φ, to the x-y coordinate system. Try constructing your own laminate, using the model below, and calculate the elastic constants for different loading angles.

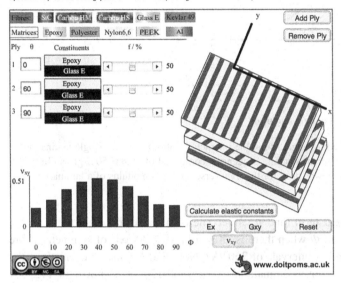

Fig. 5.3 Screenshot taken during use of the 'Stiffness of Laminates' page in the DoITPoMS TLP (Teaching and Learning Package) on Composites, accessible at www.doitpoms.ac.uk/tlplib/fibre_composites/lamina_stiffness.php.

5.2 Stresses and Distortions in Laminates

Loading of a laminate can produce complex stress distributions in the individual laminae, and also between them. These may cause various kinds of shape distortion and also, of course, create the possibility of local microstructural damage and failure. Where possible, it is important to select stacking sequences so as to minimise undesirable effects such as the distortions that are produced by the tensile–shear interactions described in Section 4.3.4.

5.2.1 'Balanced' Laminates

The tensile–shear interaction effects in the laminate as a whole depend on the set of ply orientations. The variation of the interaction compliance, η_{xyx}, with the angle Φ between the stress axis and the reference ($\phi = 0°$) direction of the laminate is shown in Fig. 5.4 for several laminates. As expected, the term becomes small for all Φ as the set of fibre directions become more evenly and more closely spaced – i.e. as the set exhibits greater *rotational symmetry* about the axis normal to the plane of the laminate.

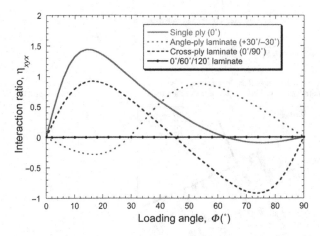

Fig. 5.4 Dependence on loading angle, Φ, of the interaction ratio, η_{xyx}, for a single lamina and for three laminates with different stacking sequences, composed of epoxy–50% glass fibres. (The equal stress expression was used for the transverse Young's modulus of a lamina.)

In fact, this tensile–shear term (and η_{xyy}) become zero – and the other terms in $\bar{S}_{ijg}$ become constant – for all Φ when there is an N-fold ($N \geq 6$) axis of symmetry (fibre directions evenly spaced at intervals of $360°/N$). Note that N must be even, since the sense of the fibre direction is immaterial. (The 0/60/120 case shown in Fig. 5.4 corresponds to $N = 6$.)

When the tensile–shear interaction terms contributed by the individual laminae all cancel each other out in this way, the laminate is often labelled as '*balanced*'. Simple cross-ply and angle-ply laminates are *not* balanced for a general loading angle, although both will be balanced when loaded at $\phi = 0°$ (i.e. parallel to one of the plies for a cross-ply or equally inclined to the $+\theta$ and $-\theta$ plies for the angle-ply case). If the plies vary in thickness, or in the volume fractions or type of fibres they contain, then even a laminate in which the stacking sequence does exhibit the necessary rotational symmetry is prone to tensile–shear distortions and computation is necessary to determine the lay-up sequence required to construct a balanced laminate. In all cases, the stacking order in which the plies are assembled does not enter into these calculations, although it may be important in determining the inter-laminar coupling stresses (see below).

5.2.2   In-plane Stresses in Individual Plies of a Laminate

The internal stresses may be subdivided into *in-plane* stresses, which can be calculated by the methods outlined below, and *inter-laminar* or *through-thickness* stresses, which can arise as a result of constraint effects. The approach to establishing in-plane stresses follows directly from the preceding treatments. Referring to Fig. 5.1, and subjecting the stack of laminae to a stress state σ_x, σ_y and τ_{xy}, the laminate strains are established from the transformed compliance matrix of the laminate

$$\begin{vmatrix} \varepsilon_x \\ \varepsilon_y \\ \gamma_{xy} \end{vmatrix} = |\bar{S}_g| \begin{vmatrix} \sigma_x \\ \sigma_y \\ \tau_{xy} \end{vmatrix} \tag{5.5}$$

Since these strains are imposed on all the plies, the strains within the kth ply (referred to as the fibre direction of the ply, i.e. the $1k$ direction) can be determined using the appropriate version of Eqn (4.50), which may be written as

$$\begin{vmatrix} \varepsilon_{1k} \\ \varepsilon_{2k} \\ \gamma_{12k} \end{vmatrix} = |T'|_{\phi=\phi_k} \begin{vmatrix} \varepsilon_x \\ \varepsilon_y \\ \gamma_{xy} \end{vmatrix} \tag{5.6}$$

The stresses in the ply, referred to as the fibre axis and transverse directions within it, are then obtained from these strains using the (on-axis) stiffness matrix for the kth ply

$$\begin{vmatrix} \sigma_{1k} \\ \sigma_{2k} \\ \tau_{12k} \end{vmatrix} = |C|_k \begin{vmatrix} \varepsilon_{1k} \\ \varepsilon_{2k} \\ \gamma_{12k} \end{vmatrix} \tag{5.7}$$

The stresses in the kth ply are thus related directly to the applied stresses by

$$\begin{vmatrix} \sigma_{1k} \\ \sigma_{2k} \\ \tau_{12k} \end{vmatrix} = |C|_k |T'|_{\phi=\phi_k} |\bar{S}_g| \begin{vmatrix} \varepsilon_{1k} \\ \varepsilon_{2k} \\ \gamma_{12k} \end{vmatrix} \tag{5.8}$$

Some results from such calculations are shown in Fig. 5.5, which gives the variation of σ_{1k} (as a ratio to the applied stress) with the angle Φ between the loading direction and the fibre direction in the 0° ply. This is shown for a single lamina and for two laminates. When more plies are present, so that the 0° ply being considered constitutes a smaller relative section, it bears a proportionately larger stress at $\Phi = 0°$. This is because the 0° ply is much stiffer than those at other orientations. Note also that the ply can be put into compression when it is oriented at a large angle to the loading direction. This is the result of a Poisson contraction effect.

This effect can also be seen in Fig. 5.6, which shows all of the stresses present in the 0° ply for a cross-ply (0/90) laminate. The compressive stress parallel to the fibres (in the 1 direction), which in this case has a magnitude of ~0.07 times the applied stress at $\Phi = 90°$, arises because the ply has a small natural Poisson contraction parallel to the fibres, but is being stressed by the larger natural contraction of the other ply (which is being loaded along its own fibre axis for $\Phi = 90°$). The nature of these stresses (transverse to the loading direction) can be seen in Fig. 5.6(b). Since there is no applied load in this direction, these two stresses must sum (as forces) to zero. It can thus be seen that stresses arising from these differential Poisson contraction effects can in some cases be surprisingly large. The details are sensitive to the Poisson ratio of fibre and matrix,

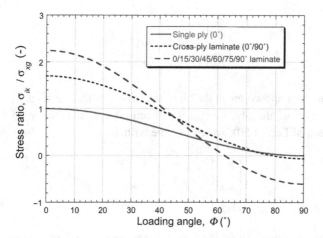

Fig. 5.5 Variation with loading angle, Φ, for a single lamina and for two laminates of epoxy–50% glass fibres, of the stress (σ_{1k}) parallel to the fibre axis in a ply initially oriented at 0° to the stress axis. The stress is plotted as a ratio to the applied stress. (The equal stress model was used for the transverse Young's modulus of a lamina.)

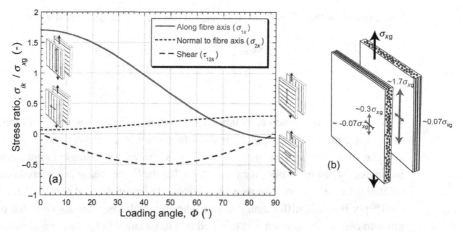

Fig. 5.6 (a) Predicted stresses within one ply of a loaded cross-ply laminate (epoxy–50% glass); and (b) a schematic of these stresses for loading parallel to one of the fibre axes.

but – since (ceramic) fibres tend to have low values (relative to both polymers and metals) – the Poisson ratio for contraction parallel to the fibre axis of a ply will always be less than that in a transverse direction and hence an effect of this type is general to most composite systems.

Being able to calculate all of the stresses within a loaded laminate is, of course, an important step towards starting to predict how it will become damaged and fail – see Chapter 8.

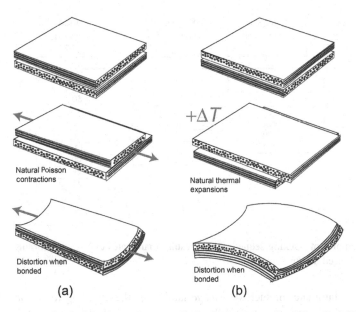

Fig. 5.7 Elastic distortions of a cross-ply laminate as a result of (a) uniaxial loading and (b) heating.

5.2.3   Coupling Stresses and Symmetric Laminates

Through-thickness (or 'coupling') stresses between the laminae are important, since they can lead to significant distortions of the laminate.[1] The general nature of these distortions can be illustrated by two examples, shown in Fig. 5.7. This refers to a cross-ply laminate, loaded mechanically or subjected to temperature change. Under uniaxial tension, as in Fig. 5.7(a), the difference in natural Poisson contractions of the two plies will cause the laminate to distort in the manner shown, and out-of-plane stresses would be needed to maintain the assembly flat. In addition to this, the transverse ply also exhibits a large through-thickness contraction as illustrated by the data in Fig. 3.7(a). A similar effect is shown in Fig. 5.7(b) for a cross-ply laminate that has been heated. Because the thermal expansion coefficients are different parallel and normal to the fibre axis (see Chapter 10), the laminate is deformed and becomes *saddle-shaped*. In this case, it is assumed that $\alpha_2 > \alpha_1$, which is expected in view of the low thermal expansion coefficients exhibited by (ceramic) fibres. The stresses that arise within such a laminate on changing the temperature can also cause microstructural damage – see Section 10.1.4.

Distortions such as these can be reduced considerably if the arrangement of the plies is symmetric about the mid-plane of the laminate, i.e. if it has a mirror plane lying in the

---

[1] These distortions mainly take the form of *curvature*, which may be different in different planes (containing the through-thickness direction). In fact, it is relatively straightforward to predict the curvature that will arise from a given in-plane *misfit strain*, such as those shown in Fig. 5.7, provided the thicknesses and elastic constants of the individual laminae are known. This topic is covered in some detail in Chapter 11. The treatment there is oriented towards coating-substrate systems, but the principles can readily be applied to (asymmetric) laminates.

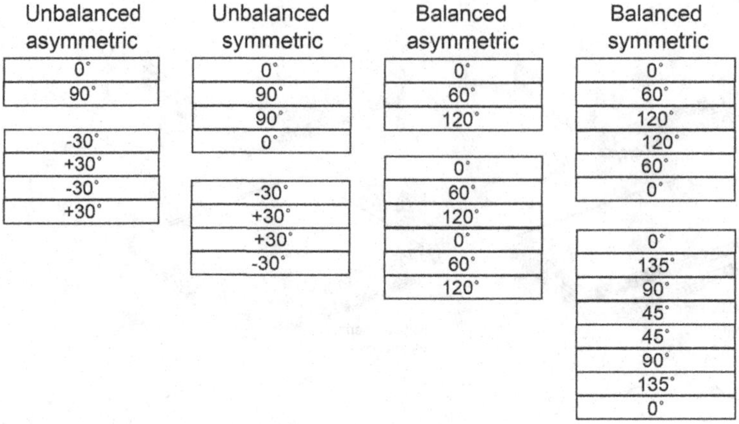

Unbalanced asymmetric	Unbalanced symmetric	Balanced asymmetric	Balanced symmetric
0°	0°	0°	0°
90°	90°	60°	60°
	90°	120°	120°
-30°	0°		120°
+30°		0°	60°
-30°		60°	0°
+30°	-30°	120°	
	+30°	0°	0°
	+30°	60°	135°
	-30°	120°	90°
			45°
			45°
			90°
			135°
			0°

Fig. 5.8 Examples of stacking sequences that result in laminates classified according to whether they are balanced and/or symmetric.

plane of the laminate. In such *symmetric* laminates, the coupling forces largely cancel out and the laminate as a whole will not distort, although there are still local stresses across the inter-laminar boundaries. In addition, the use of many thin laminae, rather than a few thick ones, minimises the distortions and leads to a reduction of the local inter-laminar stresses. The classification of laminae according to whether or not the stacking sequence is balanced and/or symmetric is illustrated in Fig. 5.8 with some examples.

There are advantages in using *balanced symmetric* stacking sequences and this is common commercial practice. However, it should be noted that a laminate is often designed in the light of information about the expected in-service stress state. For example, with tubes to be subjected to internal or external pressure, unequal biaxial tension or compression will be imposed and ply angle sequences will be chosen with this in mind. The probable mode of failure, as well as the elastic deflections, may also need to be considered (see Chapter 8). Furthermore, the type and magnitude of permissible deflections and distortions will vary widely between different applications. It is therefore rather difficult to specify an optimum stacking sequence without detailed information about the performance requirements. This highlights the important concept of designing the material and the component simultaneously – a recurrent theme when working with composites.

6 Stresses and Strains in Short Fibre and Particulate Composites

The previous three chapters cover the elastic behaviour of composites containing aligned fibres that are, in effect, infinitely long. Use of short fibres (or equiaxed particles) creates scope for using a wider range of reinforcements and more versatile processing and forming routes (see Chapter 15). There is thus interest in understanding the distribution of stresses and strains within such composites, and the consequences of this for the stiffness and other mechanical properties. In this chapter, brief outlines are given of two analytical models. In the shear lag treatment, a cylindrical (short fibre) reinforcement is assumed, with stress fields in fibre and matrix being simplified (leading to some straightforward analytical expressions). It introduces important concepts concerning load transfer mechanisms, although it is not very widely used for property prediction. The Eshelby method, on the other hand, is based on the reinforcement being ellipsoidal (anything from a sphere to a cylinder or a plate): the analysis is more rigorous, but with the penalty of greater mathematical complexity. The model is only briefly described here. Its use also introduces an important concept – that of a misfit strain, which is helpful in areas well beyond those of the mechanics of conventional composite materials.

6.1 The Shear Lag Model

A popular treatment of the effects of loading an aligned short fibre composite is the shear lag model, originally proposed by Cox [1] in 1952 and subsequently developed by others [2–4]. It centres on the transfer of tensile stress from matrix to fibre via interfacial shear stresses. The basis of the model is shown schematically in Fig. 6.1, where reference lines are drawn on a fibre and the surrounding matrix. These are initially straight and normal to the fibre axis. External loading is then applied (Fig. 6.1(b)) parallel to the fibre axis. The reference lines distort in the manner shown. Attention is concentrated on the shear distortions of the matrix close to the fibre, represented schematically in Fig. 6.1(c). The model is based on considering the shear stresses in the matrix and at the interface.

6.1.1 Stress and Strain Distributions

The basic equation of the model is obtained by setting the difference in the tensile force acting on the two end faces of a cylindrical volume element of the fibre to the shear force acting on its interfacial surface

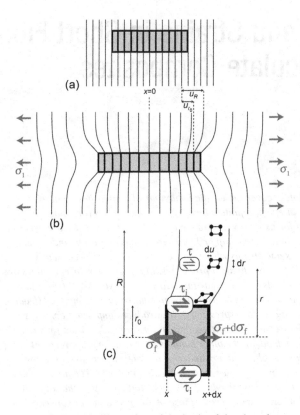

Fig. 6.1 Schematic illustration of the basis of the shear lag model, showing (a) unstressed system; (b) axial displacements, u, introduced on applying tension parallel to the fibre; and (c) variation with radial location of the shear stress and strain in the matrix.

$$-\left(\pi r_0^2\right)d\sigma_f = \tau_i(2\pi r_0 dx)$$

$$\therefore \frac{d\sigma_f}{dx} = \frac{-2\tau_i}{r_0} \tag{6.1}$$

The interfacial shear stress, τ_i, is obtained by considering how the shear stress in this direction varies within the matrix as a function of radial position. This variation is obtained by equating the shear forces on any two neighbouring annuli in the matrix

$$2\pi r_1 \tau_1 dx = 2\pi r_2 \tau_2 dx \quad \text{i.e.} \quad \frac{\tau_1}{\tau_2} = \frac{r_2}{r_1}$$

$$\therefore \tau = \tau_i \frac{r_0}{r}$$

The *displacement* of the matrix in the loading direction, u, is now considered. The shear strain at any point in the matrix can be written both as a variation in this displacement with radial position and in terms of the local shear stress and the shear modulus of the matrix, G_m

$$\gamma = \frac{\tau}{G_m} = \frac{\tau_i\left(\dfrac{r_0}{r}\right)}{G_m} \quad \text{and} \quad \gamma = \frac{du}{dr}$$

It follows that an expression can be found for the interfacial shear stress by considering the change in matrix displacement between the interface and some far-field radius, R, where the matrix strain has become effectively uniform ($du/dr \approx 0$)

$$\int_{u_{r_0}}^{u_R} du = \frac{\tau_i r_0}{G_m} \int_{r_0}^{R} \frac{dr}{r}, \quad \therefore \tau_i = \frac{(u_R - u_{r_0})G_m}{r_0 \ln \left(\frac{R}{r_0} \right)} \tag{6.2}$$

The appropriate value of R is affected by the proximity of neighbouring fibres, and hence by the fibre volume fraction, f. The exact relation depends on the precise distribution of the fibres, but this need not concern us too much, particularly since R appears in a log term. Assuming an hexagonal array of fibres, with the distance between the centres of the fibres at their closest approach being $2R$, simple geometry leads to

$$\left(\frac{R}{r_0} \right)^2 = \frac{\pi}{2f \sqrt{3}} \approx \frac{1}{f}$$

Substituting for τ_i in the basic shear lag equation now gives

$$\frac{d\sigma_f}{dx} = \frac{-2(u_R - u_{r_0})G_m}{r_0^2 \frac{1}{2} \ln \left(\frac{1}{f} \right)} \tag{6.3}$$

The displacements u_R and u_{r_0} are not known, but their differentials are related to identifiable strains. The differential of u_{r_0} is simply the axial strain in the fibre (assuming perfect interfacial adhesion and neglecting any shear strain in the fibre, which is taken to be much stiffer than the matrix)

$$\frac{du_{r_0}}{dx} = \varepsilon_f = \frac{\sigma_f}{E_f}$$

while the differential of u_R, i.e. the far-field axial strain of the matrix, can be taken as the macroscopic strain of the composite

$$\frac{du_R}{dx} \approx \varepsilon_1$$

Differentiating the expression for the gradient of stress in the fibre and substituting these two relations into the resulting equation, with the shear modulus expressed in terms of Young's modulus and Poisson ratio [$E_m = 2 G_m (1 + v_m)$], leads to

$$\frac{d^2\sigma_f}{dx^2} = \frac{n^2}{r_0^2}(\sigma_f - E_f \varepsilon_1) \tag{6.4}$$

in which n is a dimensionless constant (for a specified composite), given by

$$n = \sqrt{\frac{2E_m}{E_f(1 + v_m) \ln \left(\frac{1}{f} \right)}} \tag{6.5}$$

This is a second-order linear differential equation of a standard form, which has the solution

$$\sigma_f = E_f \varepsilon_1 + B \sinh\left[\frac{nx}{r_0}\right] + D \cosh\left[\frac{nx}{r_0}\right]$$

and, by applying the boundary condition of $\sigma_f = 0$ at $x = \pm L$ (the fibre half-length), the constants B and D can be solved to give the final expression for the variation in tensile stress along the length of the fibre

$$\sigma_f = E_f \varepsilon_1 \left\{ 1 - \cosh\left[\frac{nx}{r_0}\right] \text{sech}(ns) \right\} \tag{6.6}$$

in which s is the *aspect ratio* of the fibre ($=L/r_0$). From this expression, the variation in interfacial shear stress along the fibre length can also be found, using the basic shear lag equation (Eqn (6.1)), by differentiating and multiplying by $(-r_0/2)$

$$\tau_i = \frac{E_f n \varepsilon_1}{2} \sinh\left[\frac{nx}{r_0}\right] \text{sech}(ns) \tag{6.7}$$

6.1.2 The Stress Transfer Length (Aspect Ratio)

Eqns (6.6) and (6.7) allow predictions to be made about the stress distribution along the length of the fibre. An example is shown in Fig. 6.2. This shows the variations in fibre tensile stress and interfacial shear stress along the length of a fibre in a composite of aligned glass fibres in a polyester resin matrix (for which n has a value of about 0.22). The curves are for two different fibre aspect ratios and the composite has been subjected to a tensile strain parallel to the fibres, ε_1, of 1 millistrain (0.1%). The tensile stress is zero at the fibre ends and a maximum in the centre. The interfacial shear stress is zero in the centre and a maximum at the ends. For the high aspect ratio case ($s = 50$), the fibre is long enough for the tensile stress to build up until the fibre has a strain equal to that of the matrix and the composite. This gives rise to the plateau region of the fibre stress curve and a region of zero interfacial shear stress. (With continuous aligned fibres, all of the composite is in this equal strain condition with respect to stress in the axial direction – see Section 3.1). There are regions of the fibre near the ends that are less heavily stressed than this central plateau region, so the average fibre stress is lower than in a long fibre composite subjected to the same external load. The reinforcing efficiency decreases as the fibre length is reduced, since this increases the proportion of the total fibre length that is not fully loaded.

This behaviour leads to the concept of a *stress transfer length*, over which the strain in the fibre builds up to the plateau (matrix) value. For the case shown in Fig. 6.2, it can be seen that this length is about 30% of the distance from the end to the midpoint of the fibre for the $s = 50$ curve, which corresponds to about 7–8 fibre diameters. In fact, an estimate of this length (aspect ratio) is readily made. This can be done by recognising that, in Eqn (6.6), the equal strain condition is approached as the term in brackets gets

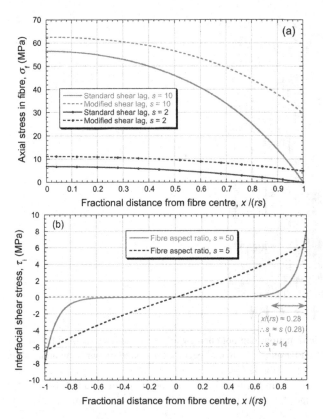

Fig. 6.2 Predicted (shear lag) variations in (a) fibre tensile stress and (b) interfacial shear stress along the axis of a glass fibre in a polyester–30% glass composite subject to an axial tensile strain of 10^{-3}, for two fibre aspect ratios.

close to unity. The stress transfer aspect ratio can thus be found by imposing this condition in the centre of the fibre ($x = 0$)

$$\cosh[0]\mathrm{sech}(ns) \ll 1$$

$$\rightarrow \frac{1}{\cosh(ns)} \ll 1 \quad (\cosh[0] = 1)$$

$$\rightarrow \cosh(ns) \approx 10 \quad (\text{say}) \quad \rightarrow ns \approx 3$$

$$\therefore s_t \approx \frac{3}{n} \tag{6.8}$$

For this case, with n having a value of ~0.22, this stress transfer aspect ratio is about 14, as shown in Fig. 6.2(b). (With a stiffer matrix, such as a metal, the value of n will be somewhat larger and so s_t will tend to be smaller, as a result of higher interfacial shear stresses.) For the low aspect ratio ($s = 5$) case shown in Fig. 6.2, the whole length of the fibre is only five diameters, so that the stress in it does not reach the plateau value. Such fibres are not providing very efficient reinforcement, because they carry much less stress than would longer fibres in the same system.

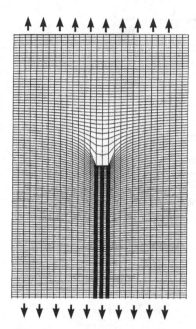

Fig. 6.3 An elastic deformation map obtained by a finite difference method [5], showing how an initially orthogonal grid around a fibre end becomes distorted on applying an axial tensile stress. (The fibre/matrix stiffness ratio is 40).

Reprinted by permission from Springer Nature: Springer Nature, Journals of Materials Science, Theoretical study of the stress transfer in single fibre composites, Termonia, Y., © 1987.

The shear lag model is qualitatively realistic. For example, outcomes of numerical modelling, such as that shown in Fig. 6.3 [5], confirm how high matrix shear strains near the end of the fibre lead to a build-up of fibre tensile strain. However, quantitative examination reveals deficiencies. The data in Fig. 6.4 were obtained by measuring the local tensile strain at different points along a fibre embedded in an epoxy matrix subjected to an external tensile load [6]. While the general appearance of the curves agrees well with the shear lag model for the higher imposed strains, there are discrepancies. Notable among these is that the fibre stress does not fall to zero at the ends. This is primarily a result of the transfer of tensile stress across the fibre ends, which is neglected in the basic model and is relatively unimportant in composites with high fibre aspect ratios. This is briefly examined in the next section.

6.1.3   Transfer of Normal Stress across Fibre Ends

Several attempts have been made [2–4] to introduce corrections for the neglect of stress transfer across the fibre ends. Any attempt to account for the effect, while retaining the attractive simplicity of the shear lag approach, must involve postulating an analytical expression for the fibre end stress σ_e. This must be an arbitrary postulate, since there is no scope within the shear lag framework for any rigorous description of stresses beyond

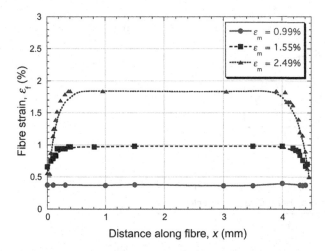

Fig. 6.4 Strains in a polydiacetylene fibre embedded in an epoxy matrix, measured by shifts in the Raman resonance spectrum [6], for three values of the macroscopic strain applied to the matrix. The fibre aspect ratio was about 200 and the fibre/matrix stiffness ratio was about 16.

the fibre end. An example is provided by the suggestion [4] that σ_e be set equal to the average of the peak fibre stress and the remote matrix stress values predicted by the standard shear lag model

$$\sigma_e = \frac{\sigma_{f0} + \sigma_{m0}}{2} \tag{6.9}$$

in which σ_{f0} is given by substituting $x = 0$ in Eqn (6.6) and σ_{m0} is taken as $E_m \varepsilon_1$ (the average matrix stress). This leads to an expression for σ_e

$$\sigma_e = \frac{\varepsilon_1 [E_f(1 - \operatorname{sec} h(ns)) + E_m]}{2} = \varepsilon_1 E'_m \tag{6.10}$$

and hence, using the new boundary conditions $\sigma_f = \sigma_e$ at $x = \pm L$ to solve Eqn (6.5), a new expression is obtained for σ_f, analogous to Eqn (6.6)

$$\sigma_f = \varepsilon_1 \left[E_f - (E_f - E'_m) \cosh\left(\frac{nx}{r_0}\right) \operatorname{sech}(ns) \right] \tag{6.11}$$

In Fig. 6.5, predictions from this equation are compared with those of the standard model (Eqn (6.6)) for polymer and metal matrix composites (PMCs and MMCs respectively). It can be seen that the predicted stresses in the fibre are significantly higher for the modified model, particularly near the fibre ends. Taking account of fibre end stress transfer naturally leads to the fibres carrying more load, particularly for short fibres. This results in an increase in the predicted stiffness of the composite (see the next section).

6.1.4 Prediction of Stiffness

The basic results of the shear lag treatment can be used to predict the elastic deformation of the composite. Consider a section of area A taken normal to the loading direction (in

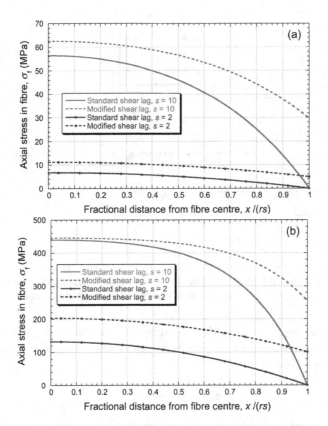

Fig. 6.5 Predicted variations in axial stress within (a) a glass fibre in a polyester–30% glass fibre composite; and (b) a SiC fibre in an Al–30% SiC fibre composite. Plots are at a composite strain of 10^{-3}, for two fibre aspect ratios ($s = 2$ and $s = 10$), according to the standard and modified shear lag models.

which all the fibres are aligned). This section intersects individual fibres at random positions along their length. The applied load can be expressed in terms of the contributions from the two components

$$\sigma_1 A = fA\bar{\sigma}_f + (1 - f)A\bar{\sigma}_m$$
$$\therefore \sigma_1 = f\bar{\sigma}_f + (1 - f)\bar{\sigma}_m \tag{6.12}$$

in which $\bar{\sigma}_f$ and $\bar{\sigma}_m$ are the volume-averaged stresses carried by fibre and matrix. This equation is often termed the *rule of averages*. The average fibre stress is evaluated from Eqn (6.6)

$$\bar{\sigma}_f = \frac{E_f \varepsilon_1}{L} \int_0^L \left[1 - \cosh\left(\frac{nx}{r_0}\right) \operatorname{sech}(ns) \right] dx$$

$$\therefore \bar{\sigma}_f = E_f \varepsilon_1 \left[1 - \frac{\tanh(ns)}{ns} \right] \tag{6.13}$$

For the matrix, it is again conventional to resort to the assumption of a uniform tensile strain equal to that imposed on the composite

$$\bar{\sigma}_m \approx E_m \varepsilon_1 \tag{6.14}$$

Combining Eqns (6.12)–(6.14) gives the stress–strain relationship for the composite

$$\sigma_1 = \varepsilon_1 \left[fE_f \left(1 - \frac{\tanh(ns)}{ns} \right) + (1-f)E_m \right] \tag{6.15}$$

The same procedure for the modified model, taking account of fibre end stress transfer, leads to

$$\sigma_1 = \varepsilon_1 \left[fE_f \left(1 - \frac{(E_f - E'_m)\tanh(ns)}{E_f ns} \right) + (1-f)E_m \right] \tag{6.16}$$

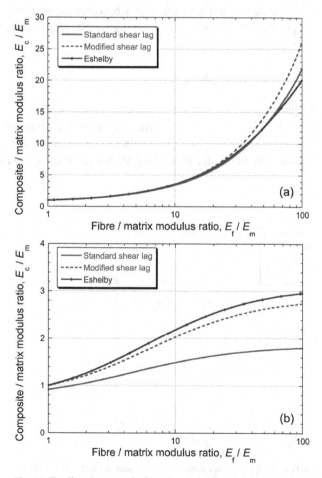

Fig. 6.6 Predicted composite/matrix modulus ratio as a function of the fibre/matrix modulus ratio, for composites with 30% reinforcement and fibre aspect ratios of (a) 30 and (b) 3. (Poisson ratios of fibre and matrix were taken as 0.2 and 0.3 respectively.)

A linear increase in stress with increasing strain is predicted in both cases. The expressions in square brackets are the predicted Young's moduli for the two models.

These equations can be tested by making comparisons with predictions from the (more rigorous) Eshelby model (see Section 6.2). For example, Fig. 6.6 shows the variation in composite Young's modulus with fibre/matrix modulus ratio for two fibre aspect ratio values. It can be seen that the standard shear lag model is inaccurate for low fibre aspect ratios. The predictions of the standard model look particularly unreliable when the fibre/matrix modulus ratio is small. This suggests that the fibre end stress modification might be particularly useful for discontinuously reinforced metal matrix composites. This is confirmed by the data in Fig. 6.7, which compares predictions from the three models with measured stiffnesses for particulate MMCs [7]. The standard shear lag model is clearly quite unsuitable for application to such materials.

It may be noted from Eqns (6.15) and (6.16) that, as expected, the stiffness approaches the limiting (rule of mixtures) value as s becomes large enough for tanh $(ns)/ns$ to become negligible ($\ll 1$). Since tanh$(ns) \sim 1$ for $ns \gtrsim 3$, and assuming that $0 \cdot 1$ can be taken as $\ll 1$

$$S_{RoM} \approx \frac{10}{n} \tag{6.17}$$

in which S_{RoM} is the fibre aspect ratio needed for the composite modulus to approach its maximum (rule of mixtures) value. As noted earlier, values of n are typically around 0.1–0.2 for polymer composites and 0.4 for those with metallic matrices. This suggests values for S_{RoM} of about 100 and 25 for PMCs and MMCs respectively. These can be

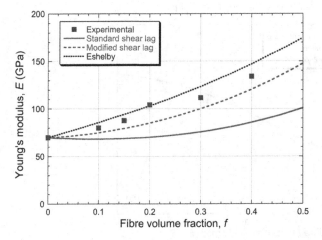

Fig. 6.7 Comparison between experimental data [7] and model predictions for the stiffness of Al–SiC (particulate) composites produced by extrusion. Since the particles were not truly equiaxed, and tended to become aligned during processing, an aspect ratio of 2 was used in the predictions.

regarded as target (minimum) aspect ratios when the main objective is to maximise the load transfer and hence the stiffness.

6.1.5 Onset of Inelastic Behaviour

Several inelastic phenomena can be promoted as the applied stress is raised. These include plastic deformation of the matrix, debonding and subsequent frictional sliding at the interface, formation of cavities or cracks in the matrix (particularly at fibre ends) and fracture of fibres. These effects change the stress distribution and hence affect the stress–strain curve. They are also related to the onset of failure and hence to the strength of the material. Detailed consideration of the factors involved in failure is presented in Chapters 8 and 9, but it is appropriate here to examine a simple extension to the basic shear lag theory designed to predict the onset of departure from elastic behaviour.

The onset of matrix plasticity or interfacial sliding is expected to occur at the fibre ends, where the shear stress is a maximum. A critical interfacial shear stress τ_{i*} can be specified for these processes. Substitution of τ_{i*} into Eqn (6.7), with $x = L$, gives the composite strain at the onset of such inelastic behaviour

$$\varepsilon_{1*} = \frac{2\tau_{i*} \coth (ns)}{nE_f} \tag{6.18}$$

This can be converted to an applied stress using Eqn (6.15), leading to the expression

$$\sigma_{1*} = \frac{2\tau_{i*}}{nE_f} \left[(fE_f + (1-f)E_m) \coth (ns) - \frac{fE_f}{ns} \right] \tag{6.19}$$

This point does not correspond to a clearly identifiable composite yield stress, since yielding (or interfacial sliding) is only taking place in a small, localised region. However, at this point the stress–strain curve will start to depart from a linear plot. As an illustration of the use of Eqn (6.18), in a typical glass fibre-reinforced polymer composite, with $\tau_{i*} = 20$ MPa, the composite strain at the onset of inelastic behaviour is about 0.6% for long fibres ($s > 30$) and about 0.3% for short fibres ($s \sim 5$).

The likelihood of fibre fracture taking place before matrix yielding or interfacial sliding can also be examined. The peak stress in the fibre at the onset of interfacial sliding or yielding is found from Eqn (6.6) by setting $x = 0$ and the composite strain to the value given by Eqn (6.18). This leads to

$$\sigma_{f0} = \frac{2\tau_{i*}}{n} \left[\coth (ns) - \text{cosech}(ns) \right] \tag{6.20}$$

Schematic plots of this relationship are shown in Fig. 6.8, which also gives an indication of the range of values expected for τ_{i*} in metallic and polymeric matrices and for the fracture stress σ_{f*} exhibited by ceramic fibres. (A distinction must be drawn between thermosetting and thermoplastic polymers: the former are brittle and tend to exhibit interfacial sliding and/or matrix micro-cracking, but not plastic yielding.) It is clear from this plot that, on increasing the load applied to either type of composite, yielding or sliding at the interface will take place before fibres start to fracture.

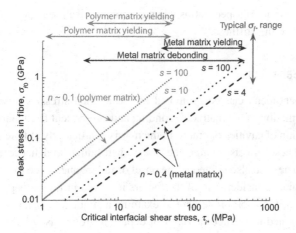

Fig. 6.8 Plots of the dependence of peak fibre stress, σ_{f0}, (at the onset of interfacial sliding or matrix yielding) on the critical shear stress for the onset of these phenomena, τ_{i*}, according to Eqn (6.20). Plots are shown for different aspect ratios, with n values typical of polymer- and metal-based composites. Also indicated are typical value ranges for fracture of fibres and for matrix yielding and interfacial debonding.

As the composite strain is increased, yielding (or sliding) spreads along the length of the fibre, raising the tensile stress in the fibre as the interfacial shear stresses increase. Fracture of fibres may then become possible and a simple treatment can be used to explore the limit of this effect. If it is assumed that the interfacial shear stress becomes uniform at τ_{i*} along the length of the fibre, then a *critical aspect ratio, s_**, can be identified, below which the fibre cannot undergo any further fracture. This corresponds to the peak (central) fibre stress just attaining its ultimate strength σ_{f*}, so that, by integrating Eqn (6.1) along the fibre half-length

$$s_* = \frac{\sigma_{f*}}{2\tau_{i*}}$$

(6.21)

A distribution of aspect ratios between s_* and $s_*/2$ is thus expected, if the composite is subjected to a large strain. In practice, the usefulness of this equation is limited, since the imposition of a large plastic strain on the matrix, without it cracking, is not common. It might, however, take place during processes such as extrusion, particularly if carried out while the matrix is relatively soft. Examples of the outcome of such processing can be seen in Fig. 6.9, which shows SEM micrographs of short fibre MMCs, containing about 15 vol% of reinforcement in Al alloy matrices, after deep etching to reveal the fibres. It can be seen that the average aspect ratio has been reduced to ~5–10 for the alumina fibres, while for the SiC whiskers, which are considerably stronger, it is ~15–20. These values are broadly consistent with Eqn (6.21), since the matrix shear yield stress during the extrusion (at about 200°C) was ~100 MPa and the fibre fracture stress is ~2 GPa for the alumina and ~5 GPa for the SiC whiskers.

Fig. 6.9 SEM micrographs of short fibre MMC composites after extrusion (and deep etching), comprising Al alloys reinforced with about 15 vol% of (a) Al$_2$O$_3$ ('Saffil') fibres and (b) SiC whiskers.

6.2 The Eshelby Method

6.2.1 History and Context

The method has its origins in some simple 'thought experiments' outlined by J. D. Eshelby [8,9] in the late 1950s. The power and versatility of the approach only became fully appreciated some time later. Various aspects of the concept are brought together in a *Collected Works* book [10]. It continues to be applied [11–14] to a wide range of situations related to the mechanics of composite materials. In fact, its usage extends beyond mechanics; for example, it can be used to model thermal and electrical conduction [15,16]. Jock Eshelby died in 1981, by which time the elegance and value of the concepts he introduced had become fully recognised.

The original study focused on an infinite matrix containing a (stiff) 'inclusion' with an ellipsoidal shape. The system is initially stress-free. The inclusion is removed and subjected to a change of size and shape (a 'stress-free' strain). The inclusion is then subjected to surface forces so as to return it to its original dimensions and replaced in the hole from whence it came. The forces are then removed, allowing the inclusion to adopt a new shape by distorting the matrix. This, in effect, is what happens on heating a matrix containing an inclusion, when the two have different coefficients of thermal expansion. (Other examples are provided by a martensitic transformation of the inclusion or uniform plastic deformation of the matrix.)

The Eshelby method involves calculating the resultant stresses in the inclusion. These turn out to be uniform throughout the inclusion, provided it has an ellipsoidal shape. This is not as severe a limitation as it sounds, since many shapes approximate to ellipsoids. In particular, a short fibre can be represented by a prolate ellipsoid having the same aspect ratio. Furthermore, while the original treatment concerned an isolated inclusion in an infinite matrix (a 'dilute' composite), the work of a number of other researchers led to adaptations that allowed the effect of neighbouring inclusions ('mean field' theories) to be incorporated in the model, so that macroscopic properties of real composites could be predicted.

Derivation of expressions for stresses and strains in composites, and hence of their elastic constants, involves some manipulation of tensors. Most of the background needed for such manipulations is provided in Chapter 4. However, the treatment given here is highly abbreviated and readers interested in the details should consult other sources [10,17,18]. Of course, as with many modelling tools, it can be used (most commonly via a computer program) without a full understanding of the internal workings.

6.2.2   A Misfitting Ellipsoid

Consider the sequence shown in Fig. 6.10(a). The initial mesh spacings of matrix and inclusion (fibre) represent unstrained material. The thicker lines of the inclusion denote its higher stiffness. For the case of differential thermal expansion on heating by ΔT, the imposed *misfit strain*, ε_{T*}, is given by

$$\varepsilon_{T*} = \Delta T(\alpha_f - \alpha_m) \tag{6.22}$$

noting that thermal expansivity, α, and strain, ε, are both second-rank tensors. Since, in most cases, $\alpha_f < \alpha_m$, a decrease in temperature produces a change in inclusion shape, relative to the matrix, as shown in Fig. 6.10(a). After replacement of the inclusion in the cavity, it adopts a new size and shape, distorting the surrounding matrix as it does so. This new size and shape represents a *constrained strain*, ε_C, relative to the original state. A key point here is that, *for the special case of an ellipsoidal shape*, this strain is uniform in all parts of the inclusion. This must also be true of the stress, which can be written

$$\sigma_f = C_f(\varepsilon_C - \varepsilon_{T*}) \tag{6.23}$$

where C_f is the stiffness tensor of the inclusion (fibre) and the term in brackets represents the net strain relative to the 'stress-free' state after ε_{T*} was imposed.

6.2.3   The Equivalent Homogeneous Ellipsoid

The essence of the Eshelby method is to consider exactly the same operations being carried out with an ellipsoid having the *same elastic constants as the matrix*. This is shown in Fig. 6.10(b). The composite is now homogeneous in terms of elastic properties. It is possible to generate exactly the same strain and stress state (in both inclusion and surrounding matrix) as arose with the real composite, provided that the imposed (stress-free) strain is the appropriate one. This is termed the *transformation strain*, ε_T. The schematic shown in Fig. 6.10(b) illustrates how choice of the appropriate value leads to the correct final stress state. The appropriate value of ε_T depends on, but differs from, the actual misfit strain, ε_{T*}. (Other terms, such as the *eigenstrain*, have been used by some authors to denote the appropriate transformation strain.)

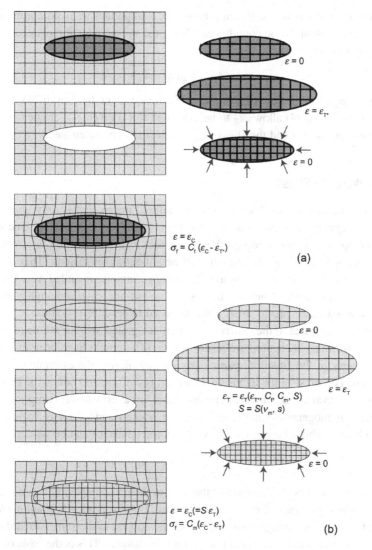

Fig. 6.10 Schematic depiction of strains during Eshelby operations of removing an inclusion from the matrix, introducing a misfit strain ε_{T*} and replacing it in the hole in the matrix. The operations are shown for (a) the actual inclusion and (b) an inclusion with the same elastic constants as the matrix, which must be subjected to a different strain, ε_T, to produce the same final stress state.

The major contribution made by Eshelby was to establish that the constrained strain, ε_C, for this *equivalent homogeneous inclusion* is uniquely related to the transformation strain by the expression

$$\varepsilon_C = S\varepsilon_T \qquad (6.24)$$

where S, the *Eshelby tensor*, is a simple function of the ellipsoid axis ratio (i.e. the aspect ratio of the fibre) and the Poisson ratio of the matrix. (The symbol S is conventionally used for the Eshelby tensor; it is important to avoid confusion with S used to represent an elastic compliance.) Now, the stress in the homogeneous inclusion can be written as

$$\sigma_f = C_m(\varepsilon_C - \varepsilon_T) \qquad (6.25)$$

This is set equal to the stress for the actual inclusion given by Eqn (6.23). Substitution for ε_C from Eqn (6.24) allows ε_T to be evaluated (for a given ε_{T*}, elastic constants and ellipsoid aspect ratio) and the stress in the inclusion to be calculated.

6.2.4   The Background Stress

The above treatment was developed for an infinite matrix and is only applicable to a 'dilute' composite, in which f is no more than, say, a few per cent. Extension to non-dilute composites requires recognition of the fact that the inclusion is affected by the matrix stress field due to the presence of neighbouring inclusions. This changes the stresses in both matrix and inclusion. The change may be regarded as a 'background stress', σ_b, superimposed on the stress state for a dilute case, although various other expressions (such as 'image stress') are also used to describe this stress.

It can be shown that, if the matrix stresses are integrated over the whole of the volume around a misfitting inclusion, the result is zero. (This might have been expected from the symmetry of the situation, particularly on considering the example of a spherical inclusion). The stress in the inclusion, however, is uniform and non-zero. This provides a means of evaluating the background stress in a non-dilute composite. For the equivalent homogeneous inclusion case, the background stress is expected to act uniformly throughout the composite, so that a force balance can be written

$$(1-f)\sigma_b + f(\sigma_f + \sigma_b) = 0 \qquad (6.26)$$

from which σ_b can be evaluated. For the case of the real composite, it is not clear how the inclusion experiences the background stress. The assumption is often made that the real and equivalent homogeneous inclusions are subject to the same strain disturbance as a result of being in a matrix containing other inclusions. This is the basis of the *mean field approximation* and detailed arguments have been put forward to support the hypothesis. It should be quite accurate, at least for volume fractions up to a practical limit for short fibre composites of around 40–50%. The term *mean stress*, $<\sigma>$, is now reserved for the volume-averaged stress in each component that contributes to this force balance

$$(1-f)\langle\sigma\rangle_m + f\langle\sigma\rangle_f = 0 \qquad (6.27)$$

The mean stresses are indicative of load transfer: the inclusion carries a mean stress of the opposite sign (and different magnitude) to that in the matrix.

From this point, some mathematical manipulations allow the mean strains of the two components, and hence of the composite, to be evaluated. For the case of a differential thermal expansion, the overall average composite strain (equal to the sum of this mean strain and the strain from the normal expansion of the matrix) then determines the composite expansivity. (Thermal expansion is examined in Section 10.1.)

6.2.5 Composite Stiffness

The analysis can be adapted to treat the case where the composite is subject to an applied stress, σ_A, and hence to predict the stiffness. The appropriate Eshelby operations are shown in Fig. 6.11. The misfit now effectively arises from the difference between the shapes that would be adopted by the inclusion (fibre) and by the cavity in the matrix,

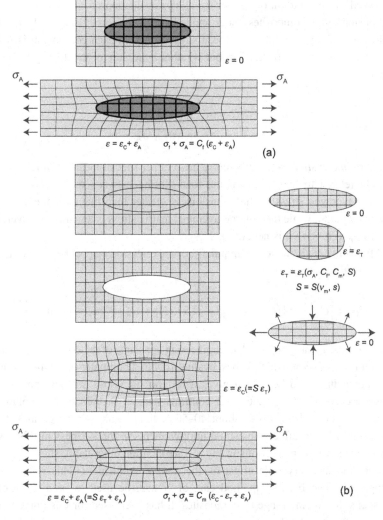

Fig. 6.11 Schematic illustration of the Eshelby operations with an applied stress σ_A for (a) actual and (b) equivalent homogeneous composites.

if the two constituents were to be independently subjected to the applied stress. For the equivalent homogeneous composite, the appropriate transformation strain, ε_T, is now dependent on σ_A, rather than on a predefined ε_{T*}. Furthermore, the resultant strain is now the sum of the constrained strain, ε_C, and a strain arising directly from the applied stress, ε_A (= $C_m^{-1} \sigma_A$). Similarly, the total stress in the inclusion is made up of σ_A and the contribution introduced by the *load transfer* between the constituents, represented by σ_f (as in the misfit strain treatment). The same technique of equating the stresses in the actual and equivalent homogeneous inclusions

$$\sigma_f + \sigma_A = C_f(\varepsilon_C + \varepsilon_A + \varepsilon_b)$$

$$= C_m(\varepsilon_C - \varepsilon_T + \varepsilon_A + \varepsilon_b) \qquad (6.28)$$

followed by substitution of $S\varepsilon_T$ for ε_C, allows evaluation of ε_T as before.

For non-dilute composites under applied stress, the force balance of Eqn (6.27) still holds, with the proviso that the mean stress, $<\sigma>$, in each component is not now the actual volume-averaged stress, $\bar{\sigma}$, but is rather a measure of the difference between this and σ_A

$$\bar{\sigma}_m = \langle\sigma\rangle_m + \sigma_A \qquad (6.29)$$

$$\bar{\sigma}_f = \langle\sigma\rangle_f + \sigma_A \qquad (6.30)$$

The term *mean stress, $<\sigma>$*, therefore has a rather special meaning in Eshelby analysis, closely related to the nature of the load-sharing between the two components: the volume-averaged stress in the inclusions is greater than that over the composite as a whole by $<\sigma>_f$, while the volume-averaged matrix stress is less than the overall average by $<\sigma>_m$ (i.e. $<\sigma>_m$ is negative).

After some mathematical manipulation, an expression can be derived for the stiffness tensor of the composite

$$C_C = \left[C_m^{-1} - f\{(C_f - C_m)[S - f(S - I)] + C_m\}^{-1}(C_f - C_m)C_m^{-1} \right]^{-1} \qquad (6.31)$$

where I is the identity tensor. The engineering constants of the composite can be derived from this expression, which is best evaluated with a computer program. Typical results are shown in Fig. 6.12, which gives axial and transverse stiffness predictions for a polymeric composite. Fig. 6.12(a) confirms that, in the practicable volume fraction range (for short fibres) up to about 40–50%, fibres with fairly high aspect ratios are needed in order to effect substantial improvements in the stiffness in this system. The transverse stiffness predictions in Fig. 6.12(b), on the other hand, show clearly that the aspect ratio has very little effect on the transverse stiffness. This is the case for all composites. The Eshelby method can, of course, be used to predict various elastic constants, for a wide range of composites. It has been used for this purpose in several places within this book.

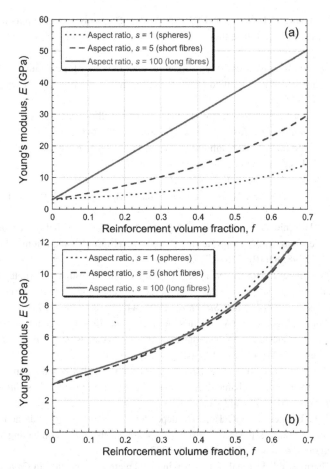

Fig. 6.12 Eshelby predictions of the Young's modulus as a function of fibre volume fraction, for glass fibres with aspect ratios of 1, 5 and 100, in an epoxy matrix for (a) axial and (b) transverse loading.

References

1. Cox, HL, The elasticity and strength of paper and other fibrous materials. *British Journal of Applied Physics* 1952; **3**: 72–79.
2. Fukuda, H and TW Chou, An advanced shear lag model applicable to discontinuous fiber composites. *Journal of Composite Materials* 1981; **15**: 79–91.
3. Nardone, VC and KM Prewo, On the strength of discontinuous silicon carbide-reinforced aluminium composites. *Scripta Metallurgica* 1986; **20**: 43–48.
4. Clyne, TW, A simple development of the shear lag theory appropriate for composites with a relatively small modulus mismatch. *Materials Science and Engineering* 1989; **A122**: 183–192.
5. Termonia, Y, Theoretical study of the stress transfer in single fibre composites. *Journal of Materials Science* 1987; **22**: 504–508.

6. Galiotis, C, RJ Young, PHJ Yeung and DN Batchelder, The study of model polydiacetylene epoxy composites: 1. The axial strain in the fiber. *Journal of Materials Science* 1984; **19**(11): 3640–3648.

7. McDanels, DL, Analysis of stress–strain, fracture, and ductility of aluminium matrix composites containing discontinuous silicon carbide reinforcement. *Metallurgical Transactions A* 1985; **16A**: 1105–1115.

8. Eshelby, JD, The determination of the elastic field of an ellipsoidal inclusion and related problems. *Proceedings of the Royal Society of London* 1957; **241**: 376–396.

9. Eshelby, JD, Elastic inclusions & inhomogeneities, in *Progress in Solid Mechanics*, Sneddon, IN and R Hill, editors. North Holland, 1961, pp. 89–140.

10. Markenscoff, X and A Gupta, eds. *Collected Works of J. D. Eshelby (The Mechanics of Defects and Inhomogeneities). Solid Mechanics and Its Applications*, 1st edition. Springer, 2006.

11. Huang, JH, YH Chiu and HK Liu, Magneto-electro-elastic Eshelby tensors for a piezoelectric–piezomagnetic composite reinforced by ellipsoidal inclusions. *Journal of Applied Physics* 1998; **83**(10): 5364–5370.

12. Giraud, A, QV Huynh, D Hoxha and D Kondo, Application of results on Eshelby tensor to the determination of effective poroelastic properties of anisotropic rocks-like composites. *International Journal of Solids and Structures* 2007; **44**(11–12): 3756–3772.

13. Lurie, S, D Volkov-Bogorodsky, A Leontiev and E Aifantis, Eshelby's inclusion problem in the gradient theory of elasticity: applications to composite materials. *International Journal of Engineering Science* 2011; **49**(12): 1517–1525.

14. Peng, X, S Tang, N Hu and J Han, Determination of the Eshelby tensor in mean-field schemes for evaluation of mechanical properties of elastoplastic composites. *International Journal of Plasticity* 2016; **76**: 147–165.

15. Le-Quang, H, G Bonnet and QC He, Size-dependent Eshelby tensor fields and effective conductivity of composites made of anisotropic phases with highly conducting imperfect interfaces. *Physical Review B* 2010; **81**(6): 064203.

16. Clyne, TW, Thermal and electrical conduction in metal matrix composites, in *Comprehensive Composite Materials II*, Clyne, T W, editor. Elsevier, 2018, pp. 188–212.

17. Withers, PJ, WM Stobbs and OB Pedersen, The application of the Eshelby method of internal stress determination for short fibre metal matrix composites. *Acta Metallurgica* 1989; **37**: 3061–3084.

18. Clyne, TW and PJ Withers, *An Introduction to Metal Matrix Composites*. Cambridge University Press, 1993.

7 The Interface Region

Previous chapters have mainly concerned the elastic behaviour of composites. Among the assumptions made in most of these treatments is that the interfacial bond is 'perfect'. This means that there is no local plasticity, debonding, cracking or sliding – in fact, no elastic or inelastic processes of any description. In practice, such phenomena may take place at or close to the interface, depending on its structure and the stresses generated there. These processes can influence the onset and nature of subsequent failure. Before treating the strength and fracture of composites (Chapters 8 and 9), it is helpful to consider the interface region in detail and examine how its response can be characterised and influenced. The meaning and measurement of bond strength are therefore outlined here. This is followed by information about the formation of interfacial bonding in various systems and the scope for its control.

7.1 Bonding Mechanisms

Several different types of mechanism may contribute to adhesion between matrix and reinforcement. Some of the main ones are depicted schematically in Fig. 7.1. These are briefly described below.

7.1.1 Adsorption and Wetting

If the surfaces of two bodies spontaneously come into intimate (atomic-scale) contact when brought close to each other (commonly with one of the bodies in liquid form), then 'wetting' is said to have taken place. Adhesion is primarily caused by van der Waals forces, although other types of bonding may reinforce these. The occurrence of wetting can be treated using simple thermodynamics, although in practice there may be chemical changes taking place that are time-dependent. Fig. 7.1(a) shows this type of contact being established for the liquid–solid case. The thermodynamic condition, often termed the *Dupré equation*, is commonly expressed in terms of surface energies, with the *work of adhesion* being a simple net sum

$$W_a = \gamma_{\text{m-air}} + \gamma_{\text{f-air}} - \gamma_{\text{m-f}} \tag{7.1}$$

Wetting is thus favoured (W_a is large) if the surface energies of the two constituents are large and their interfacial energy is small. In practice, however, a large value of the

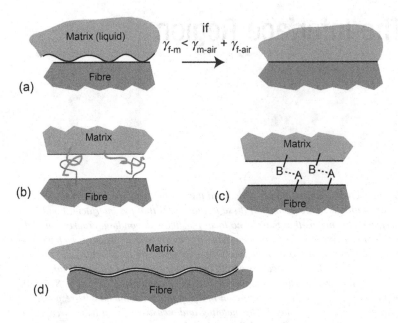

Fig. 7.1 Schematic depictions of mechanisms of interfacial bonding in composites: (a) 'wetting' (van der Waals forces); (b) diffusional processes; (c) chemical reaction; and (d) mechanical 'keying'.

liquid surface energy ($\gamma_{\text{m-air}}$) inhibits the spreading of a liquid droplet. The equilibrium wetting or contact angle, θ, is dictated by the *Young equation*, which may be expressed in the form

$$\gamma_{\text{f-air}} = \gamma_{\text{m-f}} + \gamma_{\text{m-air}} \cos \theta \tag{7.2}$$

It follows that complete wetting ($\theta = 0°$) occurs if the surface energy of the fibre is equal to or greater than the sum of the liquid (matrix) surface energy and the interface surface energy. Interface surface energies are not easy to obtain (and may be influenced by chemical reactions), but they are frequently smaller than the values for the constituents being exposed to air. The surface energies of fibres and (liquid) matrices are generally known and systems for which the former greatly exceeds the latter are likely to wet easily. For example, fibres of glass ($\gamma_{\text{f-air}} = 560$ mJ m^{-2}) and graphite ($\gamma_{\text{f-air}} = 70$ mJ m^{-2}) are readily wetted by polyester ($\gamma_{\text{m-air}} = 35$ mJ m^{-2}) and epoxy ($\gamma_{\text{m-air}} = 43$ mJ m^{-2}) resins, but polyethylene fibres ($\gamma_{\text{f-air}} = 31$ mJ m^{-2}) are not. Lack of wetting is also a problem for certain metal matrix composites (MMCs) and fibre coatings may be used to improve this.

7.1.2   Inter-Diffusion and Chemical Reactions

Various types of diffusional process can take place at the interface, with potential for promoting adhesion. For example, Fig. 7.1(b) depicts the diffusion of free chain ends at the interface between two polymers, which leads to chain entanglements and a rise in the adhesive strength. This effect is employed in some coupling agents used on fibres in thermoplastic matrices (see Section 7.3.1). Inter-diffusion can also take place in non-polymeric systems, particularly if it is accompanied by a chemical reaction. The adhesive strength is dependent on the nature of the resultant inter-atomic bonds (and also on the stresses generated by the reaction – see below).

Various types of chemical reaction may occur at the interface, either deliberately promoted or inadvertent. These can be represented, as in Fig. 7.1(c), by new A–B bonds being formed as a result of interfacial chemical reactions. These bonds may be covalent, ionic, metallic, etc. and they may be strong. There are many examples of the interfacial bond strength being raised by localised chemical reactions, although it is often observed that a progressive reaction occurs, resulting in the formation of a thick layer of a brittle product.

Carbon fibres are prone to surface reactions with organic groups. An important feature is the angle that the basal planes make on intersection with the free surface (see Section 2.1), since many reactions take place preferentially at the edges of these planes. For example, high modulus PAN-based fibres have a thick skin, with basal planes predominantly parallel to the surface. In this case, reaction takes place less readily than for carbon fibres having their basal planes inclined at large angles to the surface. Heat treatment of such fibres prior to composite fabrication can raise the bond strength, as a result of oxidation of the fibres and removal of such surface layers.

There are many instances of bond strength enhancement being stimulated by localised chemical reactions in MMCs and ceramic matrix composites (CMCs). For example, the 'Saffil' δ-alumina fibre contains a few per cent of silica, concentrated in the free surface and grain boundary regions. During exposure of this fibre to molten metals containing strong reducing agents, such as Mg, the surface is attacked [1] to the depth of silica enrichment (a few nanometres). This has been correlated with a considerable enhancement of bond strength compared with cases in which the matrix is free of Mg, presumably as a result of covalent and/or ionic bonds extending across the interface. A similar effect has been observed during the early stages of the reaction in Ti/SiC. However, in this case the reaction is a progressive one [2] that tends to result in a relatively thick layer of brittle reaction product (a mixture of Ti_5Si_3 and TiC).

7.1.3   Mechanical Keying

There may be a contribution to the strength of the interface from the surface roughness of the fibres, provided good interfacial contact has been established, as illustrated in Fig. 7.1(d). The effects are expected to be more significant under shear loading than for decohesion as a result of tensile stresses. Some improved resistance to tensile failure might result if re-entrant angles are present and there is expected to be an increase in strength under all types of loading as a consequence of the increased area of contact.

7.1.4   Electrostatic Attraction

If the surfaces carry net electrical charges of opposite sign, then a sustained adhesive force may result. This effect is utilised in certain fibre treatments, as in the deposition of coupling agents on glass fibres (see Section 7.3.1). The surface may exhibit anionic or cationic properties, depending on the oxide in the glass and the pH of the aqueous solution used to apply the coupling agents. Thus, if ionic functional silanes are used, it is expected that the cationic functional groups will be attracted to an anionic surface and vice versa. Electrostatic forces are unlikely to constitute the major adhesive bond in a composite and they can readily be reduced, for example by discharging in the presence of a strongly polar solvent, such as water.

7.1.5   Residual Stresses

The nature of the interfacial contact can be strongly affected by the presence of residual stresses. These can arise in several ways, including plastic deformation of the matrix and phase transformations involving volume changes. There are also volume changes associated with the curing of thermosetting resins. One of the most important sources of residual stress is thermal contraction during post-fabrication cooling. Since, for most composite systems, the fibre has a smaller thermal expansivity (thermal expansion coefficient) than the matrix, the resultant stresses are compressive in the fibre and tensile in the matrix. This arises because the matrix contracts onto the fibre and compresses it, as illustrated schematically in Fig. 7.2.

The stress state can be quantified for a given set of boundary conditions (including a physical representation of the composite system). For some sets, analytical solutions are available. An example of such a situation [3] is a single long fibre surrounded by a coaxial cylindrical envelope of matrix. The two radii can be chosen to correspond to a selected fibre volume fraction, although it should be appreciated that this does not correspond closely to the case of a large array of parallel fibres (since the constraint conditions at the free surface are different for the coaxial

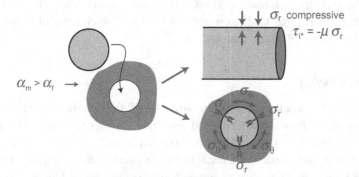

Fig. 7.2 Schematic representation of the type of residual stress state that arises as a result of differential thermal contraction during cooling of a fibre composite.

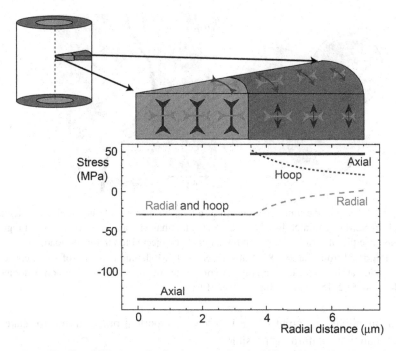

Fig. 7.3 Predicted [3] stress distribution around and within a single fibre, in a polyester–35% glass long fibre composite, from differential thermal contraction (T drop of 100 K).

model). Nevertheless, predictions from such a model, such as those shown in Fig. 7.3 (for a glass fibre surrounded by a tube of polyester resin, after cooling through 100°C), do give a good general indication of the magnitude and nature of the stresses. Of course, this is for elastic cooling: in practice, there would probably be some stress relaxation as a result of creep or viscous flow in the matrix, since stresses of the order of 50 MPa are relatively large for polymeric materials. In an MMC, on the other hand, the temperature changes during cooling are higher, and metals are usually more resistant to creep and plastic flow, so that the residual stresses are often higher.

The normal stresses across the interface (radial stresses) are usually compressive. This is particularly relevant to interfacial bonding, since this compression ensures that fibre and matrix are kept in close contact and increases the resistance to debonding and sliding. However, it should be noted that the nature of the stresses changes if the fibre volume fraction is so high that the matrix becomes broken up into isolated regions, surrounded on all sides by fibres. In this case, the matrix tends to contract away from the fibres during cooling, leading to tensile normal stress across the interface. This can make such regions vulnerable to interfacial debonding and crack propagation, since the residual stresses now tend to augment those from an applied (tensile) stress (transverse to the fibre axis), rather than counterbalancing them. This type of effect is illustrated in Fig. 7.4, which depicts how matrix isolation can occur where the local fibre volume fraction is high and also gives an example of crack propagation through such a region.

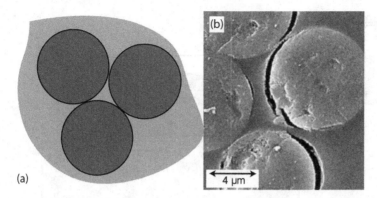

Fig. 7.4 (a) Schematic transverse section of a long fibre composite in a high-f (clustered) region; and (b) SEM micrograph [4] showing a crack that has propagated through such a region in a transverse ply of a carbon fibre–epoxy laminate (subjected to an applied load).
(b) Reprinted from Carbon, 82, Valentin, S. et al., Modelling evidence of stress concentration mitigation at the microscale in polymer composites by the addition of carbon nanotubes, 184–194., © 2015, with permission from Elsevier.

This highlights the point that it is often an important practical aim to make the fibre distribution as uniform as possible.

7.2   Experimental Measurement of Bond Strength

The nature of the interfacial bonding can affect the properties of composites in several ways, but it is not a simple matter to characterise the bond strength. This difficulty is associated both with the devising and implementation of suitable tests and with the interpretation of experimental data from them. Most measurements of bond strength involve shear debonding and sliding (often using the shear lag approach to interpret the data), with little or no attempt to change the normal stress across the interface. This is mainly because of the difficulty in applying such stresses to a cylindrical interface in a controlled manner. It might be argued that an interface exhibiting a high shear debonding stress would also be expected to offer strong resistance to a normal tensile stress. That this is not necessarily true can be seen from the fact that pronounced interfacial roughness is expected to raise the former (in shearing mode – i.e. mode II) while having little effect on the latter (in opening mode – i.e. mode I).

In relating data from interfacial tests to the macroscopic behaviour of the composite, it should therefore be borne in mind that the two cases may involve different interfacial stress states. Furthermore, some tests use artificial single fibre 'composites'. The interfaces in these specimens may differ from those in the corresponding real materials, because different manufacturing techniques are used and there are different degrees of constraint in the absence of neighbouring fibres. Finally, while interpretation of test data is commonly carried out in terms of critical stress levels, an energy-based (fracture mechanics) analysis is often more useful, at least for some purposes – see Chapter 9.

7.2.1   The Single Fibre Pull-out Test

This test is simple conceptually, but has limitations. Nevertheless, since it was originally proposed [5], it has been developed and used quite extensively [6–12]. The test is
illustrated schematically in Fig. 7.5. A single fibre, partially embedded within a matrix,
is extracted under an increasing tensile load. The load/displacement data are commonly
interpreted using an adaptation of the shear lag model. Stress distributions (normal in
the fibre and shear at the interface) are depicted in Fig. 7.5 at three different stages
during the test. Basic assumptions of the shear lag model (Section 6.1.1), such as no
shear strain in the fibre and no transfer of normal stress across the fibre end, are retained
in most treatments of this situation.

It is usually assumed that the peak in the load/displacement plot represents a
debonding event, occurring at an applied stress σ_*. The treatment then parallels that in
Section 6.1.1, but with the ratio of the radius of the matrix to that of the fibre, R/r,
retained rather than written in terms of the fibre volume fraction. The equation corresponding to Eqn (6.3) is therefore

$$\frac{d\sigma_f}{dx} = \frac{E_m(u_R - u_r)}{(1 + v_m)r^2 \ln(R/r)}$$ (7.3)

The displacement conditions are now written as

$$\frac{du_r}{dx} = \varepsilon_f, \qquad \frac{du_R}{dx} = 0$$ (7.4)

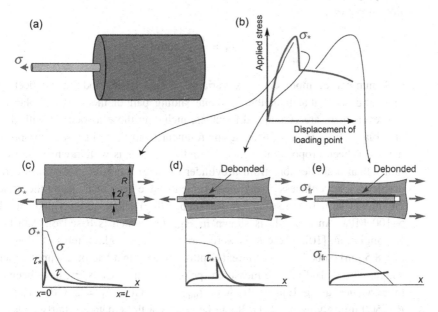

Fig. 7.5 Schematic representation of the single fibre pull-out test, showing (a) the test geometry;
(b) a stress–displacement plot and then the stress distributions (c) at the onset of debonding,
(d) during debonding and (e) during frictional pull-out.

corresponding to the interface being perfectly bonded (up until the debonding point) and the matrix remote from the interface being unstrained. The second-order linear differential equation governing the variation of σ_f along the length of the fibre is now slightly simpler than Eqn (6.7)

$$\frac{d^2\sigma_f}{dx^2} = \frac{n^2}{r^2}\sigma_f \qquad (7.5)$$

with the dimensionless constant given by

$$n = \left[\frac{E_m}{E_f(1 + \nu_m)\ln(R/r)}\right]^{1/2} \qquad (7.6)$$

Applying the boundary conditions $\sigma_f(0) = \sigma$ (where the fibre emerges from the matrix) and $\sigma_f(L) = 0$ (no stress transfer across the embedded fibre end) leads to

$$\sigma_f = \sigma\left[\frac{\sinh(n(L-x)/r)}{\sinh(nL/r)}\right] \qquad (7.7)$$

The interfacial shear stress, according to the basic equation of the shear lag model (Eqn (6.1)), then becomes

$$\tau = -\frac{r}{2}\frac{d\sigma}{dx} = \frac{n\sigma}{2}\cosh\left(\frac{n(L-x)}{r}\right)\text{cosech}\left(\frac{nL}{r}\right) \qquad (7.8)$$

On applying this at $x = 0$, the debonding shear stress, τ_* can be deduced from the peak fibre stress (σ_*)

$$\tau_* = \frac{n\sigma_*}{2}\coth\left(\frac{nL}{r}\right) \qquad (7.9)$$

A number of more complex variants of this basic model have been proposed, including some that treat the frictional sliding part of the stress–displacement plot. However, there are certainly difficulties, including those associated with the fact that an isolated fibre is in a different environment from those in a real composite. (Tests have also been proposed in which a bundle of fibres is withdrawn from a composite, rather than a single fibre from a cylinder of matrix, but this introduces certain other problems.) Nevertheless, many measurements have been made using the basic form of the model, usually yielding critical (debonding) shear stress values in the range 5–100 MPa. An example is shown in Fig. 7.6, which is based on data taken from Koyangi et al. [10]. These tests were carried out using glass fibres of diameter 17 μm ($r = 8.5$ μm) embedded to different depths in epoxy that had been cured either at room temperature or 100°C. The measured (peak) pull-out stresses, σ_*, have been converted to τ_* values using Eqn (7.9), with $E_m \sim 4.3$ GPa, $E_f = 72$ GPa, $\nu_m = 0.4$ and $R \sim 500$ μm (giving $n \sim 0.1$). It can be seen that these data are fairly consistent, with a clear indication that τ_* is higher when the matrix is cured at higher temperature (as might have been expected).

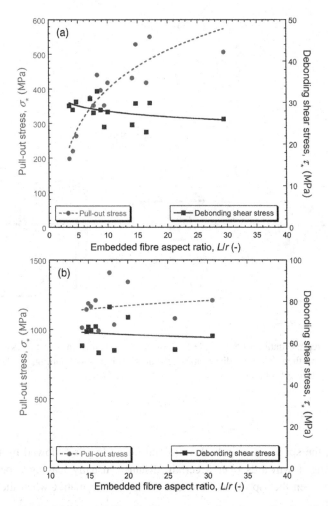

Fig. 7.6 Measured [10] pull-out stresses, and corresponding interfacial shear stresses for debonding (obtained using Eqn (7.9)), as a function of the ratio of embedded depth to fibre radius, for single glass fibres in epoxy resin that had been cured at: (a) room temperature (for two weeks) or (b) 100°C (for two hours).

7.2.2 The Single Fibre Push-out and Push-down Tests

These tests are easier to apply to pieces of actual composite than is the pull-out test, partly because of the wide availability of *instrumented indenters* ('nanoindenters'). These allow an indenter head (which can be cylindrical) to be precisely located (above a fibre) and also incorporate facilities for accurate measurement and control of (small) applied forces (and displacements). The test involves applying a compressive axial load to the top surface of an embedded fibre, until debonding occurs. In the *push-out* test, the specimen is in the form of a thin slice, with the fibre axis normal to the plane of the slice. The fibre becomes displaced so that it protrudes from the bottom of the specimen. In the

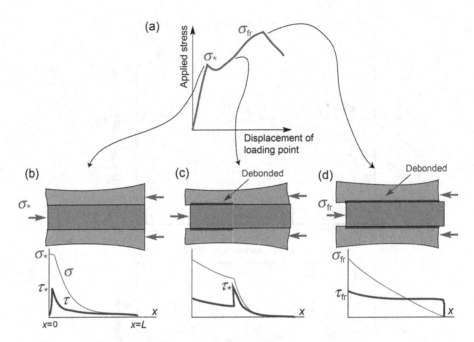

Fig. 7.7 Schematic representation of the single fibre push-out test, showing (a) a stress–displacement plot and then the stress distributions (b) at the onset of debonding, (c) during debonding and (d) during frictional push-out.

push-down test, the specimen is in bulk form and debonding is followed by the fibre frictionally sliding downwards over a certain distance, usually leaving a permanent displacement between the top of the fibre and the top of the matrix when the applied load is removed.

The test was devised [13,14] in the 1980s and has been extensively developed and used in the period since then [15–18]. Schematic illustrations of the stress distributions and load–displacement curve are shown in Fig. 7.7 for the push-out test. In contrast to the pull-out test, the Poisson effect (elastic expansion of the compressed fibre) now raises, rather than lowers, the frictional sliding stress. Analyses have been based on a shear lag approach, but finite element method (FEM) modelling has indicated that this often leads to overestimates of the peak interfacial shear stress for a given applied load. This is illustrated by the data in Fig. 7.8, which compares FEM and shear lag predictions of the shear stress distribution with experimental data from a 'macroscopic' push-out set-up, constructed from two photo-elastic resins. The peak of the shear lag curve is apparently too high by a factor of three in this case.

Since FEM work has indicated that the distribution of interfacial shear stress tends to be more uniform than the shear lag model predicts, it may be acceptable in many cases to take it as constant. A simple force balance can then be used to obtain τ_* from the stress applied to the fibre σ_*

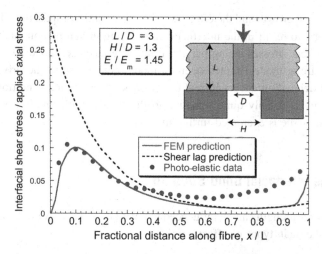

Fig. 7.8 Comparison [19] between experimental (photo-elastic) data and predictions from the shear lag model, and from FEM modelling, for the distribution of interfacial shear stress along the length of a fibre, starting from the top, during push-out of a resin 'fibre' in a matrix of a different resin.

$$\sigma_* \pi r^2 = \tau_* L(2\pi r)$$
$$\therefore \tau_* = \frac{\sigma_*}{4s} \tag{7.10}$$

where the fibre aspect ratio, s, is given by its length, L, divided by its diameter. However, FEM calculations have also highlighted the significance of thermal residual stresses, which in many polymer matrix composite (PMC) and MMC systems will strongly affect the stress distributions.

7.2.3 Other Tests

Several other tests are in use. These include the so-called 'full-fragmentation' technique. This procedure for deducing a shear strength involves embedding a single fibre in a matrix and straining the matrix in tension parallel to the fibre. The fibre fractures and fragments into a number of segments. The aspect ratios exhibited by these are measured and a value for the shear strength is inferred. Background theory for this procedure has been provided by a number of authors [20–25]. Analysis is often based on a constant τ, with the Weibull modulus (Section 2.2.2) of the fibre taken into account.

It may be noted that, for this type of test, and also for the push-out and pull-out tests, various procedures have been proposed for the extraction, not only of critical stress values, but also of interfacial fracture energies. These are certainly of considerable interest, particularly when considering the energetics of crack propagation in composites – i.e. their toughness (see Chapter 9). In fact, in general, a fracture energy value for an interface is a more fundamental measure of the resistance it offers to debonding (interfacial cracking) than is a critical stress value. There are, however, various complications in attempting to measure interfacial fracture energies, including the need to establish stable crack growth and the effect of the mode of crack propagation. These test geometries all tend to result in the interfacial crack being driven predominantly under the influence of shear stresses acting parallel to the interface (i.e. mode II). However, there is often a strong mode I (crack-opening mode) component to the driving force when interfacial debonding occurs in real composites under service conditions. Since the fracture energy is usually quite sensitive to the 'mode mix', this is a significant complication. The issue is examined further in Chapter 9.

7.3   Controlling the Interfacial Bond Strength

The interfacial bond strength can be influenced in a variety of ways. A brief summary is given below of the main types of effect involved.

7.3.1   Coupling Agents and Environmental Effects

Many coatings have been developed to improve the durability and mechanical strength of the fibre–matrix bond and these are usually termed *coupling agents* or *sizing layers* (or 'sizes'). A good example is provided by those used on glass fibres, which often suffer from problems caused by pickup of water. Some of the oxides in glass, such as SiO_2, Fe_2O_3 and Al_2O_3, form links to hydroxyl groups during contact with water and these in turn form hydrogen bonds to water molecules, so that glass picks up water very rapidly. In time, this can leach out other species in the glass, notably 'network modifiers' such as Na^+ and Ca^{2+}, to leave a weak, porous surface. The presence of the water also reduces the wettability of the fibres, potentially reducing γ from 500–600 mJ m^{-2} to something like 10–20 mJ m^{-2}. In general, the interfacial shear strength falls as polymer composites are exposed to water, although with some thermo-plastic matrices an increase has been observed [26].

Coatings that function as coupling agents are designed to eliminate the leaching effect and raise the effective γ value at least to ~40–50 mJ m^{-2}. The primary function of the coupling agent is to provide a strong chemical link between the oxide groups on the fibre surface and the polymer molecules of the resin. A wide variety of commercial coupling agents have been developed [27–30], but the principles can be illustrated by the simple example shown in Fig. 7.9. This refers to *silane coupling agents* with the general chemical formula $R–Si–X_3$. This is a multi-functional molecule that reacts at one end with the surface of the glass and at the other end with the polymer phase. The X units represent hydrolysable groups such as the ethoxy group ($–OC_2H_5$). The silane is hydrolysed to the corresponding silanol in the aqueous solution to which the fibres are exposed. These silanol molecules compete with water molecules to form hydrogen bonds with the hydroxyl groups bound to the fibre surface. When the fibres are dried, the free water is driven off and condensation reactions then occur, both at the silanol–fibre junction and between neighbouring silanol molecules. The result is a polysiloxane layer bonded to the glass surface, presenting an array of R groups to the environment.

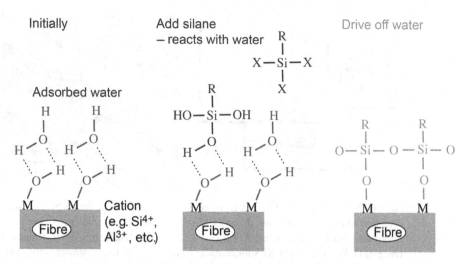

Fig. 7.9 Depiction of the action of silane coupling agents used to generate improved interfacial bonding for glass fibres in polymeric matrices. The silane reacts with adsorbed water to create a strong bond to the glass surface. The R group is one that can bond strongly to the matrix.

The surface produced is water-resistant and can also form a strong bond to a polymer matrix. If the matrix is to be a thermosetting resin, then an R group is chosen which reacts with the resin during polymerisation, thus forming a permanent link. For a thermoplastic matrix, on the other hand, all the covalent links have been formed during manufacture of the polymer. However, choice of R with a fairly short chain that can inter-diffuse with the chains of the matrix allows a strong bond to form.

It should, however, be noted how these layers are actually deposited. This is normally done immediately after the fibre has been manufactured (by spinning from a bath of molten glass), a process that is carried out at high speed. The deposition takes place via contact with an applicator roll carrying a layer of an aqueous mixture of coupling agents, lubricants, emulsifying polymers, etc. [30,31]. Pickup of the sizing typically occurs over a short contact distance (~10 mm), with a contact time of less than 0.5 ms – the drawing speed is typically ~20 m s^{-1}. Of course, production rates of this order are driven by economic factors, but one outcome is that the sizing layer is unlikely to be uniform or complete. Fig. 7.10 gives an impression [31] of typical physical dimensions and distributions of such layers. Nevertheless, they are usually effective in promoting strong and waterproof interfacial bonding. However, it is noteworthy that deposited layers may be damaged during handling (friction between fibres) and it is often important for fibres to be kept separate immediately after deposition and for composite production to take place soon afterwards.

For MMCs, it is less common for promotion of good bonding to be necessary, because some local chemical reaction often occurs naturally during fabrication. There are, however, some systems in which wetting is very poor and coatings have been used to improve this. For CMCs, although various types of coating have been developed, these are rarely designed to improve wetting or adhesion. Fibres are usually added to ceramic matrices in order to improve the toughness. Certainly, when the fibres are

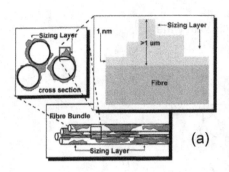
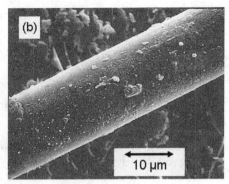

Fig. 7.10 Nature of a typical 'sizing layer' on glass fibres [31], illustrated by (a) a schematic depiction of its dimensions and (b) an SEM micrograph of a 'coated' glass fibre.
Reprinted from Composites Part A: Applied Science and Manufacturing, 30, Thomason, J. L. and Dwight, D. W., The use of XPS for characterisation of glass fibre coatings, 1401–1413, © 1999, with permission from Elsevier.

ceramic, a relatively low debonding stress is often preferred, to promote frictional sliding during fibre pull-out (see Section 9.2.3), although this may not be the case when the fibres are metallic (see Section 9.3.5).

7.3.2   Interfacial Chemical Reaction and Diffusion Barrier Coatings

Interfacial reaction can be quite extensive in certain types of composite, particularly with metallic matrices. It may occur during composite fabrication and/or under in-service conditions. There are many fibre–matrix combinations for which chemical reaction is thermodynamically favoured. Extensive reaction is usually undesirable, since it tends to promote interfacial cracking and the reaction product itself is often a brittle ceramic or intermetallic compound. Techniques for avoiding excessive reaction depend either on relatively slow reaction kinetics or on the provision of some protective layer (diffusion barrier).

Interfacial chemical reactions are of particular concern for Ti-based composites. Titanium and its alloys tend to react with most reinforcements and there is interest in their use at elevated temperatures (550–700°C). Furthermore, in the case of Ti, the surface oxide film that usually protects such reactive metals tends to dissolve in the matrix at temperatures above ~600°C. Titanium reacts with virtually all reinforcement materials during fabrication, which requires temperatures of at least about 850°C. In particular, quite substantial reaction occurs during fabrication with SiC monofilaments [2,32], which is the most promising of the available long fibre reinforcements for use in Ti.

It is common to counter this with fibre coatings, some of which are regarded as sacrificial – that is to say, they are at least partially consumed, but without impairing the properties of the fibre. For example, Fig. 7.11 shows interfacial regions in two Ti-based MMCs containing SiC monofilaments with C-rich surface coatings [33]. Excessive

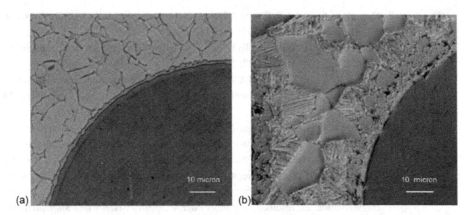

Fig. 7.11 SEM micrographs [33] of Ti-6Al-4V reinforced with SiC monofilaments after consolidation for several hours at (a) 900°C and (b) 975°C.
Reprinted from Elsevier Books, Doorbar, P. J. and Kyle-Henney, S., Comprehensive Composite Materials II, 439–463, © 2018, with permission from Elsevier.

temperature during manufacture (hot pressing of fibre–foil layers) has in one of these two cases (Fig. 7.11(b)) caused complete consumption of the coating, leading to fibre damage. In general, fewer problems of interfacial reaction arise with Al, particularly during solid state processing. Prolonged exposure of SiC to an Al melt does cause chemical reaction, but this is not a severe problem even for casting routes. Reaction problems can occur with Mg alloys, although, since Mg does not form a stable carbide, it is thermodynamically stable in contact with SiC. It does, however, tend to react with (reduce) most types of oxide reinforcement.

References

1. Cappleman, GR, JF Watts and TW Clyne, The interface region in squeeze-infiltrated composites containing delta-alumina fiber in an aluminium matrix. *Journal of Materials Science* 1985; **20**(6): 2159–2168.
2. Martineau, P, M Lahaye, R Pailler, R Naslain, M Couzi and F Cruege, SiC filament/titanium matrix composites regarded as model composites: part 2. Fibre/matrix chemical interactions at high temperatures. *Journal of Materials Science* 1984; **19**: 2749–2770.
3. Warwick, CM and TW Clyne, Development of a composite coaxial cylinder stress analysis model and its application to SiC monofilament systems. *Journal of Materials Science.* 1991; **26**: 3817–3827.
4. Romanov, VS, SV Lomov, I Verpoest and L Gorbatikh, Modelling evidence of stress concentration mitigation at the micro-scale in polymer composites by the addition of carbon nanotubes. *Carbon* 2015; **82**: 184–194.
5. Lawrence, P, Some theoretical considerations of fibre pullout from an elastic matrix. *Journal of Materials Science* 1972; **7**: 1–6.
6. Chua, PS and MR Piggott, The glass fibre–polymer interface: I. Theoretical considerations for single fibre pullout tests. *Composites Science and Technology* 1985; **22**: 33–42.

7. Desarmot, G and J-P Favre, Advances in pullout testing and data analysis. *Composites Science and Technology* 1991; **42**: 151–187.

8. DiFrancia, C, TC Ward and RO Claus, The single-fibre pull-out test 1: review and interpretation. *Composites A* 1996; **27**: 597–612.

9. Fu, SY, CY Yue, X Hu and YW Mai, Analyses of the micromechanics of stress transfer in single- and multi-fiber pull-out tests. *Composites Science and Technology* 2000; **60**(4): 569–579.

10. Koyanagi, J, H Nakatani and S Ogihara, Comparison of glass–epoxy interface strengths examined by cruciform specimen and single-fiber pull-out tests under combined stress state. *Composites Part A: Applied Science and Manufacturing* 2012; **43**(11): 1819–1827.

11. Graupner, N, J Rossler, G Ziegmann and J Mussig, Fibre/matrix adhesion of cellulose fibres in PLA, PP and MAPP: a critical review of pull-out test, microbond test and single fibre fragmentation test results. *Composites Part A: Applied Science and Manufacturing* 2014; **63**: 133–148.

12. Frikha, M, H Nouri, S Guessasma, F Roger and C Bradai, Interfacial behaviour from pull-out tests of steel and aluminium fibres in unsaturated polyester matrix. *Journal of Materials Science* 2017; **52**(24): 13829–13840.

13. Marshall, DB, An indentation method for measuring matrix–fiber frictional stresses in ceramic composites. *Journal of the American Ceramic Society* 1984; **67**(12): C259–C260.

14. Shetty, DK, Shear-lag analysis of fibre pushout indentation tests for evaluating interfacial friction stress in ceramic matrix composites. *Journal of the American Ceramic Society* 1988; **71**: C107–C109.

15. Kallas, MN, DA Koss, HT Hahn and JR Hellman, Interfacial stress state present in a 'thin slice' fiber push-out test. *Journal of Materials Science* 1992; **27**: 3821–3826.

16. Rebillat, F, J Lamon, R Naslain, E Lara-Curzio, MK Ferber and TM Besmann, Interfacial bond strength in SiC/C/SiC composite materials, as studied by single-fiber push-out tests. *Journal of the American Ceramic Society* 1998; **81**(4): 965–978.

17. Kerans, RJ, TA Parthasarathy, F Rebillat and J Lamon, Interface properties in high-strength Nicalon/C/SiC composites, as determined by rough surface analysis of fiber push-out tests. *Journal of the American Ceramic Society* 1998; **81**(7): 1881–1887.

18. Mueller, WM, J Moosburger-Will, MGR Sause, M Greisel and S Horn, Quantification of crack area in ceramic matrix composites at single-fiber push-out testing and influence of pyrocarbon fiber coating thickness on interfacial fracture toughness. *Journal of the European Ceramic Society* 2015; **35**(11): 2981–2989.

19. Kalton, AF, CM Ward-Close and TW Clyne, Development of the tensioned push-out test for study of fibre-matrix interfaces. *Composites* 1994; **25**: 637–644.

20. Curtin, WA, Exact theory of fiber fragmentation in a single-filament composite. *Journal of Materials Science* 1991; **26**(19): 5239–5253.

21. Hui, CY, SL Phoenix, M Ibnabdeljalil and RL Smith, An exact closed-form solution for fragmentation of Weibull fibers in a single filament composite with applications to fiber-reinforced ceramics. *Journal of the Mechanics and Physics of Solids* 1995; **43**(10): 1551–1585.

22. Tripathi, D and FR Jones, Single fibre fragmentation test for assessing adhesion in fibre reinforced composites. *Journal of Materials Science* 1998; **33**(1): 1–16.

23. Kim, BW and JA Nairn, Observations of fiber fracture and interfacial debonding phenomena using the fragmentation test in single fiber composites. *Journal of Composite Materials* 2002; **36**(15): 1825–1858.

24. Graciani, E, V Mantic, F Paris and J Varna, Numerical analysis of debond propagation in the single fibre fragmentation test. *Composites Science and Technology* 2009; **69**(15–16): 2514–2520.

25. Sorensen, BF, Micromechanical model of the single fiber fragmentation test. *Mechanics of Materials* 2017; **104**: 38–48.

26. Gaur, U, CT Chou and B Miller, Effect of hydrothermal aging on bond strength. *Composites* 1994; **25**(7): 609–612.

27. Almoussawi, H, EK Drown and LT Drzal, The silane sizing composite interphase. *Polymer Composites* 1993; **14**(3): 195–200.

28. Berg, J and FR Jones, The role of sizing resins, coupling agents and their blends on the formation of the interphase in glass fibre composites. *Composites Part A: Applied Science and Manufacturing* 1998; **29**(9–10): 1261–1272.

29. Tran, DT, DM Nguyen, CNH Thuc and TT Dang, Effect of coupling agents on the properties of bamboo fiber-reinforced unsaturated polyester resin composites. *Composite Interfaces* 2013; **20**(5): 343–353.

30. Dwight, DW and S Begum, Glass fiber reinforcements, in *Comprehensive Composite Materials II*, Gdoutos, EE, editor. Elsevier, 2018, pp. 243–268.

31. Thomason, JL and DW Dwight, The use of XPS for characterisation of glass fibre coatings. *Composites Part A: Applied Science and Manufacturing* 1999; **30**(12): 1401–1413.

32. Choy, KL and B Derby, Evaluation of the efficiency of TiB2 and TiC as protective coatings for SiC monofilament in titanium-based composites. *Journal of Materials Science* 1994; **29**(14): 3774–3780.

33. Doorbar, PJ and S Kyle-Henney, Development of continuously-reinforced metal matrix composites for aerospace applications, in *Comprehensive Composite Materials II*, Clyne, TW, editor. Elsevier, 2018, pp. 439–463.

8 Stress-Based Treatment of the Strength of Composites

The elastic behaviour of long and short fibre composites is described in Chapters 4–6. This involves considering the stresses in individual plies of a laminate (under an external load) and stress distributions within and around short fibres. This information is now used to explore how a composite material suffers microstructural damage, potentially leading to ultimate failure of some sort. There are two distinct aspects to these (highly important) characteristics. Firstly, there is the onset and development of microstructural damage (mainly cracking of various types) as a function of applied load. Secondly, there are the processes that cause absorption of energy within a composite material as it undergoes such failure and fracture. The latter determine the toughness of the material and are treated on a fracture mechanics basis in Chapter 9. In the present chapter, attention is concentrated on predicting how applied stresses create stress distributions within the composite and how these lead to damage and failure. The treatment is largely oriented towards long fibre composites (particularly laminates), and also towards polymer-based composites, although most of the principles apply equally to discontinuous reinforcement and other types of matrix.

8.1   Failure Modes and Strength of Long Fibre Composites

Application of an arbitrary stress state to a unidirectional lamina can lead to failure by one or more basic processes. The three most important types of failure are illustrated in Fig. 8.1. Large tensile stresses parallel to the fibres, σ_1, may cause fibre and matrix to fracture, with the fracture path approximately normal to the fibre direction. The strength is normally much lower in the transverse tension and shear modes, with the composite then fracturing on surfaces parallel to the fibre direction when appropriate σ_2 or τ_{12} stresses are applied. In these cases, fracture may occur entirely within the matrix or partly within the fibre–matrix interface and/or the fibre. To predict the strength of a lamina or laminate, values of the failure stresses σ_{1*}, σ_{2*} and τ_{12*} have to be determined.

8.1.1   Axial Tensile Strength

Understanding of failure under an applied tensile stress parallel to the fibres is relatively simple, provided that both constituents behave elastically and fail in a brittle manner. They then experience the same axial strain and hence sustain stresses in the same ratio as their Young's moduli. While it is possible to carry out various analyses, depending

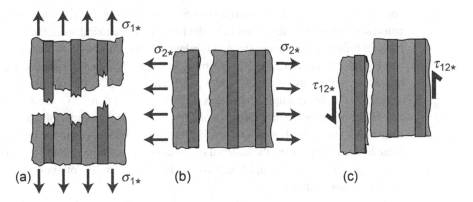

Fig. 8.1 Schematic depiction of the fracture of a unidirectional long fibre composite at critical values of (a) axial, (b) transverse and (c) shear stresses.

on which constituent has the lower fracture strain, and on assumptions made about load transfer after one of them has started to break up, it is not really appropriate to be too concerned about these details. This is because, at least in most cases, the fibre is much stronger and stiffer than the matrix. It therefore carries most of the applied load, and essentially dictates both the stiffness and the strength of the composite (under axial loading). While a rule of mixtures is reliable for the axial modulus (Eqn (3.3)), the axial strength can often be assumed to be given by

$$\sigma_{1*} \approx f\sigma_{f*} \tag{8.1}$$

This simply neglects the contribution from the matrix, but that is often a good approximation (for polymer matrix composites (PMCs)). The main point to note here is that the value of σ_{1*} is usually relatively large (compared with σ_{2*} and τ_{12*}). It can be seen in Table 2.1 (Section 2.1) that σ_{f*} values are typically of the order of 2–3 GPa, or possibly rather more. Some individual fibres will inevitably be weaker and their failure tends to trigger that of neighbours, so Eqn (8.1) tends to give something of an overestimate; nevertheless, the axial failure stress of well-aligned long fibre composites is typically in the range 500–800 MPa, which is a high strength (above the ultimate tensile stress of many metals). It is in any event a property that is best just measured experimentally, rather than attempting to predict it.

8.1.2   Transverse and Shear Strength

The above statement applies even more strongly to the values of σ_{2*} and τ_{12*} (the failure stresses in transverse tension and in shear, parallel to the fibre axis). This is entirely as expected when it is recognised that the interface plays a key role in both cases. Of course, knowing the properties of both constituents does not really give much guidance concerning the strength of the interface, which is affected by the processing conditions during production of the component. It is perhaps tempting to assume that, provided the interfacial bond is strong, these values would be dictated by properties (strength and possibly yield stress) of the matrix. However, even this is not really the case, since the constraint imposed by the presence of the fibres is likely to influence the way that these types of failure occur. In any event, it is relatively straightforward to simply measure σ_{2*} and τ_{12*} (see Section 8.2.3). The outcome, as might have been expected, is commonly that these values are relatively small. At least for PMCs, the relatively low matrix strength – see, for example, the values in Table 2.2 (Section 2.3) – ensures that typical values are of the order of a few tens of MPa (i.e. at least an order of magnitude lower than σ_{1*}). One of the key challenges in usage of composites is to ensure that, despite these low strength values, the material (component) as a whole is likely to remain elastic under typical (relatively high) applied loads. This is most commonly achieved via suitable laminate design.

8.1.3   Axial Compressive Strength: Kink Band Formation

Compressive axial failure of composites is not very common. This is largely because most composite components are designed to exploit the high tensile strength (at least parallel to the fibre axis) and often do not experience high compressive loads. However, there obviously are applications in which compressive stresses do arise – during bending of beams, for example. Also, composite components are sometimes used in situations in which the main loading is compressive – struts are sometimes made of (uniaxial) composites. The main reason for this tendency in terms of usage is simply that (axially loaded) composites are usually weaker in compression than in tension. This has long been a known feature of wood, for which the compressive strength (parallel to the grain – i.e. to the fibre axis) can be lower than that in tension by a factor of up to three. This is explored in Chapter 13, where the microstructure and properties of wood are investigated. In the meantime, some basic features of how aligned long fibre composites tend to fail under axial compression are now examined.

Compressive failure often takes place via some kind of lateral collapse of the fibres. This behaviour is related to the constraint imposed on them by the surrounding matrix, which is in turn dependent on its shear strength and that of the interface. Before examining how this occurs, however, it is important to distinguish true compressive failure from (elastic) *buckling*, which is dependent on the dimensions of the component. Of course, buckling could occur in a real component. However, it is not dependent on the compressive strength of the material, but only on its elastic properties (and the component dimensions).

True compressive failure most commonly occurs via the formation of *kink bands*. These are locations in which there has been some sort of concerted shearing, usually involving fibre fracture throughout two closely spaced, parallel planes, and often extending across the complete section of the component (when the loading is uniaxial). The micrographs shown in Fig. 8.2 illustrate the general nature of this type of failure, which has been studied and analysed over an extended period [1–5].

A feature of kink band formation that has become fairly clear is that it is more likely to occur in a region of the material where the fibres are not well-aligned with the loading

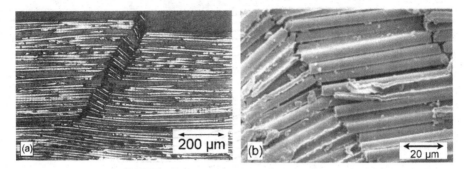

Fig. 8.2 Kink bands formed during testing of composites containing 60 vol% carbon fibre in resin: (a) optical micrograph [1] of a polished section from the compressive side of a sample tested in bending; and (b) SEM micrograph [3] of a fracture surface from a sample tested in uniaxial compression.
Reprinted by permission from Springer Nature: Springer Nature, International Journal of Fracture, Compressive response and failure of fiber reinforced unidirectional composites, Lee, S. H. and Waas, A. M., © 1987.

axis. Compressive failure of this type thus tends to occur more readily in components exhibiting a degree of 'waviness' in the fibre orientation. The creation of kink bands can be modelled as shown in Fig. 8.3. For the situation shown in Fig. 8.3(b), it can be seen using the Mohr circle (Section 4.1.4) representation of Fig. 8.3(c) that

$$\tan(2\phi) = \frac{\tau_{12*}}{(\sigma_{k*}/2)} \tag{8.2}$$

where σ_{k*} is the applied stress when the kink band forms. Since ϕ is assumed in this treatment to be a small angle, the approximation $\tan(2\phi) \sim 2\phi$ can be used, leading to

$$\sigma_{k*} = \frac{\tau_{12*}}{\phi} \tag{8.3}$$

This gives the applied stress for kink band formation in terms of the shear strength of the composite (parallel to the fibre axis) and the misalignment, ϕ, between the direction of the applied stress and the local direction of the fibre axis in the region where the band is forming. This angle is actually made up of two contributions, an initial misalignment, ϕ_0, and the rotation caused by the shear stress that is acting at that point. Thus

$$\sigma_{k*} = \frac{\tau_{12*}}{\phi} = \frac{\tau_{12*}}{\phi_0 + \left(\dfrac{\tau_{12*}}{G_{12}}\right)} = \left(\frac{1}{G_{12}} + \frac{\phi_0}{\tau_{12*}}\right)^{-1} \tag{8.4}$$

where G_{12} is the shear modulus of the composite.

The behaviour predicted by Eqn (8.4) is illustrated by the plots in Fig. 8.4 for two different shear strength values and for shear modulus data corresponding to those of an epoxy–60% carbon fibre composite. It can be seen that the compressive strength is predicted to be quite sensitive to misalignment. In principle, well-aligned material could

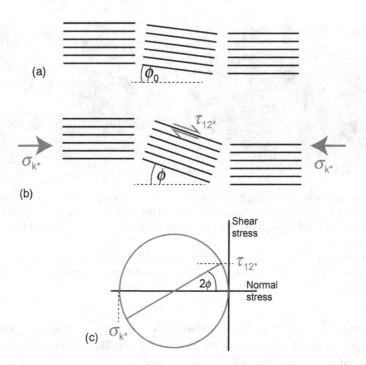

Fig. 8.3 Schematic representation of (a) fibres in a region of initial misalignment; (b) the situation at the point of formation of a kink band under compressive loading; and (c) a Mohr circle construction corresponding to that point.

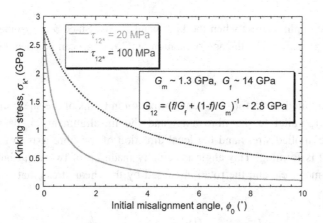

Fig. 8.4 Predicted kinking stress (Eqn (8.4)) as a function of initial misalignment angle for epoxy–60% carbon composites, with two different interfacial shear strengths.

exhibit very high strengths (~1–2 GPa in this case), although in practice it is almost impossible to avoid misalignments of at least about 1–2°, so such values are not usually observed. It is also difficult to characterise the probable misalignment in particular composite material, creating serious problems in quantitatively verifying Eqn (8.4). Nevertheless, many studies [6–9] have confirmed that composites with more severe 'fibre waviness' tend to have lower compressive strengths, and that this is commonly due to the formation of kink bands at least approximately as envisaged above.

8.1.4 Axial Compressive Strength: Other Failure Modes

Other types of compressive failure (during loading along the fibre axis) are sometimes observed. For example, kink band formation appears to be slightly less easy to stimulate in glass fibre composites than with carbon fibres. This might be associated with the lower strain to failure of carbon fibres. The tensile strengths of the two are not so different, commonly being ~3–4 GPa, but the greater stiffness of carbon fibres, ~300–500 GPa, compared with ~70 GPa for glass, means that their strains to failure are considerably lower. They thus tend to be more prone to fracture in bending, although this also depends on fibre diameter. In any event, glass fibre composites tend to be stronger in compression than those with carbon fibres, whereas the reverse is usually the case for their tensile strengths.

Linked to this trend, the failure mode in glass fibre composites is not always kink band formation. So-called 'splitting' failure [10–12] is often seen in glass fibre composites, particularly when the fibre volume fraction is relatively low ($f \sim$ 10–30%). This is the opening up of inter-fibre cracks, running parallel to the direction of applied stress. With higher volume fractions of glass fibres, it is quite common to see both types of failure mechanism operating simultaneously. This is illustrated in Fig. 8.5. In principle, there should be no normal (tensile) stress in the transverse direction, which would appear to be needed in order to open up such 'splitting' cracks. In practice, such stresses can arise in a variety of ways, particularly if there is any misalignment of the loading configuration or any form of end constraint. (A perfect uniaxial compression stress state is difficult to achieve – for example, friction on a platen-sample interface can lead to

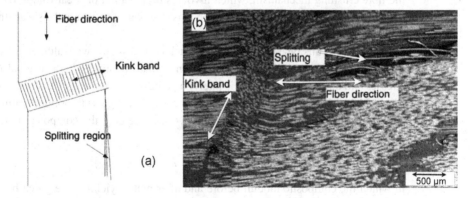

Fig. 8.5 Simultaneous operation of kink band formation and splitting during axial compression of an epoxy–60% glass fibre composite [10]: (a) schematic depiction of damage development; and (b) optical micrograph of a section through a sample after failure.

Reprinted from Composites Science and Technology, 60, Lee, S. et al., Compressive splitting response of glass-fiber reinforced unidirectional composites, 2957–2966, © 2000, with permission from Elsevier.

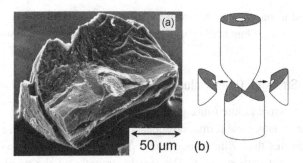

Fig. 8.6 Compressive failure of a fibre composite (Ti6Al-4V–35% SiC monofilament) with excellent fibre alignment and high transverse and shear strengths [13]: (a) a fragment of fibre extracted after failure; and (b) schematic depiction of the fibre crushing process.
Reprinted from Acta Materialia, 47, Sprowart, J. E. and Clyne, T. W., The axial compressive failure of titanium reinforced with silicon carbide monofilaments, 671–687, © 1999, with permission from Elsevier.

barrelling, and hence to development of transverse tension in the centre of the sample as the load builds up, while some types of compression test, and many situations with real components, involve even more severe constraint on the ends of the sample.) It should, of course, be noted that only relatively small transverse tensile stresses will in general be needed to cause this type of fracture – see Section 8.1.2.

A completely different kind of compressive failure is sometimes seen in well-aligned composites with high transverse and shear strengths. This involves crushing of the fibres, which is expected to require a very high compressive stress to be acting on them. An example is shown in Fig. 8.6, which shows a fragment of an SiC monofilament extracted from a metallic matric composite (MMC) based on a matrix of Ti-6Al-4V, after it had been subjected to axial compression. Also shown in Fig. 8.6 is a depiction of the fibre crushing mechanism, which involves propagation of shear (mode II) cracks at an angle of about 45° to the compression axis (the planes on which the shear stresses are at a maximum).

Modelling of this type of failure can be tackled in a simple way, although it should be recognised both that such composites often contain relatively high residual (thermal) stresses and that matrix plasticity is likely to precede fibre crushing. A plot of the type in Fig. 8.7, which shows how stresses in the system change with applied (compressive) strain, is helpful here. Neglecting matrix work hardening, the composite failure stress can be expressed as

$$\sigma_{c*} = E_{1c}\varepsilon_{cmY} + E'_{1c}(\varepsilon_{c*} - \varepsilon_{cmY}) \tag{8.5}$$

in which the composite moduli before and after matrix yielding are given by

$$E_{1c} = fE_f + (1-f)E_m \qquad E'_{1c} = fE_f$$

The strains at matrix yield, and at final failure, can be written

$$\varepsilon_{cmY} = \frac{\sigma_{mY} + \sigma_{m\Delta T}}{E_m} \qquad \varepsilon_{c*} = \frac{\sigma_{f*} + \sigma_{f\Delta T}}{E_f}$$

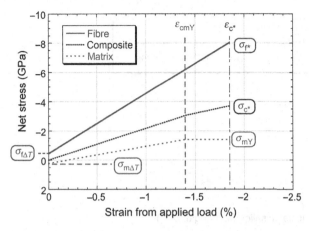

Fig. 8.7 Stresses (average axial values for fibre, matrix and composite) in a Ti-6Al-4V–35% SiC monofilament composite, as the axial strain increases during compressive loading [13]. At zero strain, these stresses are those from differential thermal contraction after production. The matrix yields at σ_{mY}. Matrix work hardening is neglected. Failure occurs when the fibre stress reaches a critical value σ_{f*}.

Substituting into Eqn (8.5), and applying a force balance $(f\sigma_{f\Delta T} + (1-f)\sigma_{m\Delta T} = 0)$

$$\sigma_{c*} = f\sigma_{f*} + (1-f)\sigma_{mY}$$

A (small) correction for the effect of misalignment then leads to

$$\sigma_{c*} = \frac{f\sigma_{f*} + (1-f)\sigma_{mY}}{\cos^2\phi_0} \tag{8.6}$$

This predicted strength is independent of the thermal residual stresses (whereas the strain at which failure occurs will depend on them).

Monofilament-reinforced composites of this type, made by a foil–fibre–foil vacuum hot pressing route, can have excellent fibre alignment (i.e. little or no 'waviness'). This allows samples to be created such that, during compression, there is a well-defined (pre-selected) misalignment angle (including a value very close to zero). The plot in Fig. 8.8 shows experimental compressive failure stress data obtained in this way [13]. Also shown in the figure are two predicted curves. One, for kink band formation, corresponds to Eqn (8.4), using appropriate values of the shear modulus, G_{12}, and shear strength τ_{12*}, for this composite. The other, for fibre crushing, is obtained using Eqn (8.6). The value of σ_{f*} used in this plot is the one giving the best fit to the experimental crushing (low ϕ_0) data. In fact, this procedure is probably the most reliable way to estimate σ_{f*}, since it is very difficult to carry out compressive tests on single fibres, even these monofilaments, without buckling affecting the results. The value obtained (~9 GPa) is about three times their tensile strength (which is easy to measure experimentally). This is certainly plausible, although accurate prediction is problematic, being dependent on the mode II fracture toughness and the flaw size distributions in individual fibres, neither of which are easy to obtain.

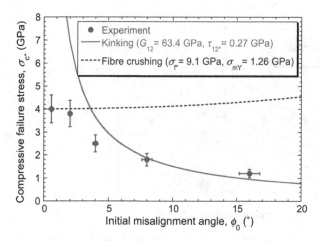

Fig. 8.8 Experimental strength data, as a function of initial angle between fibre and loading axes, during compression of Ti-6Al-4V–35% SiC specimens. Also shown are predictions for failure by kink band formation (Eqn (8.4)) and by fibre crushing (Eqn (8.6)).

8.2   Failure of Laminae under Off-Axis Loads

Laminae being subjected to an arbitrary (in-plane) stress state will undergo damage and failure via the three mechanisms (with defined values of σ_{1*}, σ_{2*} and τ_{12*}) shown in Fig. 8.1. This basic approach neglects the possibility of compressive failure, although this is taken into account in more advanced treatments – see Section 8.2.4. A number of *failure criteria* have been proposed. A key issue is whether or not the critical stress to trigger one mechanism is affected by the stresses tending to cause the others – i.e. whether there is any *interaction* between the modes of failure.

8.2.1   Maximum Stress Criterion

This is based on the simple assumption that failure occurs when a stress parallel or normal to the fibre axis reaches the appropriate critical value – that is, when one of the following is satisfied

$$\sigma_1 \geq \sigma_{1*}$$
$$\sigma_2 \geq \sigma_{2*} \qquad (8.7)$$
$$\tau_{12} \geq \tau_{12*}$$

For any applied stress state (σ_x, σ_y and τ_{xy}), the stresses parallel and transverse to the fibre axis are obtained using the transform equation Eqn (4.47)

$$\begin{vmatrix} \sigma_1 \\ \sigma_2 \\ \tau_{12} \end{vmatrix} = |T| \begin{vmatrix} \sigma_x \\ \sigma_y \\ \tau_{xy} \end{vmatrix}$$

in which the transform matrix is given by

$$|T| = \begin{vmatrix} c^2 & s^2 & 2cs \\ s^2 & c^2 & -2cs \\ -cs & cs & c^2 - s^2 \end{vmatrix}$$

with c representing $\cos\phi$ and s representing $\sin\phi$.

Monitoring of σ_1, σ_2 and τ_{12} (while the applied stress is increased) allows the onset of failure to be identified as the point at which one of the conditions in Eqn (8.7) is satisfied. Noting the form of $|T|$, and considering applied uniaxial tension, the magnitude of σ_x necessary to cause failure can be written in terms of the misalignment angle, ϕ, between stress axis and fibre axis, for each of the three failure modes

$$\sigma_{x*} = \frac{\sigma_{1*}}{\cos^2\phi} \tag{8.8}$$

$$\sigma_{x*} = \frac{\sigma_{2*}}{\sin^2\phi} \tag{8.9}$$

$$\sigma_{x*} = \frac{\tau_{12*}}{\sin\phi\cos\phi} \tag{8.10}$$

Plots of these three equations are shown in Fig. 8.9, obtained using the (typical) values of σ_{1*}, σ_{2*} and τ_{12*} indicated in the caption. The thick solid line indicates the predicted variation of the failure stress as ϕ is increased, according to the maximum stress criterion. Axial failure is in general expected only for very small misalignment angles, but the predicted transition from shear to transverse failure may occur anywhere between 20° and 50°, depending on the exact values of σ_{2*} and τ_{12*}. As with stiffness, it can be seen that the strength of a unidirectional composite tends to fall sharply as the

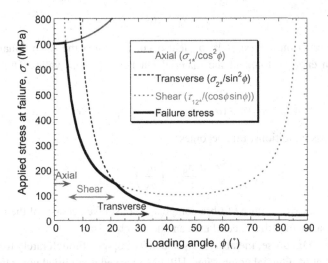

Fig. 8.9 Predicted failure stress during uniaxial loading, as a function of the loading angle, according to the maximum stress theory, for a polyester–50% glass fibre lamina (critical stress levels: $\sigma_{1*} = 700$ MPa, $\sigma_{2*} = 20$ MPa and $\tau_{12*} = 50$ MPa).

angle between fibre axis and loading axis starts to become significant ($>\sim 10°$). The reasons are also similar to the case of stiffness – i.e. as the angle increases, raising σ_2 and τ_{12}, the corresponding critical values start to be approached, in an analogous way to that in which the high compliance in these directions causes the Young's modulus to fall (Fig. 4.10(a)).

8.2.2   Tsai–Hill Criterion

The maximum stress criterion is clearly rather simplistic. In general (and not only in the context of composites), it is commonly observed that failure (fracture) occurs more readily when two modes are being significantly stimulated, compared to the situation in which either is acting in isolation. This can be rationalised on a mechanistic basis – for example, it is plausible that shear failure could occur more easily when transverse tension is simultaneously acting on the plane concerned.

In detail, this is a complex (and, of course, important) subject. There have been many proposed formulations [14–18] and various attempts to test them, including large-scale, coordinated trials [19–22]. The details of most of these are too advanced to be appropriate for coverage here. However, it may be noted that the simplest treatments are adaptations of yield criteria developed for metals. The most common yield criteria are those of *Tresca* and *von Mises*. The Tresca criterion corresponds to yield occurring when a critical value of the maximum shear stress is reached. This may be written as

$$[(\sigma_p - \sigma_q)^2 - \sigma_Y^2][(\sigma_q - \sigma_r)^2 - \sigma_Y^2][(\sigma_r - \sigma_p)^2 - \sigma_Y^2] = 0 \qquad (8.11)$$

where σ_p, σ_q and σ_r are the principal stresses and σ_Y is the yield stress under uniaxial loading. In-plane stress ($\sigma_r = 0$), Eqn (8.11) reduces to

$$\sigma_p - \sigma_q = \sigma_Y \qquad (8.12)$$

The von Mises criterion corresponds to yield occurring when the distortional (shape-changing) strain energy stored in the material reaches a critical value. This may be expressed as

$$(\sigma_p - \sigma_q)^2 + (\sigma_q - \sigma_r)^2 + (\sigma_r - \sigma_p)^2 = 2\sigma_Y^2 \qquad (8.13)$$

Under plane stress conditions, this becomes

$$\left(\frac{\sigma_p}{\sigma_Y}\right)^2 - \frac{\sigma_p \sigma_q}{\sigma_Y^2} + \left(\frac{\sigma_q}{\sigma_Y}\right)^2 = 1 \qquad (8.14)$$

Adaptation to describe failure of fibre composites must take account of their inherent anisotropy and also of differences between the mechanisms of metal yielding and composite failure. Of course, metals can also be anisotropic, although rarely to a degree comparable to that of uniaxial composites. Hill [23] derived a modified von Mises yield criterion for orthotropic metals (having three orthogonal planes of symmetry). Under plane stress, this criterion may be written as

$$\left(\frac{\sigma_1}{\sigma_{1Y}}\right)^2 + \left(\frac{\sigma_2}{\sigma_{2Y}}\right)^2 - \frac{\sigma_1\sigma_2}{\sigma_{1Y}^2} - \frac{\sigma_1\sigma_2}{\sigma_{2Y}^2} + \frac{\sigma_1\sigma_2}{\sigma_{3Y}^2} + \left(\frac{\tau_{12}}{\tau_{12Y}}\right)^2 = 1 \qquad (8.15)$$

where σ_1, σ_2 and τ_{12} are the imposed stresses, referred to the orthogonal directions in the plane, and the material properties σ_{1Y}, σ_{2Y}, σ_{3Y} and τ_{12Y} are the measured yield stresses, in tension and shear, when each is applied in isolation.

This criterion has been adapted to predict failure of unidirectional composites by replacing the yield stresses by measured failure stresses. Since such a composite is transversely isotropic, the failure stresses in the 2 and 3 directions are equal, so that the condition reduces to

$$\left(\frac{\sigma_1}{\sigma_{1*}}\right)^2 + \left(\frac{\sigma_2}{\sigma_{2*}}\right)^2 - \frac{\sigma_1\sigma_2}{\sigma_{1*}^2} + \left(\frac{\tau_{12}}{\tau_{12*}}\right)^2 = 1 \qquad (8.16)$$

This formulation was first proposed by Tsai and Azzi [14], and Eqn (8.16), commonly known as the *Tsai–Hill criterion*, is widely used in laminate analysis programs (see Section 8.3). It defines an envelope in stress space: if the stress state (σ_1, σ_2 and τ_{12}) lies outside of this envelope – i.e. if the sum of the terms on the left-hand side is equal to or greater than unity – then failure is predicted. The predominant failure mechanism is not specifically identified, although inspection of the relative magnitudes of the terms in Eqn (8.16) gives an indication of the likely contribution of the three modes. Care must be taken that the normal stresses are tensile, since the appropriate failure values are likely to be different in compression – see Sections 8.1.3 and 8.2.4.

The values of σ_1, σ_2 and τ_{12} can be obtained for a given misalignment, ϕ, between fibre axis and applied stress axis, from Eqn (4.47). For a single applied tensile stress σ_ϕ ($=\sigma_x$ at an angle ϕ to the fibre axis), substitution of these values into Eqn (8.16), and rearrangement, leads to the expression

$$\sigma_\phi = \left[\frac{\cos^2\phi(\cos^2\phi - \sin^2\phi)}{\sigma_{1*}^2} + \frac{\sin^4\phi}{\sigma_{2*}^2} + \frac{\cos^2\phi\sin^2\phi}{\tau_{12*}^2}\right]^{-1/2} \qquad (8.17)$$

This equation gives the applied stress at which failure is predicted to occur, as a function of the loading angle ϕ.

Fig. 8.10 shows plots of the Tsai–Hill and maximum stress criteria equations, obtained using the data shown in the caption for σ_{1*}, σ_{2*} and τ_{12*}, which are best-fit values for the experimental data [24] (obtained by cutting flat rectangular coupons at different angles to the fibre direction and testing them in tension). These experimental data fit rather better to the Tsai–Hill curve, notably for the specimen at $\phi = 30°$ – which was observed to fail by both shear and transverse tension.

8.2.3 Experimental Study of Off-Axis Failure

It is appropriate at this point to note that there are certain difficulties in carrying out tests designed to obtain critical stress data for different failure modes. In particular, there are several problems with the basic off-axis tensile test referred to above. One of these can

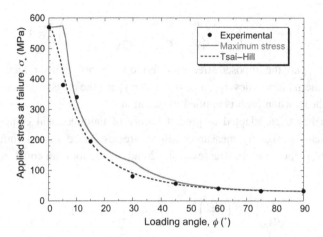

Fig. 8.10 Predicted failure stress during uniaxial loading, as a function of the loading angle, according to maximum stress and Tsai–Hill models. Also shown are experimental data [24] from off-axis tensile testing of epoxy–50% carbon fibre laminae. (The curves were obtained using critical stress levels of $\sigma_{1*} = 570$ MPa, $\sigma_{2*} = 32$ MPa and $\tau_{12*} = 56$ MPa.)

be seen in Fig. 4.12, which shows photographs of some simple 'macro-composites' (Al rods in polyurethane rubber) during this kind of testing. Tensile–shear interactions, which are particularly strong in the sample for which $\phi \sim 30°$, can cause distortion of the lamina under load, such that the constraint imposed by the gripping system introduces complex, non-uniform stresses. Furthermore, such specimens can be difficult to grip without slippage or deformation, edge effects may be significant and there is no freedom to select σ_1, σ_2 and τ_{12} independently (so as to discriminate more effectively between failure criteria).

A more versatile testing procedure is one based on tubular specimens, being tested under various combinations of applied stress. The performance of composite tube and pipe (particularly under internal pressure) is of commercial importance and this topic is covered in Section 8.4. A simple hoop-wound arrangement, with a capability for applying tension and/or torsion, is illustrated in Fig. 8.11. Any combination of σ_2 and τ_{12} can be applied, with a hoop stress (σ_1) of zero (although a tensile hoop stress can be generated if necessary, via internal pressurisation of the tube). For a thin-walled tube of diameter d and wall thickness t ($\ll d$), the shear stress in the wall is obtained from the applied torque T (in N m) via the expression[1]

$$\tau_{12} = \frac{2T}{\pi d^2 t} \tag{8.18}$$

Using a grip section with a larger diameter than the tube under test, as shown in Fig. 8.11, facilitates the application of a large torque, with less danger of slippage in the grips.

---

[1] This equation is obtained from the fact that the torque is given by the tangential force (shear stress times the sectional area on which it acts – i.e. $\pi\, d\, t$ in this case) multiplied by the radial distance from the axis ($d/2$).

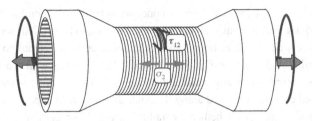

Fig. 8.11 Schematic illustration of how a hoop-wound tube is subjected to simultaneous tension and torsion in order to investigate failure mechanisms and criteria.

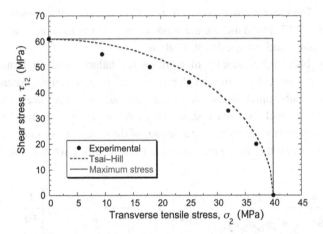

Fig. 8.12 Comparison between experimental failure data [25], obtained by combined tension/torsion testing of hoop-wound epoxy–65% glass composite tubes, and predictions from the maximum stress and Tsai–Hill criteria.

Data obtained in this way [25] are shown in Fig. 8.12, giving various combinations of σ_2 and τ_{12} that produced failure in epoxy–65% glass hoop-wound tubes. The failure envelopes for the two criteria are simple and give very different predictions for this case. The maximum stress theory gives the two lines $\sigma_2 = \sigma_{2*}$ and $\tau_{12} = \tau_{12*}$ as limits, while the Tsai–Hill condition, Eqn (8.16), reduces to

$$\left(\frac{\sigma_2}{\sigma_{2*}}\right)^2 + \left(\frac{\tau_{12}}{\tau_{12*}}\right)^2 = 1 \tag{8.19}$$

It is clear from the data in Fig. 8.12 that the Tsai–Hill criterion gives a more reliable prediction.

8.2.4 Incorporation of Compressive Loading into Failure Envelopes

The above treatment is focused on tensile and shear stresses generated within a lamina subjected to applied tensile and/or shear stress(es). This is a frequent scenario for practical use of composite materials, although these are commonly laminates, rather than single laminae – extension of the above concepts to laminates is outlined below in Section 8.3. Of course, it is not unusual for compressive loads to be applied (for example, on the compressive side of a bent beam) and these could certainly create compressive stresses both parallel and transverse to the fibre axis. However, as with tensile failure under a σ_{1*} stress, compressive failure under this kind of loading is not common (except for a unidirectional composite loaded along its axis). In other words, the value of σ_{1*} is large (relative to σ_{2*} and τ_{12*}) whether the failure is tensile or compressive. For the purposes of laminate analysis, it is therefore the values of σ_{2*} and τ_{12*} that tend to be of most significance. It is, however, worth noting that, while σ_{2*} is likely to be significantly higher in compression than in tension, compressive failure transverse to the fibre axis could still occur. Therefore, even if the σ_1 stresses are neglected, failure envelopes of the type shown in Fig. 8.12 should include the quadrant in which σ_2 has a negative sign. (Of course, shear stresses, such as τ_{12}, do not really have a sign.)

In any event, once the possibility of compressive failure (axial or transverse) is introduced, the potential level of complexity of the overall failure criterion rises considerably and a substantial number of different formulations have been put forward [15–18,26,27], some involving critical strain levels, as well as critical stress levels. It is not appropriate here to provide more than a flavour of these various formulations, which can be obtained from the following expression (sometimes termed the *Tsai–Wu criterion*)

$$\frac{\sigma_1^2}{\sigma_{1*}\sigma_{1*}'} + \frac{\sigma_2^2}{\sigma_{2*}\sigma_{2*}'} - \frac{F_{12}\sigma_1\sigma_2}{\sqrt{\sigma_{1*}\sigma_{1*}'\sigma_{2*}\sigma_{2*}'}} + \left(\frac{\tau_{12}}{\tau_{12*}}\right)^2 + \sigma_1\left[\frac{1}{\sigma_{1*}} - \frac{1}{\sigma_{1*}'}\right] + \sigma_2\left[\frac{1}{\sigma_{2*}} - \frac{1}{\sigma_{2*}'}\right] = 1$$

$$(8.20)$$

In this expression, the prime symbol indicates compression, so that σ_{1*}' is the (measured) compressive axial failure stress and σ_{2*}' is the compressive transverse failure stress. The interaction parameter, F_{12}, must be measured under biaxial loading. Of course, all such expressions are essentially empirical and the critical stress levels for failure, and any other parameters used in the criterion, must be found by experiment for the composite material concerned. It is then possible to make comparisons of the type shown in Fig. 8.12, in order to explore the validity of alternative formulations (incorporating compressive failure), although such operations can become quite complex [19–22,27]. Examples of such comparisons, using here only the Tsai–Wu criterion and experimental data from epoxy–glass and epoxy–carbon laminae [27], are shown in Fig. 8.13. These are, of course, essentially just curve-fitting exercises, but it can be seen that it is possible with such formulations to obtain quite good predictive capabilities (for specific composites).

8.3   Strength of Laminates

The strength of laminates can be predicted by an extension of the preceding sections, utilising the procedures set out in Chapter 5 for determination of the stresses in the

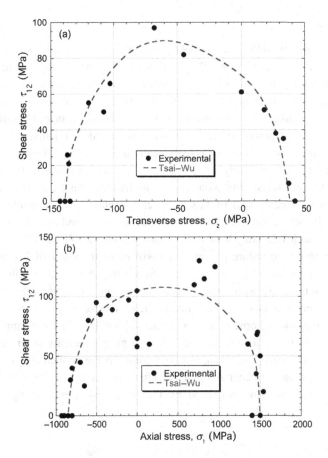

Fig. 8.13 Comparisons [27] between experimental failure data for laminae under multi-axial loading and corresponding predicted failure envelopes, according to the Tsai–Wu criterion (Eqn (8.20)), in (a) $\tau_{12} - \sigma_2$ space for an epoxy–glass fibre composite and (b) $\tau_{12} - \sigma_1$ space for an epoxy–carbon fibre composite.

component laminae. For example, Fig. 5.6 shows how the stresses within a cross-ply laminate under uniaxial tension vary with loading angle. Once these stresses are known (in terms of the applied load), an appropriate failure criterion can be applied and the onset and nature of the failure predicted. In the following sections, only the maximum stress and Tsai–Hill criteria are considered – i.e. only tensile and shear failure modes are treated, although extension of the methodology to handle other formulations (incorporating compressive failure) is at least conceptually straightforward.

It may be noted at this point that failure of an individual ply within a laminate does not necessarily mean that the component is no longer usable, as other plies may be capable of withstanding considerably greater loads without catastrophic failure. Analysis of the behaviour beyond the initial, fully elastic stage is complicated by uncertainties as to the degree to which the damaged plies continue to bear some load. Nevertheless, useful calculations can be made in this regime (although the major interest may be in the avoidance of *any* damage to the component).

8.3.1   Tensile Cracking

Consider first a cross-ply (0/90) laminate being loaded in tension along one of the fibre directions. The stresses acting in each ply of an epoxy–50% glass composite are plotted as a function of loading angle in Fig. 5.6(a) and shown schematically for the case of loading along one of the fibre directions in Fig. 5.6(b). Only transverse or axial tensile failure is possible in either ply, since no shear stresses act on the planes parallel to the fibre directions. The sequence of failure events when such a laminate is loaded progressively in this way is illustrated schematically in Fig. 8.14. It is easy to confirm that the transverse plies are likely to fail first, simply by considering the expected magnitudes of the transverse and axial strains to failure. The axial and transverse Young's moduli of a composite of this type (epoxy–glass) are, from Fig. 4.10(a), ~40 GPa and ~10 GPa respectively. The tensile strengths (σ_{1*} and σ_{2*}) are, as outlined in Sections 8.1.1 and 8.1.2, likely to be of the order of 500–800 MPa and 20–50 MPa. It follows that the strains to failure (ε_{1*} and ε_{2*}) will be of the order of 2% and 0.3% respectively, so that (since the two plies experience the same strain) failure of the transverse ply will normally occur first (despite the fact that, as shown in Fig. 8.14, the stress in the axial ply is almost six times that in the transverse ply).

Since most of the load is borne by the axial plies (about 85% for the example shown in Fig. 5.6), cracking of the transverse plies does not greatly increase the σ_1 stress and the axial plies usually remain undamaged at this point. As the applied stress increases, the next type of damage to occur is often cracking parallel to the fibres in the axial plies – see Fig. 8.14(b). This is caused by the tensile σ_2 stress, arising from the

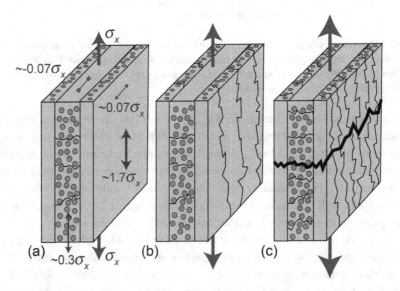

Fig. 8.14 Loading of the cross-ply laminate of Fig. 5.6 parallel to one of the fibre directions: (a) cracking of transverse plies as σ_2 reaches σ_{2*}; (b) onset of cracking parallel to fibres in axial plies as σ_2 (from inhibition of Poisson contraction) reaches σ_{2*}; and (c) final failure as σ_1 in axial plies reaches σ_{1*}.

resistance of the transverse plies to the (greater) lateral Poisson contraction of the axial plies. That an effect of this type is to be expected is clear from Fig. 4.13, where it can be seen that, while a typical Poisson ratio for an axial ply might be around 0.3, that for a transverse ply is about 0.05. (The fibres strongly inhibit the Poisson contraction under transverse loading.) It can be seen in Fig. 5.6 that these (differential Poisson contraction) stresses are about 7% of the applied stress. They are compressive in the transverse ply and tensile in the axial ply, where the cracking occurs.

For the purposes of comparing σ_1 and σ_2 in the axial plies, a limiting case is to assume that cracking of the transverse plies has made them stress-free in the loading direction, so that σ_1 in the axial plies reaches twice the applied stress – that is to say, all of the applied load is taken by the axial ply, which comprises half the thickness of the specimen. Under these circumstances, σ_1 is almost 30 times larger than σ_2 within the axial plies. However, the ratio $(\sigma_{1*}/\sigma_{2*})$ is typically (for polymer composites) in the range 30–60. Hence, transverse cracks in the axial plies are frequently observed before the final failure depicted in Fig. 8.14(c).

Effects such as these can be explored in the *Mechanics of Fibre Reinforced Composites* educational software package referred to in Section 5.1.3. One of the resources available there (www.doitpoms.ac.uk/tlplib/fibre_composites/laminate_failure.php) allows study of arbitrary cases (laminate stacking sequence and type of applied load), modelling the stresses within each ply and predicting the onset of failure (according to maximum stress and Tsai–Hill criteria). A screenshot from the page concerned is shown in Fig. 8.15.

Although the laminate depicted in Fig. 8.14(b) has not completely fractured, there is extensive micro-damage (internal cracking). The network of cracks is such that the laminate is susceptible to the passage of gases and liquids through the walls, a point of particular relevance for pressure vessels. In addition to concern about leakage, the

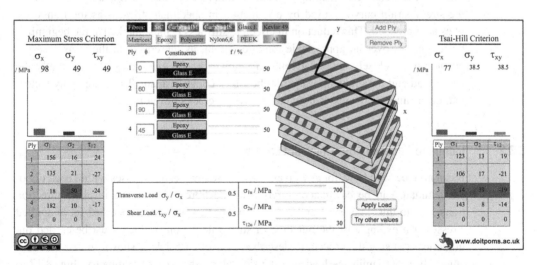

Fig. 8.15 Screenshot taken during usage of the *Failure of Laminates and the Tsai–Hill Criterion* page in the DoITPoMS TLP on Composites, accessible at www.doitpoms.ac.uk/tlplib/fibre_composites/laminate_failure.php.

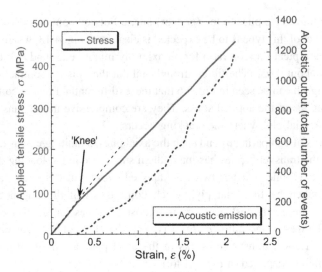

Fig. 8.16 Typical stress–strain curve and acoustic emission output of a cross-ply laminate tested in uniaxial tension parallel to one of the plies.

ingress of certain fluids may hasten final failure. For example, some composites can fracture much more readily in the presence of acids or other aggressive environments – see Section 9.3.2.

A typical stress–strain curve from testing a cross-ply laminate is shown in Fig. 8.16. Also shown on this plot is the output from acoustic emission equipment. These measurements are made with a piezo-electric transducer attached to the specimen surface, which picks up elastic waves triggered by the nucleation and growth of cracks. The onset of transverse cracking is manifest both as an acoustic signal and as a 'knee' in the stress–strain curve, caused by a reduction in specimen stiffness as the transverse plies are unloaded. This reduction in effective stiffness is most marked at the beginning, but continues over an appreciable range of strain, as the cracks in the transverse ply become more closely spaced. Eventually, however, the axial plies carry virtually the entire load and the gradient of the plot becomes constant again. There are many factors that determine the details of how the transverse cracks develop.

8.3.2 Laminate Failure under Uniaxial Loading

It can now be seen how, for any given stacking sequence, the stresses within each ply of the laminate can be calculated (referred to the fibre axis and the in-plane direction transverse to this – i.e. the 1 and 2 directions for that ply), for a specified set of applied stresses. An example of the outcome of such calculations (which must be carried out numerically, using the procedure described in Section 5.1.1) is shown in Fig. 8.17, for angle-ply ($\pm\phi$) laminates loaded in tension along the axis of symmetry. Fig. 8.17(a) shows the σ_1, σ_2 and τ_{12} stresses in one of the two plies. (For this case, symmetry dictates that those in the other ply are the same.) Using a suitable criterion, this

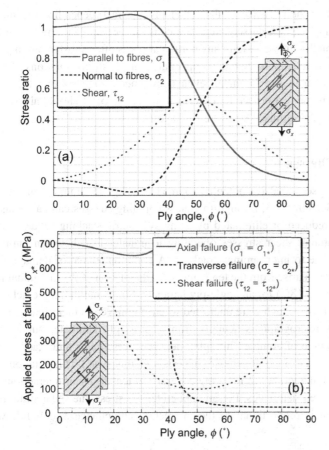

Fig. 8.17 Stresses within an angle-ply laminate of polyester–50% glass fibre, as a function of the ply angle: (a) stresses within one of the plies, as ratios to the applied stress; and (b) applied stress at failure (according to the maximum stress criterion, with $\sigma_{1*} = 700$ MPa, $\sigma_{2*} = 20$ MPa and $\tau_{12*} = 50$ MPa).

information can be used to predict the applied stress at which the ply will experience failure. This has been done in Fig. 8.17(b), using the maximum stress criterion. Since that criterion comprises three separate equations, each corresponding to a particular type of failure, the outcome is in the form of the three plots shown. As expected, for this particular loading situation, the highest strengths are obtained for low ply angles and the weakest for angles approaching 90°.

It is informative at this point to note the differences between the predictions in Fig. 8.17(b) and those in Fig. 8.9, which refers to the same composite material (with the same critical stress levels, σ_{1*}, σ_{2*} and τ_{12*}), but in that case relates to a single ply (as a function of misalignment angle). It might have been expected that, from the symmetry of the situation, these two plots would be very similar. It might even have been tempting to use the curves in Fig. 8.9, which correspond to simple analytical equations, to predict how an angle-ply laminate would fail. However, it can be seen that, while the plots bear some similarities, they are not the same. The differences can be attributed to the constraint effect of the presence of the 'other ply'. In fact, depending on the loading configuration and the stacking sequence, the constraint effect of the presence of other plies in a stack on the stresses within any particular ply is often substantial. In general, it is not acceptable to attempt to predict (using analytical transform equations) the stresses within one ply of a loaded laminate, even if symmetry suggests that this might be acceptable. The numerical procedure described in Section 5.1.1 should always be used (normally within a customised software package).

8.3.3   Inter-Laminar Stresses

Inter-laminar stresses are also a potential source of damage in laminates under load. As pointed out in Section 5.2.3, through-thickness coupling stresses can cause distortions (which may be reduced or eliminated by using symmetric stacking sequences). There are also inter-laminar shear stresses operating to transfer load between laminae, and these may give rise to inter-laminar cracking. A common source of damage is the τ_{xz} shear stresses, which arise from the tendency towards rotation of the individual laminae, as depicted in Fig. 8.18(a) for an angle-ply laminate. Some interlaminar cracks [28] are shown in Fig. 8.18(b), which were nucleated where a transverse cracks reached inter-ply interfaces. (This micrograph is actually from a cross-ply laminate, but it serves to illustrate an outcome of complex interplay between stress fields and crack patterns in highly anisotropic and inhomogeneous materials of this type.) This rotation between plies is described in Section 5.2.1, during coverage of tensile–shear interactions and the concept of 'balanced' laminates. For an applied tensile stress, the shear strain resulting from the rotation is proportional to the interaction compliance S_{16}. It follows from

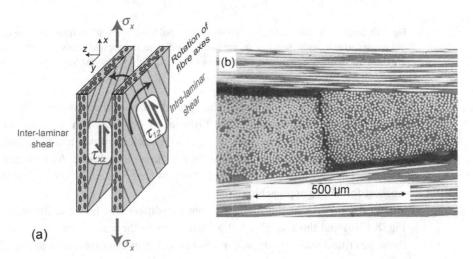

Fig. 8.18 (a) Schematic illustration of how axial tensile stressing of an angle-ply laminate generates a tendency towards rotation of the component plies, giving rise to inter-laminar shear stresses; and (b) SEM micrograph [30] showing inter-laminar cracking initiated at a transverse crack.
(b) Reprinted from Elsevier Books, 16, Larissa Gorbatikh and Stepan V. Lomov, Comprehensive Composite Materials II, 291–306, © 2018, with permission from Elsevier.

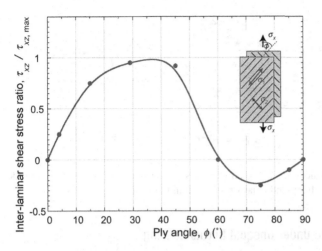

Fig. 8.19 Calculated inter-laminar shear stress as a function of ply angle, for an angle-ply carbon fibre–epoxy laminate [29].

Fig. 4.12 that, for ply angles (i.e. loading angles in Fig. 4.12) below 60°, the fibres tend to rotate towards the stress axis, while for larger angles than this they tend to rotate away from it. This is reflected in the data presented in Fig. 8.19, which shows the result of computations carried out by Pipes and Pagano [29] giving the relative magnitude of τ_{xz}, as a function of ply angle ϕ, for an angle-ply epoxy–carbon laminate.

8.4 Failure of Tubes and Pipes under Internal Pressure

In addition to the industrial significance of the failure behaviour of composite tubes under internal pressure, this mode of testing is a convenient method of generating selected stress states in a laminate, while avoiding possible complications from edge effects (see Section 8.2.3). Internal pressure acts on the cylindrical surfaces of a tube to create a hoop stress, σ_H, and an axial stress, σ_A, in the wall. Provided the wall thickness, t, is much less than the tube radius, r, these stresses are uniform within the wall. They can be obtained via simple force balances in axial and transverse directions, as illustrated in Fig. 8.20, leading to

$$\sigma_H = \frac{Pr}{t} \tag{8.21}$$

$$\sigma_A = \frac{Pr}{2t} \tag{8.22}$$

When testing is to be continued beyond initial cracking to final failure, a rubber lining is used to eliminate leakage of pressurising fluid. This loading situation – i.e. biaxial tension, with one stress equal to twice the other – is thus of considerable practical importance.

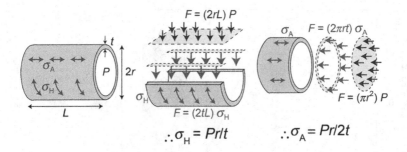

$$\therefore \sigma_{H} = Pr/t \qquad \therefore \sigma_{A} = Pr/2t$$

Fig. 8.20 Force balances applied to an internally pressurised, thin-walled tube, in order to obtain expressions for the hoop and axial stresses within the wall.

8.4.1   Laminate Failure under Unequal Biaxial Loading

Various stacking sequences can be used for composite tubes, but a common choice is to create an angle-ply laminate by alternately winding tows of (resin-impregnated) fibre at angles of $+\phi$ and $-\phi$. A clear view of the factors affecting choice of the optimum value of the ply angle ϕ is evidently needed here. Predictions are shown in Fig. 8.21 for the stresses within one of the plies, and the applied stress at failure (using the maximum stress criterion) for a polyester–glass composite, as a function of ply angle. These plots were produced in exactly the same way as Fig. 8.17, except that they now refer to a case in which the external load comprises both an applied tensile stress σ_x and a simultaneously applied σ_y stress equal to half of σ_x. This situation therefore corresponds to the stress state in the wall of an internally pressurised tube, with the hoop direction being x (i.e. $\sigma_x = \sigma_H$) and the axial direction being y ($\sigma_y = \sigma_A = \sigma_H/2$).

In contrast to the case of Fig. 8.17, the transverse stress σ_2 is now substantial for all ply angles, with the result that transverse cracking is predicted to be the dominant initial failure mode across the range of ϕ angles. This does depend on the critical stress levels σ_{2*} and τ_{12*}, but in general it is expected that σ_2 failure will occur in most cases. A more accurate prediction of failure would probably be obtained using the Tsai–Hill criterion, rather than the maximum stress condition, but again this is unlikely to alter the conclusion that transverse failure will predominate. There is nevertheless an important point to note from the plot in Fig. 8.21(b), which is that the greatest resistance to failure (highest level of σ_{x*}) will be obtained for a ply angle of about 35–40°. This kind of result is not unexpected, since having the fibres aligned rather closer to the (hoop) direction in which the larger stress is acting would appear to be a good strategy. The exact position of the peak depends on the elastic constants of the composite, and hence on the fibre and matrix types and the fibre content, but a ϕ value of around 35–38° is sometimes referred to as the 'ideal' winding angle (for a tube or pipe that is to be subjected to internal pressurisation in service).

Experimental data [31] are shown in Fig. 8.22 that broadly support this expectation. It can be seen that the failure stress is considerably higher for a ϕ value of 35° than for the other angles used in these tests. The photograph [32] in Fig. 8.23 shows a sample with this winding angle after completion of a test of this type (with a rubber lining used,

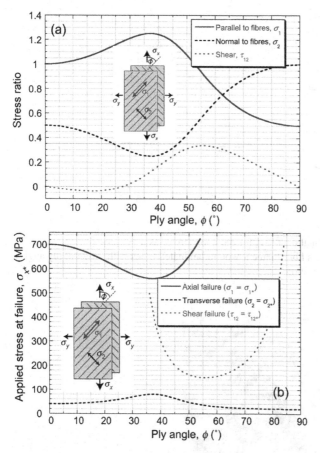

Fig. 8.21 Stresses within an angle-ply laminate of polyester–50% glass fibre, as a function of ply angle, subjected to biaxial loading, with $\sigma_x = 2\sigma_y$: (a) stresses within one of the plies, as ratios to the applied σ_x, and (b) applied stress, σ_x, at failure (according to the maximum stress criterion, with $\sigma_{1*} = 700$ MPa, $\sigma_{2*} = 20$ MPa and $\tau_{12*} = 50$ MPa).

allowing the internal pressure to be increased beyond the point where cracking has made the wall permeable to fluids). It can be seen that a crack has formed parallel to one set of fibres and final fibre fracture has then occurred in the other set. Prediction of the point of final fracture is complicated by the unloading of damaged plies. As a final topic in this chapter, a simple approach is now outlined to describe the effects of the initial damage on subsequent behaviour.

8.4.2 Netting Analysis

In netting analysis, it is assumed that only the fibres bear load, so that $\sigma_2 = \tau_{12} = 0$. Although this is not very realistic, it is broadly valid after considerable intra-laminar and inter-laminar cracking and shearing have taken place. Analysis of this case cannot be carried out rigorously, but an approximate estimate can be made as follows. For an

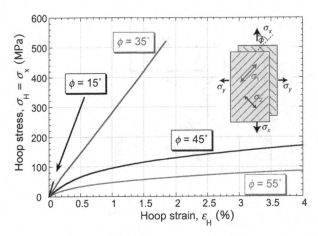

Fig. 8.22 Experimental data [31] for the hoop stress, as a function of hoop strain, for a set of polyester–50% glass fibre filament-wound laminated tubes with different ply angles, being tested by internal pressurisation.

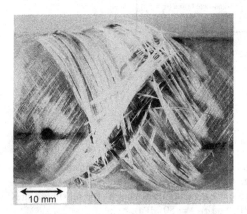

Fig. 8.23 Photograph [32] of the final failure region of an angle-ply tube ($\phi = 35°$) tested by internal pressurisation.
Reprinted from Composites, 9, D. Hull, M.J. Legg and B. Spencer, Failure of glass/polyester filament wound pipe, 17–24, © 1978, with permission from Elsevier.

angle-ply laminate, each lamina is considered independently and the ply angle ϕ taken as the angle between the fibre axis and the x-axis (i.e. the hoop direction). Using the transform equation (Eqn (4.47)) to relate applied and internal stresses, with $\sigma_x = \sigma_H$, $\sigma_y = \sigma_H/2$ and $\sigma_2 = \tau_{12} = 0$, gives

$$\sigma_H = \sigma_1 \cos^2\phi$$
$$\frac{\sigma_H}{2} = \sigma_1 \sin^2\phi \qquad (8.23)$$

There is potential for confusion here, because Eqn (4.47) apparently shows that σ_1 is equal to $\sigma_H \cos^2\phi + \sigma_H \sin^2\phi /2$, which cannot exceed σ_H. This arises because a single ply cannot in fact support the hoop and axial stresses, while maintaining $\sigma_2 = 0$, without the other ply being present to prevent failure in the transverse (2-) direction. Accepting that the treatment is simplified, it may be noted that there is a unique value of ϕ for which the equations in Eqn (8.23) both apply

$$\phi = \tan^{-1}\left(\frac{1}{\sqrt{2}}\right) \approx 35° \tag{8.24}$$

This is a simple (and general) way of obtaining the so-called 'ideal' angle for an internally pressurised tube. If the netting analysis assumptions hold, then the fibres in laminates with ply angles other than 35° must rotate towards this angle. This gives an insight into the behaviour noted in Fig. 8.22, which shows a much greater failure stress for this ply angle: in practice, rotation towards this angle for other laminates requires extensive matrix damage and this is likely to stimulate fibre fracture.

The analysis can also be used to explore the stiffness after initial damage, and the final failure stress, for the $\phi = 35°$ case. Under netting analysis conditions

$$\sigma_1 = \frac{\sigma_H}{\cos^2\phi} \approx 1.5\sigma_H \tag{8.25}$$

Given that only the fibres contribute to the stiffness ($E_1 = f E_f$), it follows that, since the axial, hoop and fibre strains are all equal in this special case, the hoop stiffness is given by

$$E_H \approx \frac{fE_f}{1.5} \tag{8.26}$$

This has a value of about 25–30 GPa for polyester–50% glass, which is in good agreement with the gradient of the $\phi = 35°$ plot in Fig. 8.22, after the initial portion; this may be compared with the value in the elastic regime of about 90–100 GPa, obtained both experimentally and from laminate analysis. Final failure occurred at $\sigma_H \sim 500$ MPa, indicating a σ_{1*} value of about 750 MPa and hence a fibre strength of about 1.5 GPa. This is about half the strength of freshly drawn glass fibres. The appearance of the final fracture site is shown in Fig. 8.23. Estimates of this type can be useful, provided the implications of the netting analysis assumptions are fully appreciated.

References

1. Parry, TV and AS Wronski, Kinking and tensile, compressive and interlaminar shear failure in carbon fiber-reinforced plastic beams tested in flexure. *Journal of Materials Science* 1981; **16**(2): 439–450.
2. Moran, PM, XH Liu and CF Shih, Kink band formation and band broadening in fiber composites under compressive loading. *Acta Metallurgica Et Materialia* 1995; **43**(8): 2943–2958.
3. Lee, SH and AM Waas, Compressive response and failure of fiber reinforced unidirectional composites. *International Journal of Fracture* 1999; **100**(3): 275–306.

4. Zidek, RAE and C Vollmecke, Analytical studies on the imperfection sensitivity and on the kink band inclination angle of unidirectional fiber composites. *Composites Part A: Applied Science and Manufacturing* 2014; **64**: 177–184.

5. Wang, Y, TL Burnett, Y Chai, C Souti, PJ Hogg and PJ Withers, X-ray computed tomography study of kink bands in unidirectional composites. *Composite Structures* 2017; **160**: 917–924.

6. Wisnom, MR, The effect of fiber waviness on the relationship between compressive and flexural strengths of unidirectional composites. *Journal of Composite Materials* 1994; **28**(1): 66–76.

7. Hsiao, HM and IM Daniel, Effect of fiber waviness on stiffness and strength reduction of unidirectional composites under compressive loading. *Composites Science and Technology* 1996; **56**(5): 581–593.

8. Elhajjar, RF and SS Shams, Compression testing of continuous fiber reinforced polymer composites with out-of-plane fiber waviness and circular notches. *Polymer Testing* 2014; **35**: 45–55.

9. Nair, SN, A Dasari, CY Yue and S Narasimalu, Failure behavior of unidirectional composites under compression loading: effect of fiber waviness. *Materials* 2017; **10**(8). DOI: 10.3390/ma10080909.

10. Lee, SH, CS Yerramalli and AM Waas, Compressive splitting response of glass-fiber reinforced unidirectional composites. *Composites Science and Technology* 2000; **60**(16): 2957–2966.

11. Yerramalli, CS and AM Waas, Compressive splitting failure of composites using modified shear lag theory. *International Journal of Fracture* 2002; **115**(1): 27–40.

12. Prabhakar, P and AM Waas, Interaction between kinking and splitting in the compressive failure of unidirectional fiber reinforced laminated composites. *Composite Structures* 2013; **98**: 85–92.

13. Spowart, JE and TW Clyne, The axial compressive failure of titanium reinforced with silicon carbide monofilaments. *Acta Mater.* 1998; **47**: 671–687.

14. Tsai, SW and VD Azzi, Strength of laminated composite materials. *AIAA Journal* 1966; **4**(2): 296–301.

15. Tsai, SW and EM Wu, General theory of strength for anisotropic materials. *Journal of Composite Materials* 1971; **5**: 58–80.

16. Liu, KS and SW Tsai, A progressive quadratic failure criterion for a laminate. *Composites Science and Technology* 1998; **58**(7): 1023–1032.

17. Gotsis, PK, CC Chamis and L Minnetyan, Prediction of composite laminate fracture: micromechanics and progressive fracture. *Composites Science and Technology* 1998; **58**(7): 1137–1149.

18. Hart-Smith, LJ, Predictions of a generalized maximum-shear-stress failure criterion for certain fibrous composite laminates. *Composites Science and Technology* 1998; **58**(7): 1179–1208.

19. Hinton, MJ and PD Soden, Predicting failure in composite laminates: the background to the exercise. *Composites Science and Technology* 1998; **58**(7): 1001–1010.

20. Soden, PD, AS Kaddour and MJ Hinton, Recommendations for designers and researchers resulting from the world-wide failure exercise. *Composites Science and Technology* 2004; **64**(3–4): 589–604.

21. Hinton, MJ, AS Kaddour and PD Soden, A comparison of the predictive capabilities of current failure theories for composite laminates, judged against experimental evidence. *Composites Science and Technology* 2002; **62**(12–13): 1725–1797.

22. Hinton, MJ, AS Kaddour and PD Soden, A further assessment of the predictive capabilities of current failure theories for composite laminates: comparison with experimental evidence. *Composites Science and Technology* 2004; **64**(3–4): 549–588.

23. Hill, R, *The Mathematical Theory of Plasticity*. Oxford University Press, 1950.

24. Sinclair, JH and CC Chamis, Fracture modes in off-axis fiber composites. *Polymer Composites* 1981; **2**(1): 45–52.

25. Knappe, W and W Schneider, The role of failure criteria in the fracture analysis of fiber/matrix composites, in *Deformation and Fracture of High Polymers*, Kausch, HH, JA Hassell and RI Jaffee, editors. Springer, 1973, pp. 543–556.

26. Puck, A and H Schurmann, Failure analysis of FRP laminates by means of physically based phenomenological models. *Composites Science and Technology* 2002; **62**(12–13): 1633–1662.

27. Kuraishi, A, SW Tsai and KKS Liu, A progressive quadratic failure criterion, part B. *Composites Science and Technology* 2002; **62**(12–13): 1683–1695.

28. Romanov, VS, SV Lomov, I Verpoest and L Gorbatikh, Modelling evidence of stress concentration mitigation at the micro-scale in polymer composites by the addition of carbon nanotubes. *Carbon* 2015; **82**: 184–194.

29. Pipes, RB and NJ Pagano, Interlaminar stresses in composite laminates under uniform axial extension. *Journal of Composite Materials* 1970; **4**: 538–548.

30. Gorbatikh, L and SV Lomov, Damage in architectured composites, in *Comprehensive Composite Materials II*, Talreja, R, editor. Elsevier, 2018, pp. 291–306.

31. Spencer, B and D Hull, Effect of winding angle on the failure of filament wound pipe. *Composites* 1978; **9**(4): 263–271.

32. Hull, D, MJ Legg and B Spencer, Failure of glass–polyester filament wound pipe. *Composites* 1978; **9**(1): 17–24.

9 Fracture Mechanics and the Toughness of Composites

The previous chapter covered factors affecting strength, in terms of the stresses at which damage and failure occur in composites. In many situations, however, it is the energy that is absorbed within the material while fracture takes place that is of prime importance. A tough material is one for which large amounts of energy are required to cause fracture. Some loading configurations, such as a component being struck by a projectile, provide only a finite amount of energy that could cause failure. In fact, there are many situations in which toughness, rather than strength, is the key property determining whether the material is suitable. In this chapter, a brief outline is given of the basics of fracture mechanics, with particular reference to the energetics of interfacial damage. This is followed by an appraisal of the sources of energy absorption in composites. Finally, progressive crack growth in composites is examined under conditions for which fast fracture is not energetically favoured (sub-critical crack growth).

9.1 Fracture Mechanics

A brief overview is presented here of some relevant background concerning fracture mechanics. Of course, a number of specialised texts are available in this area.

9.1.1 Energetics of Crack Propagation

It has long been known that stresses can become concentrated around features such as holes and cracks. In 1898 Kirsch derived the following expressions for the stress field around a circular hole in an infinite plate (Fig. 9.1)

$$\frac{\sigma_{rr}}{\sigma_0} = \frac{1}{2}\left\{1 - \left(\frac{a}{r}\right)^2\right\} + \frac{1}{2}\left\{1 - 4\left(\frac{a}{r}\right)^2 + 3\left(\frac{a}{r}\right)^4\right\}\cos 2\theta$$

$$\frac{\sigma_{\theta\theta}}{\sigma_0} = \frac{1}{2}\left\{1 + \left(\frac{a}{r}\right)^2\right\} - \frac{1}{2}\left\{1 + 3\left(\frac{a}{r}\right)^4\right\}\cos 2\theta \tag{9.1}$$

$$\frac{\sigma_{r\theta}}{\sigma_0} = \frac{1}{2}\left\{1 + 2\left(\frac{a}{r}\right)^2 - 3\left(\frac{a}{r}\right)^4\right\}\sin 2\theta$$

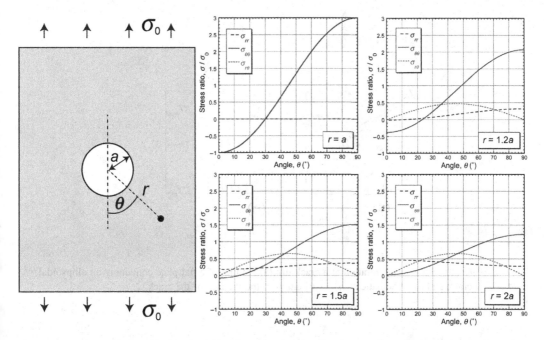

Fig. 9.1 Schematic of uniaxial loading of an infinite plate containing a circular hole and plots of the Kirsch solution (Eqn (9.1)) at four radial distances.

It can be seen in Fig. 9.1 that the maximum *stress concentration factor* has a magnitude of 3, and that this occurs, as would be expected, immediately adjacent to the hole, in the direction of the applied stress. It may also be noted that, on the scale of the hole, the stress concentration effect drops off fairly quickly with distance, with the stresses at a distance of two hole radii (at $\theta = 90°$) already starting to approach the far-field condition ($\sigma_{\theta\theta} / \sigma_0 = 1$ and $\sigma_{rr} = \sigma_{r\theta} = 0$).

Fracture mechanics has its roots in the work of Inglis (1913), who took the Kirsch treatment further by focusing on the stresses near a crack tip, which can be considerably higher than the far-field value (σ_0). He derived the following expression for the stress at the tip of an ellipsoidal crack (of the type shown in Fig. 9.2), with length c (or $2c$ if internal) and tip radius r

$$\sigma = \sigma_0 \left(1 + 2\sqrt{\frac{c}{r}} \right) \tag{9.2}$$

This confirms that a circular hole ($c = r$) has a stress concentration factor of 3. While this is physically reasonable, the case of a sharp crack ($r \to 0$) presents difficulties, in that the stress concentration, according to Eqn (9.2), becomes very large. On this basis, most components should fail, under low applied loads, at the fine surface scratches that are almost inevitably present. This is contrary to engineering experience, since most components, particularly metallic ones, are able to function even when small cracks and scratches are present.

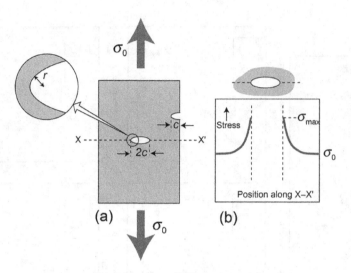

Fig. 9.2 Stress concentration at a crack tip: (a) loading of a flat plate containing an ellipsoidal crack and (b) schematic stress distribution in the vicinity of the crack.

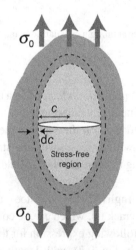

Fig. 9.3 Schematic depiction of the stress-free region shielded by a crack from an applied load.

The situation was resolved by the pioneering work of Griffith [1], who clarified in the 1920s that a crack cannot propagate unless the energy of the system is thereby decreased. The energy released during crack advance comes from stored elastic strain in the surrounding material (plus any work done by the loading system). As a crack gets longer, the volume of stress-free material 'shielded' by it from the applied stress increases – see Fig. 9.3. The driving force for crack propagation therefore increases. The strain energy stored per unit volume in (elastically) stressed material is given by

$$U = \frac{1}{2}\sigma_0\varepsilon_0 = \frac{\sigma_0^2}{2E} \tag{9.3}$$

so the energy released when the crack extends (at both ends) by dc is the product of this expression and the increase in stress-free volume. The shape of the stress-free region is not well defined, and the stress was in any event not uniform within it before crack advance, but taking the relieved area to be twice that of the circle having the crack as diameter gives a fair approximation. Thus, for a plate of thickness t, the energy released during incremental crack advance is given by

$$dW = \frac{\sigma_0^2}{2E} 2(2\pi c t \, dc) = \frac{2\sigma_0^2 \pi c t \, dc}{E} \tag{9.4}$$

A central concept in fracture mechanics is that of stored elastic strain energy being released as the crack advances. The *strain energy release rate* (crack driving force) is usually given the symbol G (not to be confused with shear modulus or Gibbs free energy). It is a 'rate' with respect to the creation of new crack area (and so has units of J m^{-2}) and does not relate to time in any way. It follows that

$$G = \frac{dW}{\text{new crack area}} = \frac{\left(2\sigma_0^2 \pi c t \, dc / E\right)}{(2t \, dc)} = \pi \left(\frac{\sigma_0^2 c}{E}\right) \tag{9.5}$$

The value of the constant (π in this case) is not well defined. It depends on specimen geometry, crack shape/orientation and loading conditions. In any event, the approximation used for the stress-free volume is simplistic. However, the dependence of G on ($\sigma^2 c/E$) is more general and has important consequences, particularly in terms of the linear dependence on crack length. It may also be noted at this point that the 'crack area' concept in this treatment refers to a 'projected area' – no attempt is made to monitor the actual crack area, which may, of course, be greater than the projected area if the crack is repeatedly undergoing minor changes of direction – being deflected, for example, by microstructural features.

In order for crack propagation to be possible, the strain energy release rate must be greater than or equal to the rate of energy absorption, expressed as energy per unit (projected) area of crack. This energy requirement is sometimes known as the *Griffith criterion*. Focusing now on an edge (surface) crack of length c, propagating inwards, for a brittle material this *fracture energy* is simply given by 2γ (where γ is the *surface energy*, with the factor of 2 arising because there are two new surfaces created when a crack forms). It can be considered as a *critical strain energy release rate*, G_c. It is a material property. It is sometimes termed the *crack resistance*. The *fracture strength* can thus be expressed as

$$G \geq G_c = 2\gamma$$

$$\therefore \pi \left(\frac{\sigma_0^2 c}{E}\right) \geq 2\gamma$$

$$\therefore \sigma_* = \left(\frac{2\gamma E}{\pi c}\right)^{1/2} \tag{9.6}$$

This equation can be used to predict the stress at which fracture will occur, for a component containing a crack of known size. However, it relates only to materials for which the fracture energy is given by 2γ – i.e. for which the energy absorbed during crack propagation is only that needed to create the new surface area. It may be noted at this point that the magnitude of 2γ does not vary very much between different materials and is always relatively low – usually $<\sim 10$ J m^{-2}. These are regarded as 'ideally brittle' materials. Some materials, including many glasses, do behave in at least approximately this way, but most materials do not – see below.

9.1.2 Mechanisms of Energy Absorption during Crack Propagation

It has long been clear that the simple Griffith condition (Eqn (9.6)) applies only to brittle materials. For metals, a fracture event, often preceded by extensive plastic flow throughout the sample, and then by localised necking, is usually very different from that in brittle materials. Apart from the possibility of energy being (permanently) absorbed by uniform plasticity before any crack growth occurs, the crack propagation process itself requires much more energy than in brittle materials. There tends to be a zone of plasticity there, irrespective of whether there has been much plastic deformation prior to the onset of fracture. Fig. 9.4 shows how this plasticity raises the radius of curvature, r, at the crack tip and reduces the stress concentration effect. (The peak stress, indicated as σ_Y in Fig. 9.4(b), is nominally the yield stress, although it will be higher than this if work hardening occurs.)

The work done during plastic deformation (which is mostly dissipated as heat) must be supplied by the crack driving force (release of elastic strain energy). It might be imagined that a simple modification could be made to Eqn (9.6), replacing 2γ by an alternative version of G_c, which takes account of these extra energy requirements (and, indeed, G_c values for tough materials such as metals are commonly greater than 2γ by

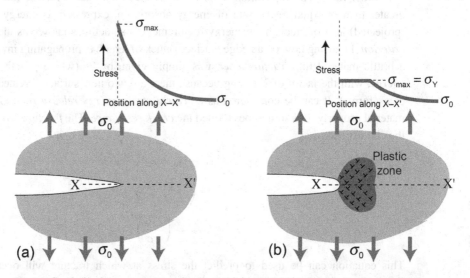

Fig. 9.4 Crack tip shapes and stress distributions for (a) brittle and (b) ductile materials.

several orders of magnitude). This apparently allows the Griffith criterion to be applied to ductile materials, but there are difficulties. It suggests that ductile materials should be as sensitive to the presence of flaws as brittle materials, although they would have higher fracture stresses for a given crack size. In fact, failure stresses are not dramatically or systematically higher for ductile materials, although they certainly fracture less readily and require much more energy input. Furthermore, ductile materials show little or no sensitivity to initial flaw size. Very ductile materials often fail by *ductile rupture* (progressive necking down to a point), with little or no crack propagation as such. The failure stress in such cases can only be predicted by analysis of the plastic flow, and Eqn (9.6) – with 2γ replaced by a much larger value of G_c – is irrelevant. However, such highly ductile materials are too soft to be useful for most purposes. Failure of engineering metals does commonly involve fracture, often after extensive plastic flow and some necking have occurred. Clearly, use of Eqn (9.6) requires care in such cases.

9.1.3   The Stress Intensity Factor

In most cases, the Inglis stress concentration equation (Eqn (9.2)) can be approximated as follows

$$\sigma_{\max} = \sigma_0 \left[1 + 2 \left(\frac{c}{r} \right)^{1/2} \right] \approx 2\sigma_0 \left(\frac{c}{r} \right)^{1/2} \tag{9.7}$$

Since a critical stress is presumably needed at the crack tip to open up the atomic planes, it follows that crack propagation is expected when

$$\sigma_0 \sqrt{c} \geq \text{critical value}$$

where the critical value is expected to be constant for a given material, but to vary between materials (since r, and probably the critical stress level as well, will differ for different materials). Around 1950, Irwin [2] proposed the concept of a *stress intensity factor, K*, such that

$$K = \sigma_0 \sqrt{\pi c} \tag{9.8}$$

The stress intensity factor, which has units of MPa $\sqrt{m}$, scales with the level of stress at the crack tip (although it does not allow the absolute value to be established). Fracture is expected when K reaches a critical value, K_c, the *critical stress intensity factor*, which is often termed the *fracture toughness*.

There are clearly parallels between K reaching a critical value, K_c, and G reaching a critical value of G_c. The magnitude of K can be considered to represent the crack driving force, analogous to G. Consider again the Griffith energy criterion

$$G \geq G_c, \quad \text{i.e.} \quad \pi \left(\frac{\sigma_0^2 c}{E} \right) \geq G_c \quad \therefore \sigma_0 \sqrt{\pi c} \geq \sqrt{E G_c} \tag{9.9}$$

It can be seen that this is similar in form to Eqn (9.8). It follows that

$$K = \sqrt{EG} \quad \text{and} \quad K_c = \sqrt{EG_c} \tag{9.10}$$

While it is reassuring to be able to treat fracture from both stress and energy viewpoints, it is not immediately apparent what advantages are conferred by using a stress intensity criterion rather than an energy-based one. However, in practice it is possible to establish stress intensity factors, and corresponding fracture toughness values, for various loading and specimen geometries, whereas this is not really possible with an energy-based approach. The 3D stress state at the crack tip, and hence the size and shape of the plastic zone, can be affected by specimen thickness and width. In fact, an important point concerning experimental measurement of G_c and K_c is that it should normally be done under conditions of *plane strain*, which effectively requires the sample to be relatively wide and also deep (so that the crack is not close to the back surface). The size of the plastic zone at the crack tip can become larger if these conditions are not maintained, raising the measured toughness.

A further issue concerning use of the stress intensity factor is that rates of *sub-critical crack growth* (progressive advance of a crack due to a corrosive environment or cyclic loading) can often be predicted from K, since it is directly related to conditions at the crack tip, whereas the strain energy release rate is a more global parameter. (Sub-critical crack growth is covered in Section 9.3.) As an example of this, it can be shown that the size of the *plastic zone* ahead of the crack tip is related to the yield stress of the material by

$$r_Y \approx \frac{1}{2\pi} \left(\frac{K}{\sigma_Y} \right)^2 \tag{9.11}$$

Similarly, the *crack opening displacement*, δ, can be expressed as

$$\delta \approx \left(\frac{K^2}{\sigma_Y E} \right) \tag{9.12}$$

Such parameters are useful when considering how energy-absorbing processes might be stimulated in composite materials, since they allow the scale of features of the crack tip to be related to the scale of the microstructure.

9.1.4   Toughness of Engineering Materials

An area of major challenge for materials scientists lies in designing materials with desirable combinations of strength and toughness, ensuring that components both retain their shape and resist fracture under service conditions. In general, the toughness of a given type of material falls as its strength (yield stress) is increased, as a consequence of reduced plasticity at the crack tip. However, large variations in toughness may be exhibited between different types of material. Some illustrative data are shown in Fig. 9.5 and Table 9.1. Pure metals, particularly fcc metals such as Al, are very tough

Table 9.1 Typical (mode I) fracture energy and fracture toughness values for various materials.

Material	Fracture energy G_c (kJ m^{-2})	Fracture toughness K_c (MPa $\sqrt{m}$)
Polymers		
Epoxy resins	0.1–0.3	0.3–0.5
Nylon	1–4	2–3
Polypropylene	5	3
Metals		
Pure Al	100–1000	100–350
Al alloy	8–30	23–45
Mild steel	100	140
Ceramics		
Soda glass	0.01	0.7
SiC	0.05	3
Concrete	0.03	0.2
Natural materials		
Woods (crack $\perp$ grain)	8–20	11–13
Woods (crack // grain)	0.5–2	0.5–1
Bone	0.6–5	2–12
Composites		
Fibreglass (glass–epoxy, planar random)	40–100	40–60
Al-based particulate MMC	2–10	15–30

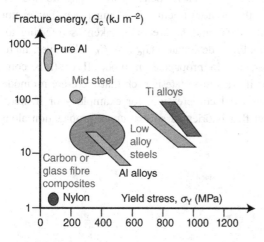

Fig. 9.5 Approximate toughness and yield stress combinations exhibited by a few selected types of materials.

($G_c \sim 200$–400 kJ m^{-2}, i.e. $K_c \sim 150$–200 MPa $\sqrt{m}$). However, their yield stresses are so low that they are not suitable for engineering applications, since they plastically deform so easily. Mild steel ($\sim$pure Fe with $\sim$0.05–0.2 wt%C) is, however, widely used. It is ductile and tough, as well as being weldable, machinable and cheap. Its yield stress (considerably higher than metals like pure Al, since it is bcc rather than fcc, and due to strengthening from the carbon) is adequate for many practical applications.

Other alloys, such as those of Ti and more complex steels, can be much stronger, while also being relatively tough. Al can be strengthened by age hardening and/or strain hardening. Toughness is thereby reduced, but is still in a range adequate for many engineering applications. In contrast, polymers and ceramics (in which dislocation motion is difficult or impossible) are much more brittle. Nylon is actually one of the tougher polymers, but its G_c value is only ~1 kJ m^{-2}, despite its relatively low strength. Ceramics are often stronger, but their toughness is in general very limited. However, it can clearly be seen that fibre composites (glass or C fibres in a polymer matrix) often have very good toughness – comparable to many metals – in addition to being relatively strong. In combination with their low density and good corrosion resistance, this clearly makes composites highly attractive. Much of the rest of this chapter is oriented towards an understanding of this apparently rather surprising characteristic.

9.1.5   Fracture Modes

The strain energy release rate needed to allow crack propagation also depends on the geometry of the stress field, relative to the crack plane and the direction of crack advance. There are three possible fracture 'modes', as illustrated in Fig. 9.6. The most common geometry of crack propagation is mode I, in which a tensile stress acts normal to the plane of the crack, so that it opens up as propagation occurs. In many cases, it is understood that this is how fracture is taking place. For example, in the treatment above, the values of G_c and K_c are often taken as referring to this case, although, strictly, these should be designated G_{Ic} and K_{Ic}. However, there are some circumstances in which cracks do propagate in mode II or (less commonly) in mode III. This might occur if the stress state is such that there are no mode I stresses acting on the crack – in uniaxial compression, for example – or if there is a (low toughness) interface present that is oriented so that crack propagation along it would

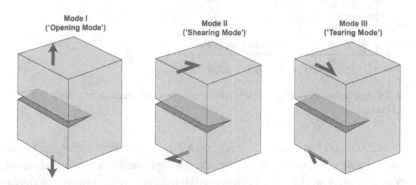

Fig. 9.6 Schematic depiction of the three crack propagation modes.

be driven only by shear stresses. In general, however, a higher driving force (G or K) is needed to cause cracks to propagate in mode II or mode III, so that the values of G_{IIc} and K_{IIc} (and also G_{IIIc} and K_{IIIc}) are expected to be significantly higher than G_{Ic} and K_{Ic} for the material concerned.

9.2 Fracture of Composite Materials

9.2.1 Interfacial Debonding

Table 9.1 shows that a composite made from glass fibres and epoxy resin can have a fracture energy comparable with those of metals ($G_{\mathrm{c}} \sim 50$ kJ m^{-2}), but also highlights that both of these constituents are themselves very brittle ($G_{\mathrm{c}} \sim 0.01$–0.1 kJ m^{-2}). This high toughness of composites is closely linked with interfacial effects. A first step in exploring this is to consider the conditions under which interfacial debonding (i.e. propagation of a crack along an interface) takes place. For a given loading configuration, such crack propagation gives rise to an energy release rate, G_{i}, in much the same way as for the case when the crack is in a homogeneous material. Also, there will be a critical value, G_{ic}, an *interfacial fracture energy*, which G_{i} must reach in order for the crack to propagate. Of course, this issue of whether interfacial debonding takes place, or at least is energetically favourable, is of wide significance. For example, not only is it central to the fracture of composite materials, but it controls the spallation of surface coatings – see Section 11.4.

Values of G_{ic} are not as readily available as G_{c} values for homogeneous materials. There are several reasons for this. First, the toughness of an interface is sensitive to the way it was produced, rather than being unique to the pair of constituents on either side. Second, interfacial cracks often propagate under *mixed mode* loading conditions. This is in contrast to a crack in a homogeneous (isotropic) material, which will always tend to advance in a direction such that the stress field at the crack tip is purely tensile (mode I) – see Fig. 9.6. An interfacial crack, however, is constrained to follow a predetermined path. Depending on the loading configuration, the stress field driving crack advance may include a significant shear stress (mode II, or, less commonly, mode III) component. In general, the energy expended in debonding the interface is greater when there is a mode II component than for the case of pure mode I loading. This complicates experimental measurement of G_{ic} and K_{ic}. Not only can it be difficult to establish the exact stress field, but it may vary with position in the specimen, particularly for fibre–matrix interfaces. The situation is further complicated if any residual stresses (e.g. see Section 10.1.1) are present.

The proportion of opening and shearing modes at the crack tip is often characterised by means of the *phase angle*, ψ (psi). This is defined in terms of the mode I and mode II stress intensity factors

$$\psi = \tan^{-1}\left(\frac{K_{\mathrm{II}}}{K_{\mathrm{I}}}\right) \tag{9.13}$$

The value of ψ is 0° for pure opening ($K_{II} = 0$) and 90° for pure shear ($K_I = 0$). The phase angle can be established for various loading arrangements, although often this calculation is not a simple one [3–5]. Several procedures have been developed for experimental measurement of G_{ic} as a function of mode mix (ψ) and outcomes are certainly available (covering a wide range of interface types and constituent materials). There have also been efforts to rationalise outcomes in terms of mechanistic contributions to the energy absorption. Assessment of the reliability of different types of test have been somewhat hampered by the large number of variables involved and the sensitivity of G_{ic} to the details of how the interface was produced. Nevertheless, the main features are fairly clear, particularly a general tendency for G_{ic} to increase quite substantially as the loading configuration moves from pure mode I ($\psi = 0°$) towards pure mode II ($\psi = 90°$). Some illustrative experimental data are shown in Fig. 9.7, covering a range of mode mix, experimental technique and interface type. The increase in G_{ic} as the loading becomes predominantly shear may be due, at least partly, to more frictional work being done immediately behind the crack tip as asperities on the crack

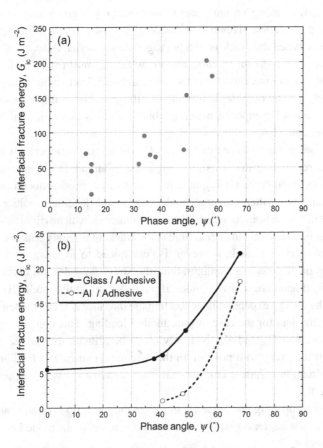

Fig. 9.7 Measured values of the interfacial fracture energy, G_{ic}, as a function of phase angle, ψ, for (a) a thin (~200 μm) epoxy layer between Al blocks, obtained using the 'Brazil Nut' test [6], and (b) interfaces between either glass or Al and a thick adhesive layer, obtained using the four-point bend delamination test [7].

flanks slide over each other. There is some evidence that the dependence on ψ may be slightly stronger when at least one of the constituents can deform plastically. The exact path taken by the crack may vary as ψ changes and this can affect the nature of the energy-absorbing processes.

9.2.2 Energy of Interfacial Debonding in Fibre Composites

A key issue concerns the phenomenon of *crack deflection*. For a composite to have a high toughness, a crack passing through the matrix must be repeatedly deflected at fibre–matrix interfaces, at least for materials based on polymer resins or ceramics. Early work by Cook and Gordon [8] on conditions for crack deflection was focused on the stress field ahead of a crack tip. It was pointed out that, when a crack approaches a fibre in a composite loaded parallel to the fibre axis, there is a transverse tensile stress ahead of the crack tip, tending to open up the interface and hence to 'blunt' and deflect the crack from entering the fibre. Since the peak value of this transverse stress is typically ~20% of the maximum axial stress, Cook and Gordon proposed that the interface will debond if its strength is less than about one-fifth of that of the matrix. However, the 'strength' of an interface is not well defined and is sensitive to the presence and distribution of flaws and to the method of loading. In view of the success of the Griffith treatment, it is clearly preferable to establish a criterion based on the energetics of crack propagation.

Energy-based crack deflection criteria have been proposed by, among others, Kendall [9,10] and He and Hutchinson [11,12]. Kendall considered two blocks bonded together and loaded in tension parallel to the interface, one of the blocks having a crack approaching the interface – see Fig. 9.8. He estimated the applied loads under which a crack would penetrate the other block or deflect along the interface, assuming that the crack requiring the lower load would predominate. This produced the following criterion for deflection

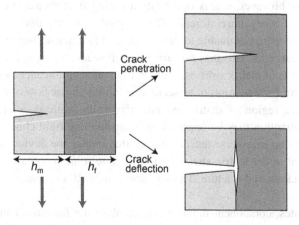

Fig. 9.8 Geometry of crack deflection at an interface (after Kendall [9,10]).

$$\frac{G_{ic}}{G_{fc}} = \left(\frac{h_m E_m + h_f E_f}{h_f E_f}\right)\left(\frac{1}{4\pi(1 - v^2)}\right) \qquad (9.14)$$

where G_{ic}, and G_{fc} are the fracture energies of interface and uncracked block (fibre), h_m and h_f are thicknesses of cracked (matrix) and uncracked (fibre) blocks, E_m and E_f are the corresponding Young's moduli and v is the Poisson ratio (taken as equal in both constituents). If h_m and h_f are taken as equal, corresponding approximately to a crack passing through the matrix between unbroken fibres in a typical composite, then this critical ratio is ~20% for E_m~E_f, falling to ~10% for $E_m \ll E_f$.

He and Hutchinson's analysis [11] is based on considering whether the penetrating or deflecting crack gives a greater net release of energy. It is more complex than Kendall's model, but also yields a critical fracture energy ratio of ~20% for the case of E_m~E_f. For $E_m \ll E_f$, however, an increase (rather than a decrease) is predicted in this ratio. There are rather few experimental data available to validate either of these models. Kendall's experimental work with rubbers did seem consistent with his predictions, but few systematic measurements have been made with systems of practical interest. It is, however, clear that the interfacial fracture energy must be appreciably lower than that of the reinforcement if matrix cracks are to be consistently deflected. Since (ceramic) fibres tend to have low fracture energies (Table 9.1), this means that interfaces of very low toughness are often required if crack deflection is essential. For ceramic matrix composites, in particular, the retention of low interfacial toughness, through processing stages that tend to promote sintering and chemical reactions, represents a major technological challenge. Indeed, the development of ceramic matrix composites (CMCs) in which the toughness is raised by repeated crack deflection has, in general, been rather unsuccessful.

Before assessing the energy associated with interfacial debonding in a fibre composite, the probable contributions from other sources should be noted. Matrix deformation (near a crack tip) is normally substantial in most metals, mainly as a result of extensive dislocation movement, while for polymers (particularly thermosets) and ceramics such a contribution is likely to be small (Table 9.1). In metallic matrix composites (MMCs), however, matrix deformation during fracture may be less than for the same material when unreinforced. This is predominantly due to increased *constraint*, so that the matrix is unable to deform as readily when surrounded by stiff and strong fibres or particles. This may be partly a result of *load transfer* (Section 1.3) reducing the magnitude of matrix stresses. Usually of greater significance, however, is the tendency for *triaxial stress states* to be set up that inhibit plastic flow of the matrix. For example, when a region of matrix extends plastically, with associated lateral contraction, this contraction may be opposed by a surrounding rigid cluster of fibres or particles. This sets up transverse tensile stresses that reduce the deviatoric (shape-changing) component of the matrix stress state. This in turn inhibits plastic flow, but may encourage cavitation and fracture. The toughness of MMCs is considered further in Section 9.2.5.

For fibre composites, component failure often involves the fracture of fibres. The contribution that this makes to the fracture energy of the material is small for most

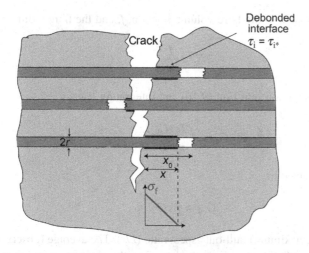

Fig. 9.9 Schematic representation of the advance of a crack in a direction normal to the fibre axis, showing interfacial debonding and fibre pull-out processes.

fibres. Typical fracture energies for fibres of glass, carbon and SiC are only a few tens of J m^{-2}. Some polymeric fibres are not completely brittle and undergo appreciable plastic deformation. For example, fracture of the cellulose fibres makes a significant contribution to the fracture energy of wood (across the grain). In such cases, fibre fracture contributes up to a few kJ m^{-2} to the overall fracture energy. Metallic fibres can in principle make even larger contributions. Thus, even at a low volume fraction, steel rods in reinforced concrete raise the toughness substantially, as well as enhancing the tensile strength. Another example is treated in some detail in Section 9.2.5. Nevertheless, for most (PMC) composites, the fibres themselves make little or no direct contribution to the overall toughness.

Interfacial energy absorption can now be examined, based on the geometry shown in Fig. 9.9. Assuming that fibres fracture at locations that are not in the macroscopic crack plane, work is done first during debonding (propagation of an interfacial crack) and subsequently during frictional pull-out. Focusing first on debonding of a single fibre, if this has occurred along a length x_0, then the associated work is given by

$$\Delta U = 2\pi r x_0 G_{ic} \qquad (9.15)$$

where r is the fibre radius. To obtain the total work of debonding for the composite, G_{cd}, this must be summed over all of the fibres intersected by the crack. An assumption is now made that there is an equal probability of all pull-out lengths between 0 and L (the maximum pull-out length). If there are N fibres per square metre, then there will be $(N\,dx_0 / L)$ per square metre with an embedded length between x_0 and $(x_0 + dx_0)$. The total work done in debonding is therefore given by

$$G_{cd} = \int_0^L \frac{N\,dx_0}{L} 2\pi r x_0 G_{ic} \qquad (9.16)$$

The value of N is related to the fibre volume fraction, f, and the fibre radius, r

$$N = \frac{f}{\pi r^2} \tag{9.17}$$

Substituting this expression into Eqn (9.16) and integrating leads to

$$G_{cd} = \frac{f 2 \pi r G_{ic}}{\pi r^2 L} \frac{L^2}{2}$$

which can be rearranged to

$$G_{cd} = f s G_{ic} \tag{9.18}$$

where s is here the (maximum) pull-out aspect ratio (L/r). The average (observed) value is expected to be about half of this. Contributions from this mechanism are usually fairly small. For example, if $s = 50$, $f = 0.5$ and $G_{ic} = 10$ J m^{-2}, then $G_{cd} = 0.25$ kJ m^{-2}. If the interfacial fracture energy is much greater than this kind of value, then the condition $G_{ic} > {\sim} G_{fc}$ is likely to apply, making it probable that the fibres will fracture in the crack plane (see Fig. 9.8), with little or no debonding. Of course, a high value of s is helpful, but it is not obvious how to promote this (and a value of 50 is fairly high). It should also be noted that this argument applies only to the issue of whether an incident matrix crack will go straight through the fibre or deflect at the interface. Subsequent propagation of an interfacial crack might occur predominantly under shear (mode II) loading, with a considerably higher G_{ic} value.

9.2.3   Energy of Fibre Pull-Out

Potentially the most significant source of fracture work for most fibre composites is interfacial frictional sliding during fibre pull-out. Depending on the interfacial roughness, contact pressure and sliding distance, this process can absorb large quantities of energy. Fracture surfaces often show evidence of extensive pull-out, with aspect ratios ranging up to several tens or even hundreds. The work done can be calculated using a similar approach to that in Section 9.2.2. Consider a fibre with a remaining embedded length of x being pulled out an increment of distance dx. The associated work is given by the product of the force acting on the fibre and the distance it moves

$$dU = 2 \pi r x \tau_{i*} \, dx \tag{9.19}$$

where τ_{i*} is the interfacial shear stress, taken here as constant along the length of the fibre. The work done in pulling this fibre out completely is therefore given by

$$\Delta U = \int_0^{x_0} 2 \pi r x \tau_{i*} \, dx = x r x_0^2 \tau_{i*} \tag{9.20}$$

The next step is a similar integration (over all of the fibres) to that used to obtain Eqn (9.18), leading to an expression for the pull-out work of fracture, G_{cp}

$$G_{cp} = \int_0^L \frac{N dx_0}{L} \pi r x_0^2 \tau_{i*}$$
(9.21)

which, using Eqn (9.17) again for N, simplifies to

$$G_{cp} = \frac{f s^2 r \tau_{i*}}{3}$$
(9.22)

This contribution to the fracture energy can be large. For example, taking $f = 0.5$, $s = 50$, $r = 5$ μm and $\tau_{i*} = 20$ MPa gives a value of ~40 kJ m^{-2}. However, it is clear that the value of s is very important: reducing it to, say, 10, brings this figure down to ~1.5 kJ m^{-2}. Fig. 9.10 shows a fracture surface [13] that might be regarded as typical, confirming that high values of s (>~100 in some cases) are in practice common. However, it might have been expected that most fibres would break in the crack plane, where the stress is highest. In fact, the observed behaviour is related to the sensitivity of the strength (of brittle materials) to the presence of flaws (see Section 2.2 and, for example, Fig. 2.5).

The effect of the variability in fibre strength, characterised by the Weibull modulus, m, is depicted schematically in Fig. 9.11. This shows how, with a deterministic (single-valued) fibre strength ($m = \infty$), the probability, P_f, of the fibre breaking in the crack plane will eventually become 100%, while remaining zero elsewhere. For a finite value of m, however, not only are high values of P_f spread over an appreciable distance on either side of the crack plane, but also it is now possible for the fibre to break almost

Fig. 9.10 SEM micrograph [13] of the fracture surface of a composite material comprising ~20 vol% of glass fibres (10 μm diameter) in a polyamide matrix.
Reprinted from Composites Part A: Applied Science and Manufacturing, 36, S.Y. Fu, B. Lauke, Y.H. Zhang and Y.-W. Mai, On the post-mortem fracture surface morphology of short fiber reinforced thermoplastics, 987–994, © 2005, with permission from Elsevier.

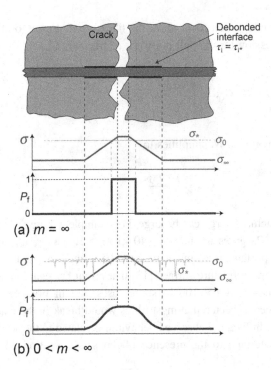

Fig. 9.11 Schematic depiction of the stress distribution, and associated probability of fracture, along a fibre bridging a matrix crack, for (a) fixed fibre strength, σ_* ($m = \infty$) and (b) strength that varies along the length of the fibre, due to the presence of flaws (finite m).

anywhere. Prediction of the fracture energy is now more complex, since account should be taken of these probabilities in calculating the contributions from different pull-out lengths. However, in practice this is not really necessary, other than to note that a low Weibull modulus tends to raise the pull-out work and that, at least under most circumstances, significant pull-out is observed in many composite systems.

A further point to note is that the pull-out energy is predicted to be greater when the fibres have a larger diameter, assuming that the fibre aspect ratio (s) is the same. This scale effect, which is well established, appears to run counter to a common perception that refinement of microstructure raises the 'strength'. Indeed, it is true that finer fibres (of brittle materials) tend to be stronger than coarser ones (because the size of the largest flaw is likely to scale with the diameter). However, the above treatment indicates that, despite their higher strength, use of finer fibres will lead to composites with lower toughness. Physically, this is due to the volume (around the crack tip) in which energy is being absorbed becoming smaller as the scale of the structure is refined. Of course, this is based on s remaining the same and there might be an expectation of higher values when the fibre is stronger. However, there are no clear grounds for such an expectation and in general composites made using very fine (nano-scale) fibres (or particles) have turned out to have relatively low toughness. This issue, and others related to fine-scale composites, are covered in Chapter 14.

9.2.4   Toughness Testing and Typical Outcomes for Polymer Composites

As a consequence of the above effects, fibre-reinforced polymer composites tend to have relatively high fracture energies, at least when fibres cross the crack plane. Of course, a fairly wide range of toughness levels can be exhibited, depending on the details of fibre content, interfacial characteristics, matrix and fibre types, etc. There is a lot of information of this type in the literature. Some limited, but representative, data [14] are shown in Fig. 9.12. This relates to uniaxial composites containing about 20 vol% of carbon fibre in an epoxy matrix, with or without additives (~10–15 vol% of rubber or alumina powder). Measured fracture energies are shown over a range of temperature. It can be seen that an increase is observed as the temperature is raised, presumably reflecting a slight tendency towards greater matrix plasticity. In general, however, the main energy-absorbing mechanisms, particularly fibre pull-out, have relatively little sensitivity to temperature. Furthermore, these changes in matrix constitution, and presumably also in the interfacial adhesion, also had relatively little effect, although the toughness with the rubberised matrix was a little lower, apparently associated with reduced pull-out [14]. Of course, many such effects will be highly specific to the system concerned.

In the context of Fig. 9.12, it should be mentioned that these data were obtained by impact (Charpy) testing, in which the energy needed to cause complete fracture of a sample is measured. This is a convenient and informative type of test, but it does have a tendency to produce higher fracture energy values than quasi-static tests, in which stable crack growth is stimulated under well-defined (plane strain, mode I) conditions. This tendency is at least partly due to the different (and changing) stress state around the crack tip during impact testing. Nevertheless, the values in Fig. 9.12 (~10–40 kJ m^{-2}),

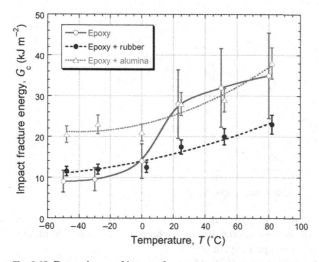

Fig. 9.12 Dependence of impact fracture energy on temperature, for three composites containing 20 vol% of carbon fibre (oriented normal to the crack plane), having an epoxy-based matrix, with or without additives [14].

obtained with composites containing only about 20 vol% fibre, do give an indication of the potential for high toughness.

Other general points about (quasi-static) fracture toughness testing include the potential for a dependence of the measured fracture energy on the crack length – often referred to as *R-curve* or *crack-bridging* effects. (The term *R*-curve is associated with a tendency to use the symbol R to represent G_c under conditions where it may vary during a test, and also to describe it as a crack growth 'resistance'.) This effect usually arises because, depending on a number of factors, frictional pull-out work may be absorbed over an appreciable distance behind the crack tip (i.e. in the 'wake' of the crack), as the crack opening displacement (Section 9.1.3) increases. This leads to more work being absorbed during propagation of longer cracks (up to a limit, above which a 'steady state' is reached). Some illustrative data are shown in Fig. 9.13, where it can be seen that G_c for this particular composite rises from ~0.3 kJ m^{-2} (at the initial notch) to ~2 kJ m^{-2} in the steady state (once the crack length has reached about 50 mm). While such effects are relevant to laboratory measurement of fracture energies, care must be taken in translating them into the performance of composites under service conditions. For example, it is not true that less energy (per unit area) will be absorbed during fracture of components with dimensions such that cracks never reach this 'steady state' length. (Complete failure will still involve the fibres being fully pulled out of their sockets across the whole of the fractured section.)

Of course, there is also interest in the toughness of laminates. Assessment tends to be more complex in this case, since there is potential for inter-laminar cracking, anisotropic behaviour, mixed-mode conditions, sensitivity to the loading geometry, etc.

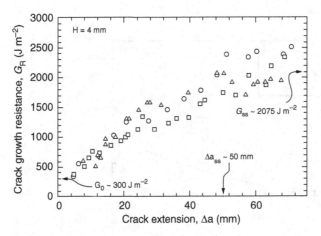

Fig. 9.13 Crack growth resistance (fracture energy) as a function of crack length extension, for an epoxy-based composite containing 60 vol.% carbon fibres, aligned normal to the crack plane [15]. The symbols refer to three repeat experiments.

Reprinted from Composites Part A: Applied Science and Manufacturing, 29, Bent F Sørensen and Torben K Jacobsen, Large-scale bridging in composites: R-curves and bridging laws, 1443–1451, © 1998, with permission from Elsevier.

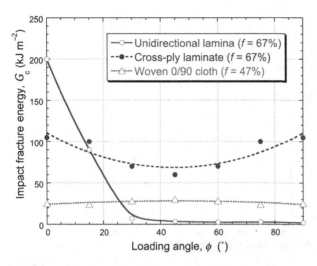

Fig. 9.14 Measured values of the fracture energy, G_c, obtained using the Charpy impact test, as a function of the loading angle, ϕ, between the axis of the test bar and the reference fibre direction, for three epoxy–glass fibre materials [16].

Nevertheless, some clear effects can be identified. For example, Fig. 9.14 shows fracture energy data, obtained using the Charpy impact test, for a unidirectional lamina and two types of laminate, all for glass fibres in epoxy. The fracture energy of the unidirectional lamina falls off sharply as the angle between the crack plane and the fibre axis is reduced (loading angle increased). This is largely because fibre pull-out becomes inhibited and fracture occurs predominantly parallel to the fibre axis. The cross-ply laminate and woven cloth material exhibit more isotropic behaviour. It is difficult to predict the dependence on loading angle in these cases, since complex interactions occur between the different plies and fibre tows. However, fracture always involves a considerable degree of interfacial debonding, fibre fracture and fibre pull-out, so that the toughness always tends to be relatively high (although, as mentioned above, impact testing often yields higher fracture energy values than quasi-static procedures). Nevertheless, long fibre laminates do often exhibit attractively high toughness levels.

9.2.5 Toughness of Metal Matrix Composites

As mentioned in Section 9.2.2, the presence of (ceramic) fibres or particles in a metal tends to reduce its toughness, mainly as a consequence of the plastic deformation (near the crack tip) being constrained. This type of effect accounts for the lower fracture energy shown in Table 9.1 for the Al-based MMC, when compared with unreinforced Al. This loss of toughness can sometimes be minimised by eliminating reinforcement clusters (Section 2.5.2) and other inhomogeneities such as pores, debonded interfaces and cracked particles or fibres. It follows that a high interfacial strength is desirable for MMCs and in most cases this is quite readily achievable (Section 7.3).

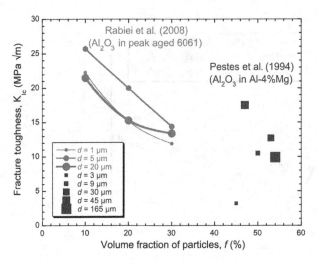

Fig. 9.15 Fracture toughness data [17,18] for Al-based MMCs containing particles of different sizes, plotted against reinforcement volume fraction.

Despite this (relatively limited) scope for optimising the toughness of (particulate-reinforced) MMCs, there is a clear tendency for it to be lower than that of the matrix concerned, and for a progressive drop to be observed as the reinforcement content is raised (and the inter-particle spacing is reduced). This is illustrated by the experimental data in Fig. 9.15, which cover the effects of particle size, as well as volume fraction. There have been many publications focused on this issue, although identification of the key effects is often hampered by the plethora of relevant variables, which include matrix hardness, particle distribution, interfacial bond strength, test procedure, etc. Nevertheless, some trends are fairly clear and understandable. There is, for example, a progressive drop in toughness as the ceramic content is raised (and constraint of matrix plasticity increases).

It should, however, be noted that this effect is not a simple or overriding one. It is clear from the data in the plot that particle size also has an influence. The expected one is a higher toughness with larger particles (raising the volume of matrix between particles that can undergo plastic deformation). The actual trend apparent in Fig. 9.15 appears to be an initial increase, but then a fall as the particle size is increased further. While results in this field can be somewhat contradictory, and are sometimes interpreted in different ways, a trend of this type has in fact been widely observed. It is usually attributed to particle cracking and/or interfacial cavitation (ahead of the crack tip) becoming more likely with larger particles, with these defects then linking up to facilitate crack propagation. It is certainly clear that cavitation at particle/matrix interfaces influences crack propagation and toughness, with particle shape, size and volume fraction all having an effect [19–21].

Modelling of such phenomena, so as to allow reliable prediction of the toughness of these composites, is not yet fully developed. Many of the early attempts were based on the Hahn and Rosenfield model [17,22]. This was not originally aimed at particulate

MMCs, but rather at the effect of inclusions (in Al alloys). These were taken to be in the approximate size range of 1–10 μm. They are often found to fracture ahead of the crack tip and it was assumed that crack extension occurs when the crack tip opening displacement is comparable with the width of the ligament separating the cavities forming at the particles. This leads to the following expression for the fracture toughness

$$K_{Ic} = \left[2\sigma_Y E \left(\frac{\pi}{6}\right)^{1/3} d \right]^{1/2} f^{-1/6} \tag{9.23}$$

where d is the particle diameter and f is the particle volume fraction. This does predict that the toughness goes down as the volume fraction is raised and goes up as the particle size is increased, broadly in line with experiment. The magnitude is usually somewhat under-predicted; explanations for this, and possible modifications, have been suggested [17,23], although in general the number of microstructural variables is such that toughness prediction is unlikely to be very accurate. As a general observation, it may be noted that, while relatively low toughness is certainly an issue with many particulate MMCs, products are being developed that offer attractions such as enhanced stiffness, hardness and wear resistance, and reduced thermal expansivity, in combination with acceptable toughness levels. Furthermore, cermets, which may be regarded as high volume fraction particulate MMCs, have been very successful over an extended period – see Section 9.2.6.

9.2.6 Toughness of Cermets

The term 'cermet' is simply a combination of 'ceramic' and 'metal', so that various materials incorporating both could in principle be encompassed. In practice, it normally [24] refers to an assembly of ceramic particles bonded together by a small proportion of a metallic phase. There is a strong case for regarding cermets as a special class of MMCs. While they could be considered as ceramics that have been toughened by the presence of a small proportion of metal, the metallic phase often forms some sort of partially interlinked network, so that they are in effect particulate-reinforced MMC with a very high proportion of ceramic particles. It may also be noted that the terms 'hardmetal' and 'cemented carbide' are commonly used. The differences between these are sometimes unclear and largely arise from historical origins [24]. The literature on cermets is large and advances continue to be made on their processing and formulation [25–28].

Cermets are normally produced by blending of ceramic and metallic powders, followed by liquid-phase sintering. Typically, the ceramic particles are ~1–10 μm in diameter. Blending is usually followed by cold isostatic pressing, or injection moulding, to give the required shape, and then by holding at a suitable temperature under vacuum, inert gas or hydrogen. The binder, which is usually a metal with a high melting point and low chemical reactivity (such as Co), is commonly in a strain-hardened state [29] as a consequence of the high triaxial constraint and the plastic deformation imposed during

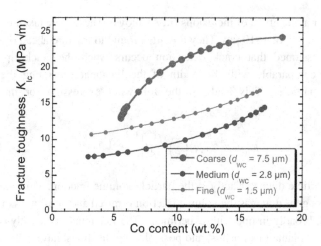

Fig. 9.16 Dependence on Co content, and the size of the WC particles, of the fracture toughness of sintered WC–Co cermets [30].

cooling. Nevertheless, it does confer a substantial increase in toughness, as well as facilitating the consolidation process. This is illustrated by Fig. 9.16, where it can be seen that the toughness [30–33] rises with increasing Co content (decreasing 'reinforcement' level). Of course, this trend is consistent with that in 'conventional' MMCs, but it can be seen that the actual toughness values are relatively high and certainly do not appear to lie on the same trend line as the data in Fig. 9.15 – the values of f in Fig. 9.16 are all over 80%.

It is of interest to consider why cermets are so tough – they are certainly tougher than most other materials with comparable stiffness ($E \sim 600$ GPa) and hardness ($H_V \sim 10$–20 GPa); these values are, as expected, similar to those of monolithic engineering ceramics – see Fig. 1.2 in Section 1.2. It seems likely that the relatively small amount of metal present is effective in conferring good toughness at least partly because it is very uniformly distributed. Also, it appears that cavitation takes place less readily in cermets than with higher metal content MMCs, probably because stress concentration effects are less pronounced. Finally, it is also possible that prolonged industrial experience with cermets has promoted better optimisation of processing conditions, etc., so that characteristics such as interfacial bond strength, homogeneity, avoidance of porosity, etc. may be superior. In fact, it seems likely that research into optimisation of (lower ceramic content) particulate MMCs could benefit from improved familiarity with the extensive experience and expertise associated with cermet technology.

For applications such as cutting tools, the behaviour at high temperature is clearly important. The binder will tend to soften on heating, but this will not necessarily impair the hardness very much, particularly at relatively low binder levels, since it is heavily constrained by the presence of the surrounding particles. Cermets thus exhibit excellent hot hardness (markedly superior to high-speed-tool steels) in the range 700–1000°C. However, at temperatures above this, grain boundary sliding of the carbides starts to occur, leading to extensive plastic deformation. This is pronounced in WC–Co cermets. The TiC cermets, commonly having TiCN–Mo–Ni formulations, are more resistant to this effect and are thus commonly used in high-speed cutting operations that may raise the temperature of the tool above 1000°C. For such applications, their exceptional hot hardness offsets the rather lower hardness and toughness these systems exhibit at room temperature when compared with WC–Co. Of course, many other (more complex) cermet formulations are also available.

9.2.7 Toughness of a Metal Fibre-Reinforced Ceramic Composite

There have been many attempts to produce CMCs with relatively high toughness, mostly with limited success. Much work has been oriented towards the incorporation of ceramic fibres into a ceramic matrix, with the intention being to promote crack deflection and pull-out in a similar manner to that in PMCs. Unfortunately, this has often required processing at high temperatures, such that the interfaces become strongly bonded, reducing the incidence of fibre pull-out (see Section 9.2.2). While there are commercial products of this type, notably carbon–carbon composites used in applications such as aircraft brakes (Section 16.5.1), such material is not in widespread use.

A more promising approach, at least in many respects, is based on the introduction of a network of *metallic fibres* into a ceramic matrix. The objective here is to ensure that fracture always involves extensive plastic deformation of the fibres, as well as pull-out work, requiring large amounts of energy. There are in fact some commercial products based on this type of structure, such as the one marketed under the tradename 'Fiberstone'. A typical microstructure [34] is shown in Fig. 9.17. The fibres are usually about 0.5 mm in diameter, although finer fibres can be used. They are usually made of

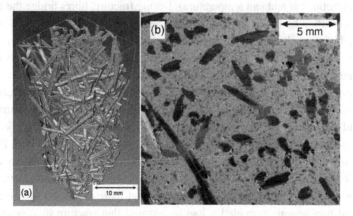

Fig. 9.17 Microstructure [34] of a metal fibre-reinforced ceramic composite: (a) tomographic reconstruction of the fibre architecture; and (b) low-magnification optical micrograph of a polished section.

Reprinted from Composites Science and Technology, 71, S.R. Pemberton, E.K. Oberg, J. Dean, D. Tsarouchas, A.E. Markaki, L. Marston and T.W. Clyne, The fracture energy of metal fibre reinforced ceramic composites (MFCs), 266–275, © 2011, with permission from Elsevier.

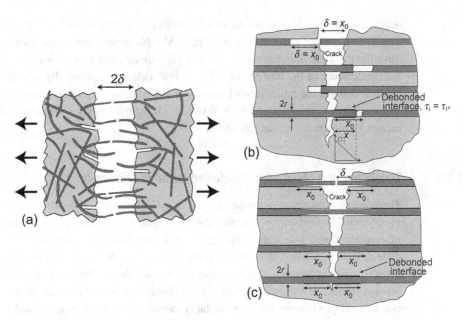

Fig. 9.18 Schematic [34] of Fiberstone fracture: (a) overall geometry; (b) debonding, possibly fracture, and then frictional pull-out; and (c) debonding, plastic deformation and then fracture. Reprinted from Composites Science and Technology, 71, S.R. Pemberton, E.K. Oberg, J. Dean, D. Tsarouchas, A.E. Markaki, L. Marston and T.W. Clyne, The fracture energy of metal fibre reinforced ceramic composites (MFCs), 266–275, © 2011, with permission from Elsevier.

stainless steel, while the matrix is predominantly alumina (plus a small proportion of cementitious material – components are manufactured by infiltration of a ceramic slurry into an assembly of fibres in a mould, with consolidation then taking place largely via hydration reactions at ambient temperature). During fracture, fibres bridge the crack and energy is absorbed by both frictional pull-out and plastic deformation, as illustrated in Fig. 9.18. These mechanisms are expected to dominate the work of fracture.

The potential for energy absorption via fibre plasticity in such materials has been highlighted repeatedly [35–38], with particular attention having been drawn by Ashby and co-workers [35] to the concept of the constraint imposed by the surrounding matrix having an influence on the volume of fibre in which plastic deformation can occur, and hence on its effective ductility (and on the total plastic work). This is clearly dependent on the interfacial bond strength, which thus affects both fibre pull-out and fibre plasticity. There have been various observations, over an extended period [39–44], concerning the extent and nature of fibre pull-out and plastic deformation in composites of this type. It seems clear that there is at least the potential for pronounced toughening, provided it can be ensured that fracture involves extensive plastic deformation of fibres. The simple model [34] described below allows prediction of the fracture energy from information about how fibre deformation and rupture occurs during crack propagation.

The composite work of fracture, G_{cnet}, is taken as the sum of contributions from pull-out, G_{cpo}, and plastic deformation, G_{cfd}, of fibres bridging the crack plane, assuming that a fraction g of them undergo the former (and $(1 - g)$ the latter). The equation for the pull-out work of fracture (Eqn (9.22)) can be used, but the relationship between N and f depends on the fibre orientation distribution. Eqn (9.22) refers to a set of aligned fibres (normal to the crack plane). As can be seen in Fig. 9.17(a), this material has an approximately isotropic (random) distribution of fibre orientations. It can be shown [45] that, for this case, N is reduced by a factor of 2, leading to:

$$G_{cpo} = \frac{gfs^2 r \tau_{i*}}{6}$$ (9.24)

where s is the ratio of δ (maximum fibre length extending beyond the crack plane – see Fig. 9.18) to the fibre radius, r.

The value of G_{cfd} is obtained assuming interfacial debonding to a distance x_0 from the crack plane (Fig. 9.18(c)), treating the work done as if it were that in a tensile test with an original fibre length $2x_0$.

$$G_{cfd} = (1 - g)2x_0 N U_{fd} = (1 - g)2x_0 \left(\frac{f}{2\pi r^2}\right) W_{fd} \pi r^2 = (1 - g)x_0 f W_{fd}$$ (9.25)

where U_{fd} and W_{fd} are the work of deformation of the fibre, expressed respectively per unit length (J m^{-1}) and per unit volume (J m^{-3}). The latter is the area under the stress–strain curve of the fibre during a tensile test. The value of δ is in this case given by the product of x_0 and ε_* (fibre strain at failure), leading to

$$G_{cfd} = (1 - g)\left(\frac{\delta}{\varepsilon_*}\right)f W_{fd} = (1 - g)\frac{s r f W_{fd}}{\varepsilon_*}$$ (9.26)

The work of fracture can now be estimated by summing the energy absorbed via both processes, assuming that a fraction g of the fibres bridging the crack plane undergo pull-out and the remainder $(1 - g)$ undergo plastic deformation and rupture

$$G_{cnet} = g G_{cpo} + (1 - g)G_{cfd} = g\frac{fs_{po}^2 r \tau_{i*}}{2} + (1 - g)\frac{fs_{fd} r W_{fd}}{\varepsilon_*}$$ (9.27)

For a given (maximum) fibre protrusion aspect ratio, s ($= \delta/r$), both pull-out and plastic deformation contributions increase linearly with absolute scale (fibre diameter), again reflecting the tendency for the toughness of composites to be greater with a coarser scale of structure. This tendency is apparent in the comparison shown in Fig. 9.19 between predicted and observed fracture energy levels, for composites containing either fine ($r = 25$ μm) or coarse ($r = 250$ μm) fibres. There is, in general, good agreement between predicted and measured data. It is also worth noting that, for the coarse fibre composite, both observed and predicted fracture energies are in a range corresponding to a genuinely tough material. In a separate study [46], it has been shown that the model can be used to account for the effect of (severe) heat treatment on the toughness of these materials, with use of an oxidation-resistant stainless steel fibre giving markedly better performance. These materials are usable [47] at temperatures well over 1000°C, despite the presence of a metallic constituent.

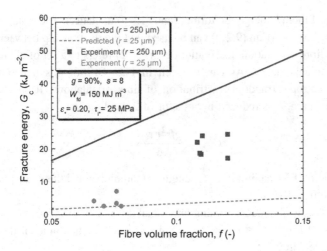

Fig. 9.19 Experimental [34] fracture energy values of 'Fibrestone', as a function of fibre fraction, and predictions obtained using Eqn (9.27), for fine and coarse fibres.
Reprinted from Composites Science and Technology, 71, S.R. Pemberton, E.K. Oberg, J. Dean, D. Tsarouchas, A.E. Markaki, L. Marston and T.W. Clyne, The fracture energy of metal fibre reinforced ceramic composites (MFCs), 266–275, © 2011, with permission from Elsevier.

There are, of course, several simplifications incorporated in Eqn (9.27), including the one that all fibres are assumed to undergo either pull-out or plastic rupture, but not both. In practice, the majority may well experience both, at least to some degree, depending, of course, on the fibre strength and ductility, and the interfacial bond strength. Further-more, while some of the parameter values, including g, are difficult to estimate, none of them are arbitrarily adjustable – i.e. they all represent physically meaningful effects. In general, it is clear that ceramic-based composites of this type do have considerable potential in terms of offering the usual attractions of ceramics (thermal and chemical stability, hardness, stiffness, wear resistance, etc.) in combination with good toughness (and also versatile and economic processing characteristics). Their usage is likely to increase in the future.

9.3   Sub-Critical Crack Growth in Composites

When the rate of energy release (driving force) during crack propagation is lower than the critical value, spontaneous fast fracture does not occur. Under some circumstances, however, an existing crack may advance slowly under this driving force. Since crack growth leads to an increased driving force (for the same applied load), this process is likely to lead to an accelerating rate of damage, culminating in conditions for fast fracture being satisfied. There are two common situations in which such sub-critical crack growth tends to occur. First, if the applied load is fluctuating in some way, local conditions at the crack tip may be such that a small advance occurs during each cycle. Second, the penetration of a corrosive fluid to the crack tip region may lower the local toughness and allow crack advance at a rate determined by the fluid penetration kinetics or chemical interaction effects. In both cases, as indeed with fast fracture, the presence or absence of an initial flaw, which allows the process to initiate, is likely to be of considerable importance.

9.3.1   Fatigue Crack Growth

For metals, fatigue failure is an important topic that has been the subject of detailed investigation over many years. Analysis is commonly carried out in terms of the difference in stress intensity factor between the maximum and minimum applied load (ΔK). This is because, while the maximum value, K_{max}, dictates when fast fracture will occur, the cyclic dissipation of energy is dependent on ΔK. It is, however, common to also quote the *stress ratio*, R ($= K_{min}/K_{max}$), which enables the magnitudes of the K values to be established for a given ΔK. The resistance of a material to crack extension is given in terms of the crack growth rate per loading cycle (dc/dN). At intermediate ΔK, the crack growth rate usually conforms to the Paris–Erdogan relation [48]

$$\frac{dc}{dN} = \beta \Delta K^n \tag{9.28}$$

where β is a constant. Hence, a plot of crack growth rate (m/cycle) against ΔK, with log scales, gives a straight line in the Paris regime, with a gradient equal to n. At low stress intensities, there is a threshold, ΔK_{th}, below which no crack growth occurs. The crack growth rate usually accelerates as the level for fast fracture, K_c, is approached.

An alternative way of presenting fatigue data is in the form of S/N_f curves, showing the number of cycles to failure (N_f) as a function of the stress amplitude (S). Many materials show rapid crack growth (low N_f) when the stress amplitude is high, a central portion of decreasing S with rising N_f, corresponding to the Paris regime, and a *fatigue limit*, which is a stress amplitude below which failure does not occur, even after large numbers of cycles. This corresponds to a stress intensity factor below ΔK_{th}.

It is instructive to examine how the presence of particulate reinforcement affects the fatigue behaviour of a metal. Some illustrative data are shown in Fig. 9.20 for an Al alloy with and without 20 vol% of SiC particulate. It can be seen that, depending on the value of R, ΔK_{th} is around 1–3 MPa√m for the alloy and 2–5 MPa√m for the MMC. Several explanations for this increase have been proposed [49,50], including crack deflection at interfaces and a reduction in slip band formation due to the particles. This beneficial effect of the reinforcement in inhibiting the onset of fatigue cracks is a useful feature of such MMCs. However, MMCs are less tough than unreinforced metals, mainly as a result of the constraint imposed on matrix plasticity (Section 9.2.5). The Paris regime is usually relatively short and the exponent n is often around 5–6, which is higher (particularly at high stress ratios) than those typical of unreinforced systems (~3–4). Final fast fracture of the composite is usually initiated at lower ΔK values than for the unreinforced metal. In practical terms, this means that MMCs can offer improvements in fatigue performance compared with metals, providing that applied stress levels

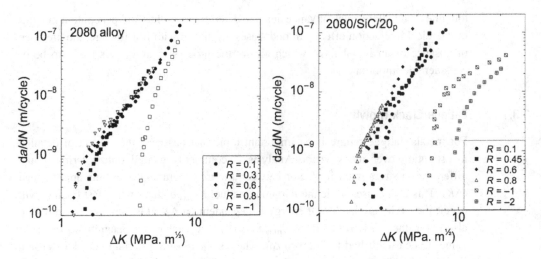

Fig. 9.20 Fatigue crack growth data [50] for 2080 Al alloy, with and without 20 vol% of SiC particulate (8 μm diameter), for several different values of the stress ratio, R.
Reprinted from International Journal of Fatigue, 32, N. Chawla and V.V. Ganesh, Fatigue crack growth of SiC particle reinforced metal matrix composites, 856–863, © 2010, with permission from Elsevier.

are low and/or flaw sizes are kept small. Control over processing of MMCs so as to eliminate inhomogeneities may be important to ensure this.

While particulate reinforcement usually produces relatively minor modifications to the fatigue behaviour of the matrix, the presence of long fibres has a more pronounced effect. This is particularly true for polymer composites, in which it is often the type and orientation distribution of the fibres that is of most significance. The propagation of cracks through the matrix and along the interfaces usually dictates how fatigue progresses, but this is strongly influenced by how the fibres affect the stress distribution. The failure strain of the matrix is also important. A further difference from the particulate case lies in the distribution of damage. It is common in long fibre composites for matrix and interfacial micro-cracks to form at many locations throughout the specimen. Fibre bridging across matrix cracks often occurs, reducing the stress intensity at the crack tip. In contrast to this, fatigue crack growth in monolithic and particle-reinforced materials usually involves a single dominant crack with a well-defined length.

While the axial fatigue resistance of long fibre composites tends to be very good, particularly with high-stiffness fibres, performance is usually inferior for laminates or under off-axis loading. This is illustrated by the plots [51] in Fig. 9.21, which are for glass-reinforced polyester. The cross-ply and woven cloth laminates fail at lower loads than the unidirectional material and show little evidence of a fatigue limit stress value being identifiable. Damage to the transversely oriented regions starts at low applied loads (see Section 8.1.2), transferring extra load and eventually causing cracks to propagate into the axial regions. Nevertheless, the fatigue resistance of such materials compares quite well with that of many metals. Finally, the chopped strand mat and dough moulding compound show relatively poor fatigue resistance. In these materials,

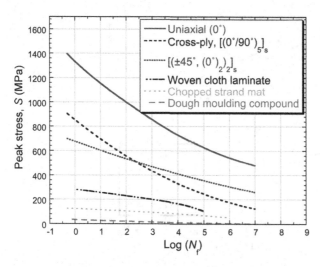

Fig. 9.21 Experimental S/N_f plots [51] showing the number of cycles to failure during fatigue loading of glass fibre–polyester composites with various fibre distributions, as a function of the peak applied stress. The stress ratio, R, was 0.1 in all cases.

fibres are misaligned and often have relatively low aspect ratios, particularly for the dough moulding compound.

A further point worthy of note with respect to the fatigue of composites is that the behaviour is often sensitive to the absolute values of the stresses being applied, rather than just the ΔK range. In particular, the introduction of compressive stresses usually reduces the resistance to fatigue. This is largely due to the axially aligned fibres having poor resistance to buckling (see Section 8.1.3). This results in damage to the fibres and the surrounding matrix and also allows larger stresses to bear on neighbouring, transversely oriented regions, accelerating their degradation. This is illustrated by the data [52] in Fig. 9.22, which show that the fatigue resistance of carbon fibre laminates becomes poor for negative R values. (It should, however, be noted that the effect is exaggerated by plotting the peak stress, rather than the stress difference, which is larger for the lower R values.)

A further point concerns the frequency of cycling. For metals, this usually has little effect, but in polymer composites a higher frequency often hastens failure. This is partly because of the viscoelastic response of polymers. Matrix damage is more likely if the strain is imposed rapidly, allowing no time for creep and stress relaxation. A second effect arises from the low thermal conductivity of polymers, particularly if the fibres are also poor conductors (glass, Kevlar, etc.). Such composites tend to increase in temperature during fatigue loading as a result of difficulties in dissipating the heat generated locally by damage and viscoelastic deformation. This is accentuated at high cycling frequencies. Since the strength of the matrix falls with increasing temperature, this tends to accelerate failure.

It should finally be emphasised that the details of how (polymer) composite laminates perform under fatigue loading are both complex and technologically important – particularly in view of their extensive and increasing usage in the aerospace industry.

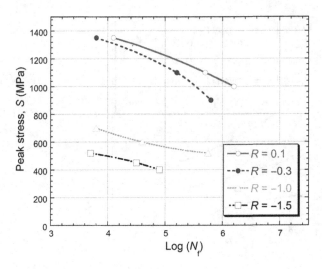

Fig. 9.22 Experimental S/N_f data [52] for a carbon fibre laminate with a $([(\pm45°, 0°_2)_2]_s$ stacking sequence, showing the effect of changing the stress ratio, R.

The effects described above are no more than the tip of a very large iceberg. Consultation of more detailed literature [53–56] is recommended if anything more than a cursory appreciation of the area is required. In particular, certain hybrid materials with layered polymer laminae/metal sheet structures (such as GLARE) are of considerable interest for airframe structures, with their fatigue properties [57–59] being of central importance.

9.3.2   Stress Corrosion Cracking

Stress corrosion cracking is the term given to sub-critical crack growth that occurs as a result of the effect of a corrosive fluid reaching the advancing crack tip. The micromechanisms responsible for this lowering of the local toughness of the material vary widely between different materials and environments. For example, in Al-based systems, the presence of moist or salt-laden air causes an acceleration of fatigue crack growth of 5–10 times. There are indications [60,61] that such effects can be even stronger in the corresponding particulate MMCs. This may be due to the presence of high-stress regions, and interfaces, in the vicinity of particles, where chemical attack is accentuated.

In the case of Al, stress corrosion cracking is usually due to the evolution of (atomic) hydrogen at the crack tip, which is promoted by the presence of various liquids. This tends to cause embrittlement by impeding dislocation motion and/or creating microvoids containing molecular hydrogen. In MMCs, a further factor is introduced in the form of strong traps for hydrogen at the matrix/reinforcement interface, promoting cracking. It has been shown [62] that, while the strength levels of MMCs were unaffected by testing in a hydrogen environment, the ductility was sharply reduced, irrespective of matrix ageing. This was attributed to strong trapping of hydrogen at the interface, where voids (not seen in air-tested specimens) were observed.

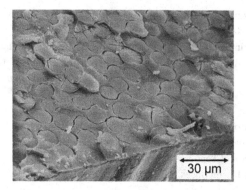

Fig. 9.23 SEM micrograph of the fracture surface [65] of a glass fibre–polyester resin lamina tested in axial tension in the presence of dilute hydrochloric acid.
Reprinted from Micromechanisms of crack growth in composite materials under corrosive environments, Hogg, P. J. and Hull, D., Materials Science and Technology, 1980, Taylor & Francis, reprinted by permission of the publisher (Taylor & Francis Ltd http://www .tandfonline.com).

For polymer composites, the behaviour of the matrix is often quite sensitive to the presence of fluids. For example, the properties of many polymers are modified by absorption of moisture. In many cases, this penetration occurs by diffusion through the matrix and is not confined to access along the crack. Various features of such absorption have been identified [63,64]. Commonly, water absorption raises the toughness and strain to failure of the matrix. Thus, the fatigue resistance of glass-reinforced epoxy can be raised by prior boiling in water [51]. However, water uptake also tends to promote interfacial debonding, impairing stiffness and strength, particularly under shear and transverse loading. Some fluids sharply degrade the strength of the fibres. An example of this can be seen in Fig. 9.23, which shows the fracture surface of a polyester–60% glass composite tested in the presence of dilute hydrochloric acid [65]. This attacks the fibre surface at the crack tip, reducing its strength considerably. Not only does this impair the strength of the composite, but also the toughness and fatigue resistance are sharply reduced, since the fibres all break in the crack plane and the pull-out work (Section 9.2.4) is virtually zero. The danger of such effects must always be borne in mind when glass-reinforced polymer composites are to be used under load in chemically aggressive environments, particularly mineral acids [66].

References

1. Griffith, AA, The phenomena of rupture and flow in solids. *Philosophical Transactions of the Royal Society A* 1920; **221**: 163–198.
2. Irwin, GR, Fracture dynamics, in *Fracturing of Metals*. ASM, 1948, pp. 147–166.
3. Rice, JR, Elastic fracture mechanics concepts for interfacial cracks. *Journal of Applied Mechanics* 1988; **55**: 98–103.

4. Evans, AG and JW Hutchinson, Effects of non-planarity on the mixed mode fracture resistance of bimaterial interfaces. *Acta Metallurgica* 1989; **37**(3): 909–916.

5. Wang, J-S and Z Suo, Experimental determination of interfacial fracture toughness curves using Brazil-Nut-sandwiches. *Acta Metallurgica et Materialia* 1990; **38**(7): 1279–1290.

6. Akisanya, AR and NA Fleck, Brittle fracture of adhesive joints. *International Journal of Fracture* 1992; **58**: 93–114.

7. Cao, HC and AG Evans, An experimental study of the fracture resistance of bimaterial interfaces. *Mechanics of Materials* 1989; **7**: 295–304.

8. Cook, J and JE Gordon, A mechanism for the control of crack propagation in all-brittle systems. *Proceedings of the Royal Society A* 1964; **282**: 508–520.

9. Kendall, K, Control of cracks by interfaces in composites. *Proceedings of the Royal Society of London* 1975; **341**: 409–428.

10. Kendall, K, Transition between cohesive and interfacial failure in a laminate. *Proceedings of the Royal Society A* 1975; **344**: 287–302.

11. He, M-Y and JW Hutchinson, Crack deflection at an interface between dissimilar elastic materials. *International Journal of Solids and Structures* 1989; **25**: 1053–1067.

12. He, M-Y and JW Hutchinson, Kinking of a crack out of an interface. *Journal of Applied Mechanics* 1989; **56**: 270–78.

13. Fu, SY, B Lauke, YH Zhang and YW Mai, On the post-mortem fracture surface morphology of short fiber reinforced thermoplastics. *Composites Part A: Applied Science and Manufacturing* 2005; **36**(7): 987–994.

14. Kim, JK, C Baillie, J Poh and YW Mai, Fracture-toughness of CFRP with modified epoxy-resin matrices. *Composites Science and Technology* 1992; **43**(3): 283–297.

15. Sorensen, BF and TK Jacobsen, Large-scale bridging in composites: R-curves and bridging laws. *Composites Part A: Applied Science and Manufacturing* 1998; **29**(11): 1443–1451.

16. Harris, B, *Engineering Composite Materials*. Institute of Metals, 1980.

17. Rabiei, A, L Vendra and T Kishi, Fracture behavior of particle reinforced metal matrix composites. *Composites Part A: Applied Science and Manufacturing* 2008; **39**(2): 294–300.

18. Pestes, RH, SV Kamat and JP Hirth, Fracture toughness of Al-4%Mg/Al$_2$O$_3$p composites. *Materials Science and Engineering A: Structural Materials, Properties, Microstructure and Processing* 1994; **189**(1–2): 9–14.

19. Whitehouse, AF and TW Clyne, Critical stress criteria for cavitation in MMCs. *Acta Metallurgica et Materialia* 1995; **43**: 2107–2114.

20. Song, SG, N Shi, GT Gray and JA Roberts, Reinforcement shape effects on the fracture behavior and ductility of particulate-reinforced 6061-Al matrix composites. *Metallurgical and Materials Transactions A: Physical Metallurgy and Materials Science* 1996; **27**(11): 3739–3746.

21. Derrien, K, D Baptiste, D Guedra-Degeorges and J Foulquier, Multiscale modeling of the damaged plastic behavior and failure of Al SiCp composites. *International Journal of Plasticity* 1999; **15**(6): 667–685.

22. Hahn, GT and AR Rosenfield, Metallurgical factors affecting fracture toughness of Al alloys. *Metallurgical Transactions A* 1975; **6**: 653–670.

23. Garrison, WM and NR Moody, Ductile fracture. *Journal of Physics and Chemistry of Solids* 1987; **48**(11): 1035–1074.

24. Mari, D, Cermets and hardmetals, in *Encyclopaedia of Materials, Vol. X: Composite Materials*, Mortensen, A, editor. Elsevier, 2001.

25. Venkateswaran, T, B Basu, GB Raju and DY Kim, Densification and properties of transition metal borides-based cermets via spark plasma sintering. *Journal of the European Ceramic Society* 2006; **26**(13): 2431–2440.

26. Kim, HC, IJ Shon, JK Yoon, JM Doh and ZA Munir, Rapid sintering of ultrafine WC–Ni cermets. *International Journal of Refractory Metals & Hard Materials* 2006; **24**(6): 427–431.

27. Zhang, SC, GE Hilmas and WG Fahrenholtz, Zirconium carbide–tungsten cermets prepared by in situ reaction sintering. *Journal of the American Ceramic Society* 2007; **90**(7): 2296–2296.

28. Deorsola, FA, D Vallauri, GAO Villalba and B De Benedetti, Densification of ultrafine WC–12Co cermets by pressure assisted fast electric sintering. *International Journal of Refractory Metals & Hard Materials* 2010; **28**(2): 254–259.

29. Sigl, LS and HF Fischmeister, On the fracture toughness of cemented carbides. *Acta Metallurgica* 1988; **36**(4): 887–897.

30. Lueth, RC, Determination of fracture toughness parameters for tungsten carbide–cobalt alloys, in *Fracture Mechanics of Ceramics*, Bradt, RC, editor. Plenum Press, 1974, pp. 791–806.

31. Pickens, JR and J Gurland, The fracture toughness of WC–Co alloys measured on single-edge notched beam specimens precracked by electron discharge machining. *Materials Science and Engineering* 1978; **33**: 135–142.

32. Bouaouadja, N, M Hamidiuche, H Osmani and G Fantozzi, Fracture toughness of WC–Co cemented carbides at room temperature. *Journal of Materials Science Letters* 1994; **13**: 17–19.

33. Ravichandran, KS, Fracture toughness of two phase WC–Co cermets. *Acta Metallurgica et Materialia* 1994; **42**: 143–150.

34. Pemberton, SR, EK Oberg, J Dean, D Tsarouchas, AE Markaki, L Marston and TW Clyne, The fracture energy of metal fibre reinforced ceramic composites (MFCs). *Composites Science and Technology* 2011; **71**(3): 266–275.

35. Ashby, MF, FJ Blunt and M Bannister, Flow characteristics of highly constrained metal wires. *Acta Metallurgica* 1989; **37**(7): 1847–1857.

36. Kazmin, VI, ST Mileiko and VV Tvardovsky, Strength of ceramic matrix–metal fibre composites. *Composites Science and Technology* 1990; **38**: 69–84.

37. Kaute, DAW, HR Shercliff and MF Ashby, Delamination, fibre bridging and toughness of ceramic matrix composites. *Acta Metallurgica et Materialia* 1993; **41**(7): 1959–1970.

38. Vekinis, G, E Sofianopoulos and WJ Tomlinson, Alumina toughened with short nickel fibres. *Acta Metallurgica* 1997; **45**(11): 4651–4661.

39. Simpson, LA and A Wasylysh, Fracture energy of Al2O3 containing Mo fibers. *Journal of the American Ceramic Society* 1971; **54**(1): 56.

40. Hing, P and G Groves, The strength and fracture toughness of polycrystalline magnesium oxide containing metallic particles and fibres. *Journal of Materials Science* 1972; **7**(4): 427–434.

41. Morton, J and GW Groves, The effect of metal wires on the fracture of a brittle-matrix composite. *Journal of Materials Science* 1976; **11**: 617–622.

42. Shah, SP and C Ouyang, Mechanical behaviour of fiber-reinforced cement-based composites. *Journal of the American Ceramic Society* 1991; **74**(11): 2727–2738, 2947–2953.

43. Donald, IW and BL Metcalfe, The preparation, properties and applications of some glass-coated metal filaments prepared by the Taylor-wire process. *Journal of Materials Science* 1996; **31**: 1139–1149.

44. Cailleux, E, T Cutard and G Bernhart, Pullout of steel fibres from a refractory castable: experiment and modelling. *Mechanics of Materials* 2005; **37**(4): 427–445.

45. Underwood, EE, *Quantitative Stereology*. Addison-Wesley Publishing Company, 1970.

46. Lam, SK and TW Clyne, Toughness of metal fibre/ceramic matrix composites (MFCs) after severe heat treatments. *Materials Science and Technology* 2014; **30**(10): 1135–1141.

47. Clyne, TW and LW Marston, Metal fibre-reinforced ceramic composites and their industrial usage, in *Comprehensive Composite Materials II*, Clyne, TW, editor. Elsevier, 2018, pp. 464–481.

48. Paris, P and F Erdogan, A critical analysis of crack propagation laws. *Journal of Basic Engineering* 1963; **85**: 528–534.

49. Christman, T and S Suresh, Effects of SiC reinforcement and aging treatment on fatigue crack growth in an Al-SiC composite. *Materials Science and Engineering* 1988; **102A**: 211–216.

50. Chawla, N and VV Ganesh, Fatigue crack growth of SiC particle reinforced metal matrix composites. *International Journal of Fatigue* 2010; **32**(5): 856–863.

51. Harris, B, Fatigue of glass fibre composites, in *Handbook of Polymer Fibre Composites*, Jones, FR, editor. Longman Green, 1994, pp. 309–316.

52. Lee, JA, DP Almond and B Harris, The use of neural networks for the prediction of fatigue lives of composite materials. *Composites Part A: Applied Science and Manufacturing* 1999; **30**(10): 1159–1169.

53. Gamstedt, EK and BA Sjogren, Micromechanisms in tension–compression fatigue of composite laminates containing transverse plies. *Composites Science and Technology* 1999; **59**(2): 167–178.

54. Payan, J and C Hochard, Damage modelling of laminated carbon/epoxy composites under static and fatigue loadings. *International Journal of Fatigue* 2002; **24**(2–4): 299–306.

55. Quaresimin, M, L Susmel and R Talreja, Fatigue behaviour and life assessment of composite laminates under multiaxial loadings. *International Journal of Fatigue* 2010; **32**(1): 2–16.

56. Movahedi-Rad, AV, T Keller and AP Vassilopoulos, Fatigue damage in angle-ply GFRP laminates under tension–tension fatigue. *International Journal of Fatigue* 2018; **109**: 60–69.

57. Wu, GC and JM Yang, The mechanical behavior of GLARE laminates for aircraft structures. *JOM* 2005; **57**(1): 72–79.

58. Kotik, HG and JEP Ipina, Short-beam shear fatigue behavior of fiber metal laminate (GLARE). *International Journal of Fatigue* 2017; **95**: 236–242.

59. Li, HG, YW Xu, XG Hua, C Liu and J Tao, Bending failure mechanism and flexural properties of GLARE laminates with different stacking sequences. *Composite Structures* 2018; **187**: 354–363.

60. Singh, PM and JJ Lewandowski, Effects of heat treatment on stress-corrosion cracking of a discontinuously reinforced aluminium (DRA) 7XXX alloy during slow strain rate testing. *Scripta Metallurgica et Materialia* 1995; **33**(9): 1393–1399.

61. Singh, PM and JJ Lewandowski, Stress corrosion cracking of discontinuously reinforced aluminium (DRA) alloy 2014 during slow strain rate testing. *Journal of Materials Science Letters* 1996; **15**(6): 490–493.

62. You, CP, M Dollar, AW Thompson and IM Bernstein, Microstructure-property relationships and hydrogen effects in a particulate-reinforced aluminium composite. *Metallurgical Transactions A: Physical Metallurgy and Materials Science* 1991; **22**(10): 2445–2450.

63. Shen, C-H and GS Springer, Moisture absorption and desorption of composite materials. *Journal of Composite Materials* 1976; **10**: 2–20.

64. Kumosa, L, B Benedikt, D Armentrout and M Kumosa, Moisture absorption properties of unidirectional glass/polymer composites used in composite (non-ceramic) insulators. *Composites Part A: Applied Science and Manufacturing* 2004; **35**(9): 1049–1063.

65. Hogg, PJ and D Hull, Micromechanisms of crack growth in composite materials under corrosive environments. *Metal Science* 1980; **14**(8–9): 441–449.

66. Price, JN and D Hull, Effect of matrix toughness on crack propagation during stress-corrosion of glass reinforced composites. *Composites Science and Technology* 1987; **28**(3): 193–210.

10 Thermal Effects in Composites

The behaviour of composite materials is often sensitive to changes in temperature. This arises for two main reasons. Firstly, the response of the matrix to an applied load is often temperature-dependent; and secondly, changes in temperature can cause internal stresses to be set up as a result of differential thermal contraction and expansion of the two constituents. These stresses affect the thermal expansivity (expansion coefficient) of the composite. Furthermore, significant stresses are normally present in the material at ambient temperatures, since it has in most cases been cooled at the end of the fabrication process. Changes in internal stress state on altering the temperature can be substantial and may influence the response of the material to an applied load. Thermal cycling can thus have strong effects on, for example, creep characteristics. Finally, the thermal conductivity of composite materials is of interest, since many applications and processing procedures involve heat flow of some type. This property can be predicted from the conductivities of the constituents, although the situation may be complicated by poor thermal contact across the interfaces.

10.1 Thermal Expansion and Thermal Stresses

10.1.1 Thermal Expansivity Data for Reinforcements and Matrices

There is commonly a substantial difference between the thermal expansivities of matrix and reinforcement. This is clear from a glance at Fig. 10.1, which shows some approximate values (as a function of temperature) for a few representative materials (plastics, metals and ceramics). Noting the log scale of this plot, it can be seen that the differences are often large. For example, while glass (fibres), which are mainly silica, have a value of ~1 microstrain K^{-1}, that for an epoxy (matrix) is ~50–80 microstrain K^{-1}.

It may be noted at this point that there is a relationship, or at least a trend, linking the expansivity of a material to its stiffness (Young's modulus). This is illustrated in Fig. 10.2, which qualitatively depicts potential wells representing the equilibrium separation between neighbouring atoms in a solid. Thermal expansion arises because of the asymmetry of this well, so that the average spacing between atoms tends to rise as the amplitude of thermal vibration increases (Fig. 10.2(a)). Stiffness depends on the restoring force arising if the separation is changed, and hence on the average gradient either side of the minimum. Materials with strong bonds, such as ceramics, have deep wells, and hence a tendency to exhibit both a high stiffness and a low thermal expansivity. Polymers, on the other hand, have weak inter-atomic bonds (shallow

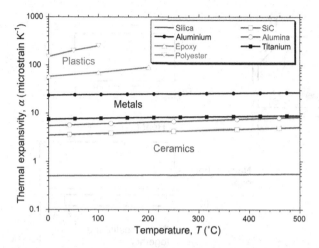

Fig. 10.1 Approximate values of thermal expansion coefficients (expansivities) for various materials over a range of temperature.

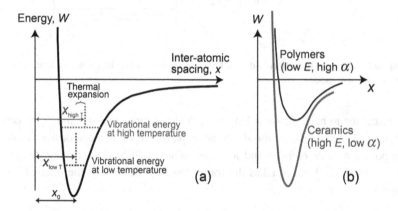

Fig. 10.2 Schematic representation of the effect of inter-atomic separation on the energy (Lennard–Jones potential), showing (a) how thermal expansion arises and (b) typical shapes of the curve for materials exhibiting inter-atomic bonding that is weak (polymers) or strong (ceramics).

potential wells) and so tend to have low stiffness and high expansivity (assuming that these are being controlled by forces between atoms in neighbouring chains and not between atoms along the length of a set of aligned chains).

10.1.2 Axial Expansivity of Long Fibre Composites

Derivation of a reliable expression for the axial expansivity of a long fibre composite, in terms of those of fibre and matrix, can be carried out using the slab model (Section 3.1). The basis for this is shown in Fig. 10.3. The response of the material to a temperature increase is considered in two stages – first, a 'free' expansion of the two constituents (as if they were not bonded together) and second imposition of the condition that they are

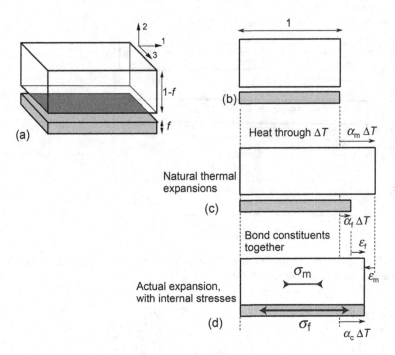

Fig. 10.3 Schematic depiction of thermal expansion in the fibre direction of a long fibre composite, using the slab model.

constrained to have the same length (in the axial direction – i.e. the '1' direction in the figure). The expression for the axial expansivity of the composite is obtained by imposing a *force balance* and a *strain balance*. Since there is no externally applied force, the force balance takes the form of setting the sum of the internal forces equal to zero

$$(1-f)\sigma_m + f\sigma_f = 0$$
$$\therefore (1-f)E_m\varepsilon_m + fE_f\varepsilon_f = 0 \tag{10.1}$$

in which these strains are the ones that arise by imposing the condition that the lengths of the constituents must be the same (in the axial direction). The strain balance is set by requiring the *misfit strain* (between the stress-free dimensions of the two constituents) to be accommodated as the difference between the strains in Eqn (10.1), although, taking account of their different signs, this has a magnitude equal to their sum

$$(\alpha_m - \alpha_f)\Delta T = \varepsilon_f - \varepsilon_m \tag{10.2}$$

The misfit strain is often written as $\Delta\alpha\,\Delta T$, although care needs to be taken with signs. Also, while these are being treated as scalars for the moment, both $\Delta\alpha$ and the misfit strain itself are in fact second-rank tensors.

An expression for the composite expansivity is now obtained by expressing it as the sum of the 'free' expansion of the matrix and the matrix strain resulting from the internal stresses

$$\alpha_c \Delta T = \alpha_m \Delta T + \varepsilon_m \tag{10.3}$$

The value of this matrix strain is obtained by combining the force and strain balance equations

$$(1-f)E_m\varepsilon_m + fE_f[\varepsilon_m + (\alpha_m - \alpha_f)\Delta T] = 0$$

$$\therefore \varepsilon_m = \frac{-fE_f(\alpha_m - \alpha_f)\Delta T}{(1-f)E_m + fE_f} \tag{10.4}$$

Substituting this into Eqn (10.3) it follows that the composite expansivity is given by

$$\alpha_c = \alpha_m - \frac{fE_f(\alpha_m - \alpha_f)}{(1-f)E_m + fE_f} \tag{10.5}$$

This expression can be rearranged into a form that makes its nature a little clearer

$$\alpha_{c,ax} = \frac{\alpha_m(1-f)E_m + \alpha_f fE_f}{(1-f)E_m + fE_f} \tag{10.6}$$

It can now be seen that this is a weighted mean formulation, with the weighting being carried out, not only by volume fraction, but also by the stiffness of the constituents. This is expected, since a stiffer constituent will tend to have a stronger influence on the final dimensions of the composite than a more compliant constituent. Since the axial force balance in a long fibre composite is accurately reflected in the slab model (Section 3.1), this prediction is expected to be reliable.

A simple conclusion may be noted here, which is that, just by considering the two contributions to the final dimensions in this way, it becomes clear that the presence of voids (porosity) in a material will have no effect on its thermal expansivity: since they effectively have zero stiffness, they will not create any internal stresses or associated strains. It is a well-established observation that porous materials, including foams, have the same expansivity as the corresponding fully dense material.

10.1.3   Transverse Expansivities

The expansivity in the transverse direction, and the values for short fibre and particulate composites, are more difficult to establish, since the stresses and strains vary with position. Nevertheless, as with the transverse stiffness, some useful approximations can be made. Bowles and Tompkins [1] provide a review of relevant models. One of the most successful for the transverse expansivity of long fibre composites is that due to Schapery [2], who used an energy-based approach to obtain the expression

$$\alpha_{c,tr} = \alpha_m(1-f)(1+\nu_m) + \alpha_f f(1+\nu_f) - \alpha_{c,ax}\nu_{12c} \tag{10.7}$$

in which the axial expansivity, $\alpha_{c,ax}$, is that given by Eqn (10.4) and the Poisson ratio ν_{12c} is obtained from a simple rule of mixtures between those of the constituents (Section 3.3.2).

The predicted dependence of these two expansivities on fibre content, given by Eqns (10.6) and (10.7), is shown in Fig. 10.4 for a polymer matrix composite. Also shown are

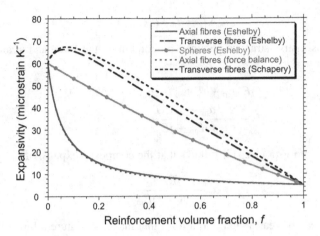

Fig. 10.4 Predicted dependence of thermal expansivity on reinforcement content for glass fibres and spheres in an epoxy matrix, according to the force balance (Eqn (10.6)), Schapery (Eqn (10.7)) and Eshelby (Eqn (10.8)) models. The curves were obtained using the property data in Tables 2.1 and 2.2.

predictions obtained using the Eshelby method (Section 6.2). This technique involves calculating the volume-averaged stresses in both constituents as a result of an imposed misfit strain, so that it is readily adapted for the prediction of expansivities. The equation obtained [3] is

$$\alpha_c = \alpha_m - f\{(C_m - C_f)[S - f(S - I)] - C_m\}^{-1}C_f(\alpha_f - \alpha_m) \tag{10.8}$$

where the stiffness, C, and expansivity, α, are here both tensors, as are the Eshelby tensor, S (which depends on the aspect ratio of the fibre) and the identity tensor, I. It may be noted that, as in the derivation of the axial expansivity (Section 10.1.2), the outcome is expressed as the matrix expansivity plus the change in it induced by the internal stresses. While the Eshelby formulation is not exact – there are no analytical expressions giving exact solutions for these composite properties – the ('mean field') approximations involved in their derivation are in general considered to be quite reliable, at least up to reinforcement contents of the order of 50% – see Section 6.2.4. It can be seen that the simpler expressions represented in Fig. 10.4 (to which one could add a linear rule of mixtures for the spheres – i.e. for equiaxed particles) give predictions that agree reasonably well with the Eshelby curves and hence can be regarded as fairly accurate (particularly the force balance equation).

Among the features of interest in Fig. 10.4 is that the transverse expansivity of a (long fibre) composite tends to rise initially as the fibre content increases. This occurs because, on heating the composite, axial expansion of the matrix is strongly inhibited by the presence of the fibres and the resultant axial compression of the matrix generates a Poisson expansion in the transverse direction, which more than compensates for the reduction effected in the normal way by the presence of the fibres, at least for low fibre contents.

Such plots are useful, since they allow the tailoring of expansivity via selection of constituents and reinforcement contents. However, it is important to note that these predictions are based on elastic behaviour. The associated internal stresses may become large, particularly if the temperature changes involved are substantial, and under these circumstances the matrix is likely to undergo plastic flow, or creep, which will alter the dimensional changes exhibited by the composite and make them more difficult to predict. This is examined in more detail below.

10.1.4   Thermal Stresses

The previously mentioned treatments can also be used to obtain information about the internal stresses created as a result of temperature change. For example, the expression derived in Section 10.1.2 for the axial strain in the matrix of a long fibre composite (Eqn (10.4)) is readily converted to a stress

$$\sigma_m = \frac{-fE_f E_m(\alpha_m - \alpha_f)\Delta T}{(1-f)E_m + fE_f} \tag{10.9}$$

Predicted curves obtained using this equation are shown in Fig. 10.5, with a temperature change of +100 K and some property data representative of polymer and metal matrix systems. (It should be noted that this treatment is based on the slab model, giving a uniform stress in the matrix, whereas in practice it will depend on radial distance – see below; however, with typical fibre volume fractions, this variation will be small.) It is immediately clear that stresses generated in this way can be relatively large ($>\sim$100 MPa), even with moderate temperature changes. In the case of a long fibre composite, transverse stresses will be much lower, so that there will tend to be a large deviatoric (shape-changing) stress, which is likely to provoke inelastic behaviour such

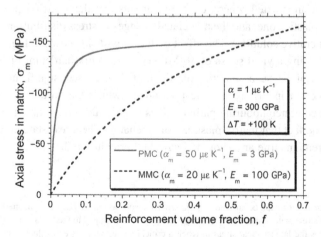

Fig. 10.5 Predicted dependence on reinforcement content of the axial matrix stress within a long fibre composite, arising from a temperature rise of 100 K, obtained using Eqn (10.9), with expansivity and stiffness data representative of polymer and metal matrix systems.

as plastic deformation or creep. This is a general feature of the behaviour of composite systems when significant temperature changes take place.

This will also tend to occur with other reinforcement geometries, as can be seen by considering the case of a sphere. For an isolated sphere in an infinite matrix, the stress field is the same as that produced by a spherical bubble of radius R under a pressure P (= a hydrostatic stress, σ_H, of $-P$). There is a simple analytical solution to this case, which was first obtained in 1852 by Lamé [4], who showed that the radial and hoop (tangential) stresses in the matrix are given by:

$$\sigma_r = \frac{-PR^3}{r^3}, \quad \sigma_\theta = \frac{PR^3}{2r^3} \tag{10.10}$$

The pressure in a spherical particle as a result of a misfit strain $\Delta\alpha\,\Delta T$ can also be expressed analytically [5]

$$P = \frac{4G_m(1+v_m)}{3(1-v_m)}\frac{K_p}{K_m}\Delta\alpha\Delta T\left[\frac{(1+v_m)}{3(1-v_m)}\left(\frac{K_p}{K_m}-1\right)+1\right]^{-1} \tag{10.11}$$

where G, v and K are the shear modulus, Poisson ratio and bulk modulus, respectively. The result of applying these equations to an alumina sphere in an aluminium matrix, after cooling through 100 K, can be seen in Fig. 10.6(a). This shows that a compressive stress is developed in the radial direction, while the hoop stress in the matrix is tensile (and the stress state in the particle is pure hydrostatic compression). This is broadly the type of stress field that would tend to be present in such a composite at room temperature, since its production would normally involve cooling from an elevated temperature (although the likely effect of plasticity should be noted – see below).

The matrix stress state thus incorporates a deviatoric (shape-changing) component, particularly in the region close to the particle. This may well cause yielding and plastic deformation.[1] For an isolated sphere, equations are also available [5] for prediction of the extent of the plastic zone, and the associated changes in stress distribution (although they do require iterative solution). The effect on the stress field shown in Fig. 10.6(a) of the (Al) matrix having a yield stress of 40 MPa (with no work hardening) is shown in Fig. 10.6(b). It can be seen that the plastic flow has caused significant changes, including a reduction in the hydrostatic pressure to which the particle is subjected. While the presence of neighbouring particles does modify the behaviour somewhat in a real composite, the fact that matrix plasticity often changes the stress state predicted for the elastic case remains true and should always be borne in mind.

---

[1] Yielding and plastic deformation occur when a *yield criterion*, such as that of *Tresca*, is satisfied. These take account of the full stress state, rather than just setting the largest stress equal to the (uniaxial) yield stress. In fact, Tresca requires the largest shear stress to reach a critical value, which is equivalent to the difference between the largest and smallest principal stresses being equal to the yield stress. It can be seen that this is satisfied in Fig. 10.6(b). The other common criterion is that of *von Mises*, which involves all three principal stresses, but reduces to Tresca in certain cases, including that in Fig. 10.6(b).

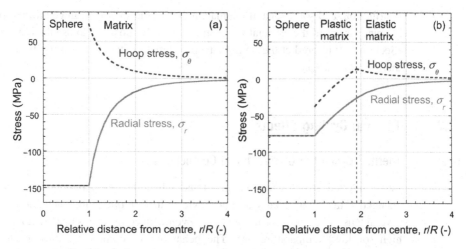

Fig. 10.6 Predicted variation of stresses with distance from the centre of a spherical α-alumina particle in an infinite Al matrix, after cooling by 100 K, obtained using the formulations of Lee et al. [5]. (Property data were taken from Tables 2.1 and 2.2, with Eqn (4.35) being used to obtain the bulk moduli.) The plots are for: (a) the elastic case and (b) after plastic deformation, with a matrix yield stress of 40 MPa and no work hardening.

It may be noted at this point that, if the system had been heated instead of cooled, or if it was cooled and the sphere had a higher expansivity than the matrix, then this operation would produce a stress state within the sphere of pure hydrostatic tension. Such a stress state is difficult to generate in a body during conventional testing (whereas pure hydrostatic compression is readily induced by simply immersing the body in a pressurised fluid). Of course, it would only occur if the interface were sufficiently strong to resist debonding under the radial tensile stress. In any event, for both hydrostatic tension and hydrostatic compression, there is no deviatoric stress and hence no possibility of plastic deformation (although the body could fracture under hydrostatic tension).

The elastic stress field for an infinitely long single fibre surrounded by a cylindrical matrix (of finite radial extent) can also be obtained analytically [6]. A typical result is shown in Fig. 7.3, for a glass–polyester composite after cooling through 100 K. This stress field shows similarities to the case of the sphere, although, as might have been expected, the axial stresses are more prominent for the fibre case. It may be noted here that the mismatch in expansivity, $\Delta\alpha$, is usually larger in polymer matrix composites (PMCs) than in metallic matrix composites (MMCs), tending to create larger misfit strains. However, this is often offset by the lower matrix stiffness, so that, as can be seen in Fig. 10.5, the resultant matrix stresses (for a given temperature change) are often of broadly similar magnitude in the two cases. Of course, polymers often have low yield stresses, and are susceptible to creep (Section 10.3.1), so large stresses are unlikely to be sustained in them.

In a short fibre composite, the stress field is more complex, particularly when the effect of neighbouring fibres is considered. In this case, recourse to numerical methods is often necessary to predict the stress field. However, the general nature of the stress distribution, and some idea of the magnitudes of the stresses, can be inferred from these simpler cases and the associated equations. It is also possible to use the Eshelby method (Section 6.2) to predict the volume-averaged matrix stresses resulting from differential thermal contraction.

10.2　Thermal Cycling Effects

10.2.1　Thermal Cycling of Unidirectional Composites

Large internal stresses can be generated when the temperature of a composite is changed. This often occurs during service, since temperature fluctuations of at least several tens of degrees Celsius are likely even with components that are not designed for high (or low) temperature use. The behaviour of composites during such thermal excursions are therefore of practical importance. Composites may respond to the associated internal stress changes in an inelastic manner. For example, dilatometry (length) measurements made on composites often exhibit significant hysteresis (i.e. the heating and cooling curves do not coincide).

A schematic representation of some of the effects of thermal cycling is presented in Fig. 10.7. This shows changes in axial matrix stress and in composite strain as the temperature is changed (over a relatively large range), for a matrix prone to yielding (e.g. a metal or a thermoplastic). It should be clarified that, while the composite

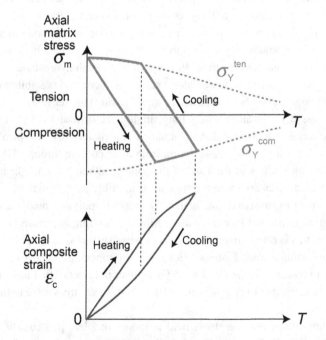

Fig. 10.7 Schematic depiction of the variations in matrix stress and in axial strain of the specimen during thermal cycling of an aligned long fibre composite.

strain shown here is thermal expansion, the matrix stress is an internal one (arising from the misfit strain), which must be balanced (as a force) by the stress in the fibres. It should also be noted that it is being assumed here that yielding and plasticity occur when the axial matrix stress reaches the (uniaxial) yield stress, with no work hardening. As indicated in Section 10.1.4, criteria for yielding should be based on the full stress state (i.e. all three principal stresses). However, in this case the transverse matrix stresses are relatively low, so this approximation is acceptable for present illustrative purposes.

The matrix is taken to have an initial residual tensile stress equal to its yield stress (because of the cooling cycle). On heating, this stress falls, becomes compressive and eventually causes yielding in compression. A period of progressive plastic flow then follows, with the stress being limited to the yield stress (for the temperature concerned). On cooling, the matrix stress becomes tensile, causing yield at some point, before returning to its starting condition. The dilatometry traces (axial strain/temperature plots) are thus predicted to show hysteresis, but no net dimensional change.

Experimental data for long fibre composites [7,8] are broadly consistent with this view. A simple analysis of the dilatometer trace, proposed by Masutti et al. [9], can be used to estimate the changing axial matrix stress in a long fibre composite. Assuming no interfacial sliding, fibre yielding, etc., the axial strain of the composite must be equal to that of the fibres, which can be expressed as the sum of their natural thermal expansion and their elastic strain

$$\varepsilon_{1c} = \varepsilon_{1f} = \alpha_f \Delta T + \frac{\sigma_{1f} - \nu_f(\sigma_{2f} + \sigma_{3f})}{E_f} \tag{10.12}$$

Since the radial and hoop stresses in the fibre (σ_{2f} and σ_{3f}) are relatively small compared with the axial stress (see Fig. 7.3) and the Poisson ratio of ceramic fibres is normally quite low (~0.2), the contribution from the transverse fibre stresses can be neglected. The axial force balance

$$f\sigma_{1f} + (1 - f)\sigma_{1m} = 0 \tag{10.13}$$

can then be used to find the axial stress in the matrix.

$$\sigma_{1m} = \frac{f}{(1 - f)} E_f(\alpha_f \Delta T - \varepsilon_{1c}) \tag{10.14}$$

Hence σ_{1m} is proportional to the difference, $\Delta\varepsilon$, between the natural thermal strain of the fibre and the measured strain of the composite. The initial thermal stress in the matrix can be deduced by taking the composite up to high temperature (~0.8 T_m), where the matrix stress will become very small, and running the fibre thermal expansion line back from this region. This is illustrated in Fig. 10.8, which shows an experimental dilatometer trace [9] and the corresponding deduced matrix stress history. The latter is seen to be broadly of the form shown in Fig. 10.7. This simple procedure is a convenient way of studying the initial matrix stress, as well as the high-temperature characteristics. For example, Masutti et al. [9] showed that a quench in liquid nitrogen followed by a return to room temperature generated a compressive

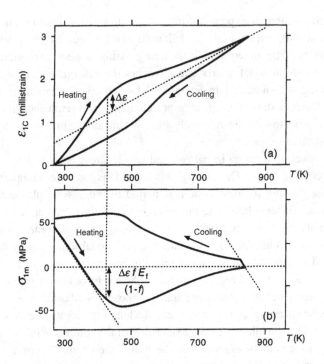

Fig. 10.8 (a) Experimental strain history [9] during thermal cycling and (b) deduced variation of matrix axial stress for an Al-3%Mg/30%SiC long fibre composite. The matrix stress is taken, Eqn (10.14), as being proportional to the difference between the measured composite strain and the natural thermal expansion of the fibres, which can be obtained by extrapolation of the high-temperature data. The initial portions of the stress history curve, at the start of heating and of cooling, have gradients corresponding to elastic behaviour, indicated by the dotted lines.

residual stress of ~25 MPa in an Al–SiC composite, which may be compared with a tensile stress of ~50 MPa after heating and cooling back to room temperature. This highlights the fact that the thermal history of a composite sample can strongly influence the residual stresses within it.

The behaviour of short fibre (and particulate) composites during thermal cycling is rather similar to that described above, but there is more scope for various stress relaxation processes to operate. This is illustrated by the neutron diffraction data [10] presented in Fig. 10.9, which shows the changing axial (lattice) strains (and hence stresses), in both fibre and matrix, for an Al matrix reinforced by SiC whiskers (short fibres). The matrix stress development is similar to those shown in Figs 10.7 and 10.8, although the observation of a significant drop in the final residual stress over an eight-hour period at room temperature is suggestive of continuing stress relaxation (creep) processes. The scope for such stress relaxation varies markedly with the type of matrix, the fibre aspect ratio (and possibly its absolute length) and the temperature range. When stress relaxation cannot occur easily, and the matrix is relatively brittle, then matrix micro-cracking is likely. This is more common with laminates than with unidirectional composites – see Section 10.2.2.

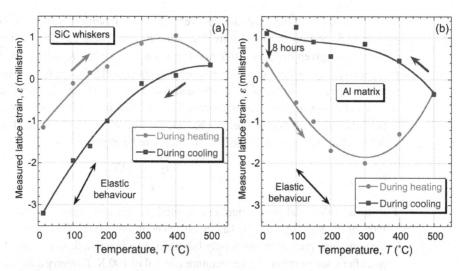

Fig. 10.9 Neutron diffraction data [10] for an Al-5vol%SiC whisker (short fibre) composite, showing lattice strains (and hence stresses) within matrix and reinforcement during unloaded thermal cycling. (111) peaks were used for both constituents. The gradients shown are calculated values for elastic behaviour, assuming a fibre aspect ratio of 10.

10.2.2 Thermal Cycling of Laminates

Problems associated with thermal stresses can be severe with laminates. Thermal misfit strains now arise not only between fibre and matrix, but also between the individual plies of the laminate. For example, since a lamina usually has a much larger expansion coefficient in the transverse direction than in the axial direction, heating of a cross-ply laminate will lead to the transverse expansion of each ply being strongly inhibited by the presence of the other ply. This is useful in the sense that it will make the dimensional changes both less pronounced and more isotropic, but it does lead to internal stresses in the laminate. These can cause the laminate to distort, so that it is no longer flat; an illustration of the type of distortion that can occur is shown in Fig. 5.7 (Section 5.2.3). However, even if such distortions do not occur, the stresses that arise on changing the temperature can cause microstructural damage and impairment of properties.

Thermal stresses within a laminate can be calculated using the numerical techniques outlined in Chapter 5. For the special case of a cross-ply laminate, a simple analytical approach can be used. The stresses and strains parallel and normal to the fibres in each ply can be found using the method outlined in Section 10.1.2. Fig. 10.3 is again applicable, but the two constituents are now the two plies (one oriented axially and the other transversely), rather than the matrix and the fibres. The same equations result, but the meaning of the parameters changes slightly. For example, taking the transverse ply as the 'matrix' and the axial one as the 'fibres', a modified version of Eqn (10.4) now gives the elastic strain in the transverse ply on heating through a temperature interval ΔT

$$\varepsilon_2 = \frac{-E_1(\alpha_2 - \alpha_1)\Delta T}{E_2 + E_1} \tag{10.15}$$

where E_1 and E_2 are the axial and transverse Young's moduli (given by Eqn (3.3) and by Eqn (3.6) or Eqn (3.7) respectively), while α_1 and α_2 are the axial and transverse expansivities (given by Eqns (10.6) and (10.7) respectively). The volume fraction, f, of the two constituents does not appear, since it is equal to 0.5 for a cross-ply laminate in which the plies are of equal thickness and therefore cancels out in Eqn (10.4). This strain can be converted directly to a stress, since it arises from the two plies being forced to have the same length (and does not include the free thermal expansion). The stress to which the transverse ply is subjected can therefore be expressed as

$$\sigma_2 = \frac{-E_1 E_2 (\alpha_2 - \alpha_1)\Delta T}{E_2 + E_1} \tag{10.16}$$

The stress in the axial ply is equal in magnitude to this and opposite in sign.

Plots obtained using Eqn (10.16) are presented in Fig. 10.10. This shows the stresses in the transverse plies of two cross-ply laminates, as a function of the volume fraction of fibres, after cooling through a temperature interval of 100 K. Property data for the fibres and the matrix were taken from Tables 2.1 and 2.2. Several features are of interest in this figure. A peak is observed on increasing the fibre content. This is expected, since the stresses are due to the anisotropy in α and the difference between α_1 and α_2 has a peak at some intermediate value of f (see Fig. 10.4). For the carbon fibre composite, the stress does not fall to zero as f approaches 100%. This is because the carbon fibres are inherently anisotropic in thermal expansivity (see Table 2.1); they show zero (or negative) expansion axially on heating, but expand in the transverse (radial) directions.

Levels of thermal stress are of interest for prediction of damage development. For a fibre content of ~40–70%, it can be seen from Fig. 10.10 that stresses of ~25–30 MPa are expected (from a –100 K temperature change, which is of the approximate order expected during production of the material, when account is taken of a tendency for

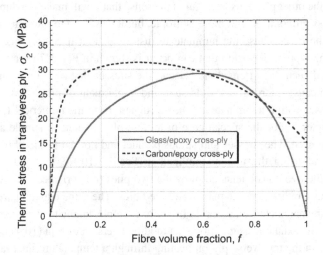

Fig. 10.10 Predicted dependence on fibre content of the thermal stress in the transverse direction within each ply of two cross-ply laminates, arising from a temperature decrease of 100 K. The plots were obtained from Eqn (10.16), using property data in Tables 2.1 and 2.2.

stress relaxation to occur during cooling). This is in the range that might be expected for transverse failure stress for typical laminae (e.g. see Fig. 8.10). It thus appears to follow that cooling (from an initially stress-free state) through a temperature interval of this order may cause damage in such a cross-ply laminate, particularly if it takes place over a temperature range in which stress relaxation is expected to be limited (e.g. below room temperature).

While this is very broadly consistent with experimental observations, the requirement under these circumstances is not really to ensure that the stress imposed by these thermal misfit strains should be below the (transverse) strength of the laminate, but rather that it should be able to sustain a particular transverse strain without damage. This is just one example of many situations in which it is *strain tolerance*, rather than strength (expressed as a maximum stress level), that is of prime importance for durability. There is thus considerable interest in the development of polymer formulations and manufacturing methods giving high transverse strains to failure, particularly for components likely to be subjected to large temperature fluctuations.

Even when such matrices are used, however, severe thermal cycling is often observed to cause networks of transverse cracks in laminates. Examples [11] are shown in Fig. 10.11, which relates to carbon fibre cross-ply laminates. While the matrices concerned are both relatively tough and designed for high-temperature use, it can be seen that dense networks of cracks were created, particularly after large numbers of

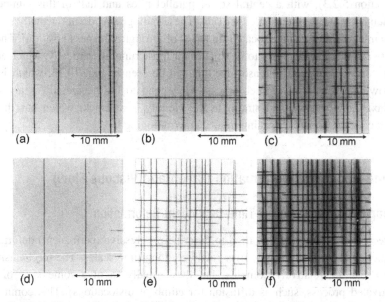

Fig. 10.11 X-ray radiographs [11] of carbon fibre cross-ply laminates, having a matrix of either (a)–(c) a thermoplastic, PEEK (polyether ether ketone) or (d)–(f) a thermosetting resin, G-40. These samples were thermally cycled between +150°C and −196°C, with the number of cycles being (a) 100, (b) 500, (c) 1000, (d) 100, (e) 500 and (f) 900.

Reprinted from Composites Part B: Engineering, 34, Satoshi Kobayashi, Kazuhiro Terada and Nobuo Takeda, Evaluation of long-term durability in high temperature resistant CFRP laminates under thermal fatigue loading, 753–759, © 2003, with permission from Elsevier.

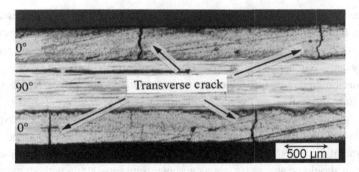

Fig. 10.12 Optical micrograph [11] of a section from a carbon fibre cross-ply laminate, with a PEEK matrix, after 200 thermal cycles between +150°C and –196°C.
Reprinted from Composites Part B: Engineering, 34, Satoshi Kobayashi, Kazuhiro Terada and Nobuo Takeda, Evaluation of long-term durability in high temperature resistant CFRP laminates under thermal fatigue loading, 753–759, © 2003, with permission from Elsevier.

cycles. This is an example of sub-critical crack growth (Section 9.3) and there are points in common between these conditions and conventional (mechanical) fatigue loading (Section 9.3.1). Of course, it should be recognised that the thermal cycling concerned in this case (between +150°C and –196°C) was severe (for a polymer matrix).

A further point to note about Fig. 10.11 is that the laminates were all symmetrical (Section 5.2.3), with a central set of parallel plies and half of this number of plies oriented at 90° to them on either side. These radiographs do not actually reveal where the transverse cracks are located (in terms of through-thickness position). The spacings between the cracks appear to be smaller for those running vertically, which are those in the 'outside' plies. The reason for this can be seen in Fig. 10.12, which is a section showing a small set of such cracks. They tend to be 'staggered' as a result of the relaxation of stress (and strain) in the vicinity of a crack, which has the effect of reducing the apparent spacing in an X-ray radiograph.

10.3   Time-Dependent Deformation (Creep and Viscous Flow)

10.3.1   Simple Representation of Time-Dependent Deformation

Creep is a term commonly used to describe progressive (permanent) deformation of a material under load, usually over relatively long time scales. The mechanisms responsible can be relatively complex, but they usually involve some kind of thermally activated process, such as diffusion (or climb of dislocations). This commonly leads to a strong (Arrhenius) dependence of the strain rate on temperature. Furthermore, it is commonly observed that, after an initial transient (*'primary' creep*), the strain rate becomes stable for an extended period (*'secondary' creep*). Broadly, primary creep corresponds to the setting up of some kind of microstructural balance, which is then maintained during the quasi-steady state of secondary creep (before the onset of microstructural damage, such as void formation, when a final tertiary regime is entered).

For example, this balance might be one in which dislocations are surmounting obstacles at a rate dictated by diffusional processes. Interest commonly focuses on the strain rate during secondary creep, although properties such as the creep rupture strain may also be of concern. The steady-state strain rate is usually expressed as

$$\frac{d\varepsilon_{cr}}{dt} = A\sigma^n \exp\left(\frac{-Q}{RT}\right) \tag{10.17}$$

where A is a constant, σ is the applied stress, n is the stress exponent and Q is an activation energy. In practice, primary creep sometimes extends over a substantial fraction of the period of interest. Also, the transition between primary and secondary regimes may not be very well defined. Formulations that cover both can therefore be helpful. One example (often termed the *Miller–Norton law*) gives the creep strain as a function of time

$$\varepsilon_{cr} = \frac{C\sigma^n t^{m+1}}{m+1} \exp\left(\frac{-Q}{RT}\right) \tag{10.18}$$

where C is a constant and m is a parameter controlling the way that the transition between the two regimes takes place. Plots showing the general shape of predicted curves, and the effects of changes in σ, n and m, are presented in Fig. 10.13. As expected, raising σ and, particularly, n give sharp increases in strain rate. The sign of m is always negative: a lower magnitude delays the transition to the secondary (linear) regime, while a higher magnitude accelerates it (although, strictly, the strain rate never becomes exactly constant with this formulation).

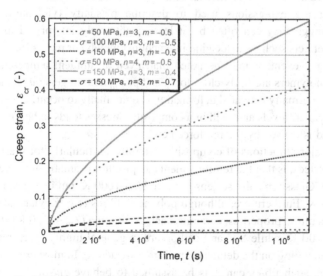

Fig. 10.13 Plots produced using the Miller–Norton law (Eqn (10.18)), giving the creep strain as a function of time, for different values of the applied stress, σ, the stress exponent, n, and the parameter controlling the transition from primary to secondary regimes, m. The value of $C\exp(-Q/RT)$ has been taken to be 10^{-10} MPa^{-n} s$^{-(m+1)}$.

Of course, there is also a high (exponential) sensitivity to temperature. As a very broad and general observation, creep tends to become significant only at temperatures above about 0.5 T_m (i.e. 50% of the melting temperature in kelvin). For most metals, this translates into temperatures of at least a few hundred degrees Celsius, although there are a few, such as Pb and Sn, for which ambient temperature is around 0.5 T_m. For most ceramics, on the other hand, creep is only likely to become significant at high temperatures (see below), often beyond the usable range for polymers and metals, so it can often be assumed that the reinforcement undergoes no creep (or plastic deformation) in composites.

Polymers are, of course, rather more complex. Thermosetting resins do not have well-defined melting temperatures, but they tend to become chemically and mechanically degraded when subjected to moderate temperature increases (~100–200 K) above ambient. They are usually fairly resistant to creep at room temperature. Thermoplastic polymers, on the other hand, have a fairly well-defined glass transition (amorphous) temperature, T_g, and/or a melting (crystalline) temperature, T_m, above which they become (viscous) liquids. They are not normally usable in this form, but many polymers are partially crystalline and some of these are used in the range between T_g and T_m. In this case, the structure consists of (small) crystallites surrounded by viscous liquid and such material is often observed to conform to mathematical formulations representing viscoelastic behaviour.

It should, however, be appreciated that such formulations, and the equations used to describe creep, are simply empirical relationships. While some sort of rationalisation is sometimes possible – for example, the values of n and Q in Eqn (10.17) may in some cases be correlated with identifiable atomic-scale processes – these equations are not closely tied to microstructure or mechanisms. Of course, a similar statement can be made regarding the expressions used to describe plasticity (yielding and work hardening), although they can often be at least qualitatively interpreted in terms of microstructural effects such as dislocation interactions (metals) or realignment of chains (polymers). In any event, it may be noted that, even at ambient temperature, most (thermoplastic) polymers are fairly close to T_g and/or T_m and, in general, some kind of time-dependent, thermally activated deformation is quite likely to occur. Of course, this does depend on stress levels and, within a composite, stresses tends to be low (because much of the load is carried by the reinforcement).

One of the major attractions of composite materials, particularly for thermoplastic and metallic matrices, is that the reinforcement can produce dramatic improvements in creep resistance. Glass typically softens at about 900–1000 K, but is resistant to creep up to about 600 K. Furthermore, although polymeric fibres (e.g. Kevlar) do not have good high-temperature properties, ceramics such as alumina (T_m ~2300 K) show little creep below ~1100 K, while SiC and carbon are in general highly creep-resistant to over 1200 K, depending on the details of their microstructure. In most creep situations with composites, such fibres can thus be assumed to behave elastically. The creep behaviour of the composite as a whole depends on load partitioning and constraint, which is in turn dependent on geometrical factors and, in some cases, on the nature of the interface.

10.3.2   Axial Creep of Long Fibre Composites

Provided that the fibres remain elastic, with no interfacial sliding, treatment of this case is straightforward. The initial strain, generated immediately when the composite is loaded, is the applied stress divided by the Young's modulus of the composite, given by a rule of mixtures (Section 3.1)

$$\varepsilon_0 = \frac{\sigma}{fE_f + (1-f)E_m} \tag{10.19}$$

During creep of the matrix, the applied stress is progressively transferred to the fibres. As the fibres strain elastically, the stress in them increases. The limit comes when the fibres carry all of the applied load. At this point, the strain of the fibres, and hence of the composite, is given by

$$\varepsilon_\infty = \frac{\sigma}{fE_f} \tag{10.20}$$

The strain approaches this value asymptotically, since the rate of creep of the matrix falls off as the stress in it decreases, and a true steady state is never set up. From an engineering point of view, this situation appears attractive, since the creep strain should never exceed a predetermined level (which in general is small), assuming that the fibres do not break.

However, in practice the behaviour is likely to be more complex. Among the more comprehensive experimental studies is that of Schwenker and Eylon [12], who examined the Ti-6Al-4V-36%SiC monofilament system in some detail, at various temperatures and applied stress levels. Some of their strain-time data are plotted in Fig. 10.14(a), together with corresponding ε_0 and ε_∞ values (obtained using information provided [12] about the effect of temperature on stiffness). It can clearly be seen that, while the concept of these being bounds for the strain range is not so far from reality, particularly for ε_0, it is not accurately correct. Actual strains at long times are appreciably above the limit of Eqn (10.20), particularly at the higher temperatures. This is almost certainly due to interfacial sliding in this case (see also Section 10.3.3), although extensive shear deformation of the matrix close to the fibres could produce a similar effect. Of course, there may be cases where neither occurs.

In addition to this complication, reliable prediction of the creep strain evolution is not easy, at least using analytical formulations. Some such models have been developed, an example being that of McLean [13], who showed that, for a matrix conforming to Eqn (10.17), the evolution of strain with time can be expressed as:

$$\varepsilon = \varepsilon_\infty - (\varepsilon_\infty - \varepsilon_0)\left[1 + \left(1 - \frac{\varepsilon_0}{\varepsilon_\infty}\right)^{n-1} \frac{(n-1)A\sigma^n \exp(-Q/RT)t}{\varepsilon_\infty\left(1 + \frac{fE_f}{(1-f)E_m}\right)(1-f)^n}\right]^{-1/(n-1)} \tag{10.21}$$

However, this also is subject to limitations that may not apply in practice. The predictions shown in Fig. 10.14(b) were obtained, for one of the temperatures concerned, using information [12] about the (secondary) creep rate of the matrix in this case ($n = 7.1$ and

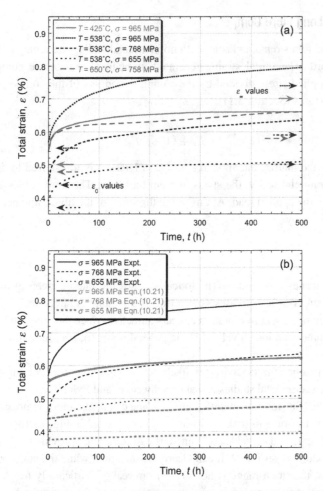

Fig. 10.14 Experimental data [12] for strain-time during axial loading of a Ti-6Al-4V/36%SiC monofilament composite, showing comparisons with: (a) initial and final strain levels, predicted using Eqns (10.19) and (10.20), and (b) complete curves for $T = 538°C$, predicted using Eqn (10.21).

$A\exp(-Q/(RT)) = 2.1\ 10^{-23}$ MPa^{-n} s^{-1}). It can be seen that use of this equation leads to substantial over-prediction of the time needed for the limiting strain to be approached. This is probably due to neglect of the primary regime, during which creep tends to be considerably faster, often making a substantial contribution to the overall strain. The presence of thermal residual stresses (Section 10.1.4) could also have an effect, although these tend to become small if the composite is held at elevated temperature.

10.3.3   Creep under Transverse Load and in Discontinuously Reinforced Composites

The situation is different if the composite is loaded transversely, or if the reinforcement is discontinuous (short fibres or particles). The stress field is more complex, and less

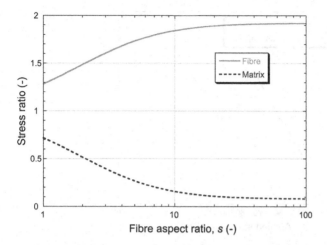

Fig. 10.15 Predicted dependence on fibre aspect ratio of the volume-averaged stress in each constituent, as a ratio to the applied stress, for axial loading of polyester–50% glass composites, obtained using the Eshelby model (Section 6.2.5).

amenable to simple analysis, than for axial loading of long fibre composites. Deformation is expected to be progressive and steady-state (constant strain rate) creep is often established. Creep rates are strongly dependent on the creep characteristics of the matrix. However, the extent to which the reinforcement relieves the matrix of load is important. For example, in a short fibre composite under axial load, the load partitioning depends strongly on the fibre aspect ratio. This is illustrated by the data in Fig. 10.15, which shows volume-averaged stresses in matrix and fibre for elastic loading of a polyester–50% glass composite, as a function of fibre aspect ratio. In this case, the stress in the matrix is reduced by a factor of about 5 as the fibre aspect ratio rises from 1 to 10.

Experimental data for creep of short fibre composites under axial load confirm a strong sensitivity to fibre aspect ratio. An example is shown in Fig. 10.16, which gives minimum (steady-state) creep rates [14] for an Al alloy reinforced with SiC particles (aspect ratio ~1) or with SiC whiskers (aspect ratio ~5). The creep rates of the whisker material are lower by about two orders of magnitude, at any given stress level, even though the reinforcement content is lower. In many cases, however, the observed behaviour is not only being influenced by the degree of elastic load transfer. During creep, the stress in the matrix is also dependent on the degree to which *stress relaxation* occurs. This term refers here to unloading of the matrix (mainly by diffusive processes) via changes in its stress-free shape (so as to transfer stress to the reinforcement). Such effects are likely to depend on reinforcement scale, as well as aspect ratio. Since the whisker diameter (~500 nm) is considerably smaller than that of the particles (~20 μm), load transfer of this type may occur more readily. It is in any event clear that the substantially greater creep resistance of the fibrous material is at least mainly due to load partitioning effects. As pointed out by Nieh [14], inter-reinforcement distances are too large to inhibit directly the motion of dislocations, even in the whisker-reinforced material.

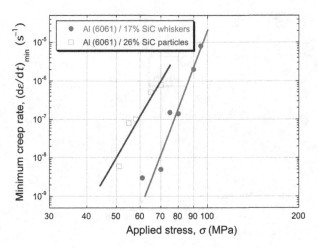

Fig. 10.16 Experimental steady-state creep rate data [14] for isothermal creep at 561 K of Al alloy (6061) MMCs, reinforced with either 26 vol% of SiC particles or 17 vol% of SiC whiskers.

Another factor that is sometimes important in creep of discontinuously reinforced composites is the effect of thermal residual stress. When the temperature is changed, there is a change in the level of thermal stress in the matrix (Section 10.1.4). This may augment or reduce the stresses from the applied load and hence reduce or enhance the creep rate. Anomalously high values of the activation energy and the stress exponent for creep, often observed for MMCs, have been explained on this basis [15,16]. Furthermore, while thermal residual stresses tend to become relaxed, and hence are often of minor significance during isothermal creep, they can be of prime importance during *thermal cycling*, when they are continuously regenerated. Under these circumstances, the creep behaviour can differ substantially from that expected on the basis of the dependence of creep rate on temperature observed during isothermal testing.

The effects outlined above are illustrated by the plots in Fig. 10.17, which give strain history data obtained [17,18] during thermal cycling creep of two short fibre ($s = 5$) Al-based MMCs (produced by squeeze infiltration, followed by extrusion). The microstructures (fibre characteristics and matrix grain structures) of these two materials can be seen in Fig. 10.18. The predicted plots in Fig. 10.17 were obtained by using the Eshelby model (Section 6.2.5) to predict the stresses in the matrix from the applied load and as a consequence of the changes in temperature. The changing strain rates were then obtained from the known dependence of matrix creep rate on stress and temperature. Also shown, for the Al-10%Al$_2$O$_3$ composite, are the measured and modelled isothermal creep rates of the composite at the diffusional mean temperature of the thermal cycle [19]. This is the mean temperature of the imposed cycle, weighted for the temperature dependence of the creep rate. It can be seen that the average strain rate during cycling is considerably greater than the isothermal rate, reflecting the influence of the thermal residual stresses, with rapid creep during the periods when high matrix stresses are generated. It is a cause for concern that such thermal cycling can accelerate the creep of MMCs, since there are many service conditions that involve this.

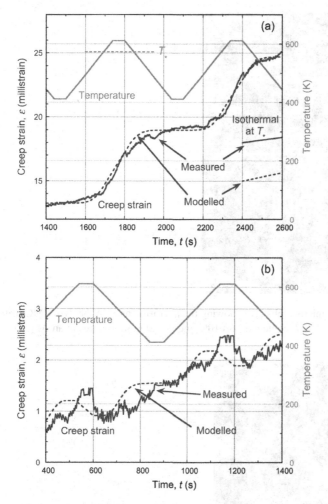

Fig. 10.17 Comparison [17,18] between measured and modelled creep strain histories during thermal cycling of 2618 Al containing: (a) 10% and (b) 20% Saffil (short Al_2O_3 fibre), under 20 MPa applied load (in the direction of fibre alignment). Data are also shown in (a) for the corresponding isothermal creep at the diffusional mean temperature.

Finally, the interface sometimes plays an important role during creep of discontinuously reinforced materials. An example is provided by the transverse loading of long fibre-reinforced titanium alloys, which has been examined in some detail [20,21]. Much of this type of material has been based on SiC monofilaments with a graphitic surface coating. This protects them from handling damage and chemical attack during fabrication, but does not allow strong interfacial bonding. At ambient temperature, fibre–matrix cohesion remains good as a result of the radial compressive thermal residual stress (Section 10.1.4). At elevated temperature (>500°C), however, the radial stress tends to become tensile, allowing the interface to be opened up by moderate applied loads. (This poor bonding probably also plays a role in the straining beyond the expected limit under axial loading that is described in Section 10.3.2.) If creep is also

Fig. 10.18 Optical micrographs [18] showing the microstructures of the two MMC materials referred to in Fig. 10.17, containing (a) 10% and (b) 20% Saffil fibres in an Al (2618) matrix.

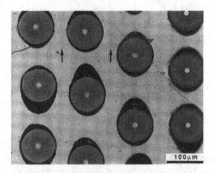

Fig. 10.19 SEM micrograph [22] of a polished section near the fracture surface of a Ti-6Al-4V/ 30%SiC composite, after rupture under transverse loading at a stress of 25 MPa with thermal cycling (between 400°C and 700°C, at 100°C per minute). The direction of loading is indicated by the arrows.

enhanced by thermal cycling, large voids can develop quite rapidly at the interface. This is illustrated by Fig. 10.19, which shows a micrograph [22] of such a composite after transverse loading under thermal cycling conditions. Such large interfacial voids lead to high (stage 3) creep rates and rapid onset of creep rupture. Outcomes such as these have inhibited the use of this type of composite in many of the applications for which they were originally envisaged.

It should, however, be appreciated that this type of effect is less significant for polymer-based composites, which are normally used over much more limited temperature ranges. While relatively large misfit strains can be generated in them (as a consequence of the large differences in expansivity), the higher fibre contents and pronounced stress relaxation in the matrix (transferring load to the fibres) usually means that thermal cycling effects are not strong and creep effects in general are usually small, at least in laminates.

10.4 Thermal Conduction

High thermal conductivity is useful in improving the resistance of materials to thermal shock and avoiding the development of 'hot spots' during localised heating. In other situations, low thermal conductivity can be beneficial in providing thermal insulation. In any event, there are many applications in which the conductivity (possibly electrical, as well as thermal) of a component needs to be known (including any anisotropy). Before considering conductivity levels in fibres, matrices and composites, it is useful to review briefly the mechanisms of thermal conduction.

10.4.1 Heat Transfer Mechanisms and Typical Material Conductivities

The main heat transfer mechanisms within a material are normally the transmission of phonons (lattice vibrations) and/or free electrons (if present). Both of these *carriers* have a certain *mean free path*, λ, between collisions (energy exchange events) and an average velocity v. The *thermal conductivity*, K, is related to these parameters by a simple equation derived from kinetic theory

$$K = \frac{1}{3}cv\lambda \tag{10.22}$$

where c is the volume-specific heat of the carrier concerned. Both metals (electron-dominated) and non-metals (phonons only) have zero conductivity at a temperature of 0 K (where c becomes zero in both cases), followed by a sharp rise to a peak and then a gradual fall as the temperature is progressively increased. The rise reflects the increase in c towards a plateau value ($\sim 3Nk$, where N is Avogadro's number and k is Boltzmann's constant, with all the vibrational modes of all the molecules active). The fall is caused by the decreasing λ as the greater amplitude of lattice vibration causes more scattering of the carriers. The maximum (low-temperature) value of λ is dictated by atomic-scale defects for electrons and by the physical dimensions of the specimen for phonons.

The average carrier velocity is insensitive to temperature in both cases. The phonon velocity (speed of sound) is high in light, stiff materials. The mean free path of a phonon is structure-sensitive and can be very large in pure specimens of high perfection and large grain size. Single crystals of materials like diamond and SiC thus exhibit very high thermal conductivities, as do some recently developed carbon fibres [23,24] – including some derived from carbon nanotube structures [25,26]. With the exception of such cases, metals have the highest conductivities, because electrons usually have a much larger mean free path than phonons. This can, however, be substantially reduced by the

Table 10.1 Thermal conductivity data [29–34] for a range of materials.

Material	Thermal conductivity K (W m^{-1} K^{-1})		Material	Thermal conductivity K (W m^{-1} K^{-1})	
	At 300 K	At 900 K		At 300 K	At 900 K
Diamond	600	–	Al	~220	~180
Graphite (parallel to c-axis)	355	–	SiC (single crystal)	~100–300	~70 – 200
Graphite (normal to c-axis)	89	–	SiC (polycrystal)	~20–100	~10 – 40
Carbon fibres (various)	10 – 1,000	–	Ti	~18	~18
Ag	425	325	Ti-6Al-4V	~8	~12
Cu	400	340	E-glass fibre	1.3	
Cu-2%Ag	~390	~340	Epoxy resin	0.2–0.5	–
Cu-35%Zn	~100	~180	Polyester resin	0.2–0.24	–
Cu-40%Ni	~20	~40	Nylon	0.2–0.25	–

presence of solute atoms and various defects that cause electron scattering. It may also be noted that there are some metals and alloys in which the electron availability or mobility is relatively low, such that phonon transport may make a significant contribution to the overall heat transfer. For example, this is often the case [27,28] for titanium and its alloys. Polymers have no free electrons and low stiffness (low phonon velocity), so conductivities are low.

These trends are all apparent in the data [29–34] shown in Table 10.1. The wide range of values listed in some cases, particularly for certain ceramics (including carbon fibres), arises from the high sensitivity of the conductivity to the presence of impurities, grain boundaries and other defects. Thus, while there is in principle scope for enhancement of thermal conductivity of a metal by incorporation of ceramic (e.g. SiC in Ti), the sensitivity to size and microstructure for non-metals and the possibility of interfacial thermal resistance means that care should be exercised in examining data such as these. On the other hand, PMCs are clearly expected to have substantially higher conductivities than the corresponding matrices.

It can be shown [31] that, in materials with free electrons, the ratio of thermal conductivity to electrical conductivity is expected to be constant at a given temperature (*Wiedemann–Franz law*) and the ratio $K/(\sigma T)$, sometimes called the *Lorentz number*, is also a constant, when comparing values obtained at different temperatures. It is predicted to be given by $3k^2/e^2$, which has a value of about 22 W nΩ K^{-2}. Some of the room-temperature data presented in Table 10.1 are incorporated into Fig. 10.20, which shows thermal conductivity plotted against electrical resistivity for pure metals and for various grades of carbon fibre [24].

A general trend of higher thermal conductivity being associated with lower electrical resistivity is apparent for both types of material, but they clearly lie on different curves. The value of the Lorentz number corresponding to the data for the metals is in the approximate range of 20–30 W nΩ K^{-2} in all cases, consistent with basic theory for free electron materials. However, in the case of the carbon fibres, the Lorentz numbers (~500–5000 W nΩ K^{-2}) vary substantially and are considerably higher than the value

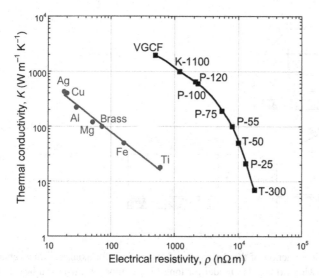

Fig. 10.20 Room-temperature data for metals and for carbon fibres, plotted as thermal conductivity against electrical resistivity (after Ting and Lake [24]). VGCF, vapour-grown fibre.

predicted by theory, particularly for the higher-conductivity products. This suggests that thermal conduction in these carbon fibres is not dominated by electrons acting as carriers. The turbostratic structure of carbon fibre allows electrons to flow within the π orbitals parallel to the graphitic basal planes and high degrees of alignment of these planes along the fibre axis can lead to relatively high electrical conductivities. However, the anomalously high Lorentz numbers indicate that phonons are carrying much of the thermal energy in these materials, particularly for the high-conductivity fibres, in which the degree of crystalline perfection is high. Thus, while the location of the T-300 fibre on the plot is not very far from an extrapolation of the line for the metals (corresponding to free electron carriers being dominant), the vapour-grown fibre (VGCF), which has a high degree of crystalline perfection, allowing extensive transfer of heat via phonons, has a much higher thermal conductivity than would a metal with a similar electrical resistivity.

It follows from data such as these that there is considerable scope for controlling the thermal conductivities of composites by tailoring the type, content and morphology of the reinforcement. There are, however, a number of complications. These include the possibility of interfacial resistance being present and also of the reinforcement conductivity being anisotropic. For example, the thermal conductivities of carbon fibres are expected to be considerably lower in directions transverse to the fibre axis, although it can be quite difficult to measure such properties reliably.

10.4.2 Models for the Conductivity of Composites

The basic (1D) steady-state heat flow equation may be written as

$$q = -K\frac{\partial T}{\partial x} \tag{10.23}$$

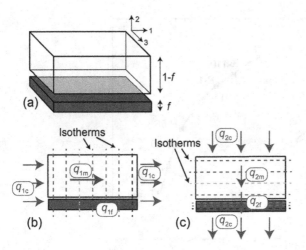

Fig. 10.21 Schematic depiction of the derivation of expressions for axial and transverse conductivities, showing (a) the slab model for long fibre composites, (b) axial heat flow and (c) transverse heat flow (for a case in which the conductivity of the fibre is less than that of the matrix).

where q is the heat flux (W m^{-2}) arising from a thermal gradient $\partial T/\partial x$ (K m^{-1}) in a material of thermal conductivity K (W m^{-1} K^{-1}). It is important to distinguish K from the *thermal diffusivity a* (=K/c), the parameter determining the rate at which a material approaches thermal equilibrium, which appears in the unsteady diffusion equation

$$\frac{\partial T}{\partial t} = a\frac{\partial^2 T}{\partial x^2} \tag{10.24}$$

Prediction of the conductivity of a composite requires assumptions about the flow of heat through the constituents, i.e. the shape of the isotherms. Many reviews [35–39] are available covering the effective conductivity of composites. For heat flow in long fibre composites, the slab model can be used. Axial and transverse heat flows are represented schematically in Fig. 10.21. For the axial (long fibre) case, the thermal gradient (spacing between the isotherms) is the same in each constituent. The total heat flux in the axial (1-) direction is given by the sum of the flows through fibre and matrix

$$q_{1c} = fq_{1f} + (1-f)q_{1m} \tag{10.25}$$

These can be written in terms of conductivities and thermal gradients

$$K_{1c}\frac{\partial T}{\partial x} = fK_f\frac{\partial T}{\partial x} + (1-f)K_m\frac{\partial T}{\partial x} \tag{10.26}$$

so that the composite conductivity is given by a simple rule of mixtures

$$K_{1c} = fK_f + (1-f)K_m \tag{10.27}$$

This can also be expressed as

$$\frac{K_{1c}}{K_m} = f\beta + (1-f) \tag{10.28}$$

where β is the ratio of the conductivity of the reinforcement to that of the matrix. This expression should give a reliable value for the axial conductivity of a long fibre composite (and also of a layered structure).

When treating transverse conduction in long fibre composites (and behaviour in the presence of short fibre, particulate and plate-like reinforcement), rigorous analytical expressions are not obtainable. This is a consequence of the complex thermal field around each reinforcing body, particularly when their content is high enough to ensure that the associated disturbances to the field cannot be taken as independent of each other. The spatial distribution of the reinforcement within the matrix can thus affect the conductivity, since it will influence the nature of the interaction effects, although in practice the dependence on this factor is expected to be weak.

The simplest treatment of transverse conductivity, based on the slab model (i.e. a layered arrangement), is implemented by equating the heat fluxes through the two constituents (Fig. 10.21(c))

$$q_{2c} = K_{2c}\frac{\partial T_c}{\partial x} = q_{2f} = K_f\frac{\partial T_f}{\partial x} = q_{2m} = K_m\frac{\partial T_m}{\partial x} \tag{10.29}$$

The thermal gradients in the two constituents are related to the overall average value by the expression

$$\frac{\partial T_c}{\partial x} = f\frac{\partial T_f}{\partial x} + (1-f)\frac{\partial T_m}{\partial x} \tag{10.30}$$

leading to the following expression for the transverse conductivity

$$K_{2c} = \left(\frac{f}{K_f} + \frac{(1-f)}{K_m}\right)^{-1} \tag{10.31}$$

This can also be written as

$$\frac{K_{2c}}{K_m} = \left(\frac{f}{\beta} + (1-f)\right)^{-1} \tag{10.32}$$

This is analogous to the conventional equal stress ('Reuss') expression for the transverse stiffness (Section 3.2). As in the derivation of that expression, the assumption that the two constituents lie 'in series' with each other leads to this expression being of poor accuracy for fibre composites, although it is accurate for a layered structure. It represents a lower bound for fibres, but becomes very unreliable when the fibres have a low conductivity; for insulating fibres, the composite is predicted to have zero conductivity even when the fibre volume fraction is small.

A number of more reliable models are available for transverse conduction in long fibre composites. Some of these were developed a considerable time ago. For example, in 1892 Rayleigh [40] examined the effect of a rectangular array of fibres on the flux of electrical current in the transverse direction and derived the expression

$$\frac{K_{2c}}{K_m} = \frac{(1-f)+\beta(1+f)}{(1+f)+\beta(1-f)} \tag{10.33}$$

Other expressions have also been derived. These include that obtained by Davies [41], who applied effective medium theory to a random array of parallel fibres, leading to an expression representing the solution of a quadratic

$$\frac{K_{2c}}{K_m} = \left(\frac{1}{2}-f\right)(1-\beta) + \left\{\frac{(1-2f)^2(\beta-1)^2}{4}+\beta\right\}^{1/2} \tag{10.34}$$

Predictions obtained using Eqns (10.32)–(10.34) are shown in Fig. 10.22 for a matrix with 25% of the conductivity of the fibres and vice versa. The Rayleigh expression is expected to be reliable, provided the error introduced by neglect of interfacial resistance is small. This will tend to be the case for low-conductivity fibres ($\beta = 0.25$), but not necessarily with a relatively high ratio (such as $\beta = 4$). The inverse rule of mixtures expression (Eqn (10.32)) will always tend to give an underestimate. It can be seen that the Davies model predictions are close to those of Rayleigh for low-conductivity fibres, but above them with more conductive reinforcement.

It is also worth noting that the case of a dispersed set of spherical particles within a matrix has been analysed. This problem was addressed in 1873 by Maxwell [42], who showed that a solution is given by

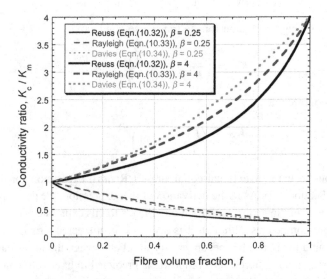

Fig. 10.22 Plots of predicted conductivity ratio against fibre volume fraction, obtained using three analytical equations for the transverse conductivity of long fibre composites, shown for the cases of fibres having 25% and 400% of the conductivity of the matrix.

$$\frac{K_c}{K_m} = \frac{2(1-f) + \beta(1+2f)}{(2+f) + \beta(1-f)} \qquad (10.35)$$

It can be seen that this is similar to the Rayleigh expression (Eqn (10.33)).

The Eshelby approach (Section 6.2) can be adapted to predict the conductivity of composites [3,43–46]. The problem is mathematically simpler than that for stresses and strains, but the physical operations that the mathematical steps represent are less clear. A visualisation such as that shown in Fig. 10.23 can be helpful. This shows isotherms in

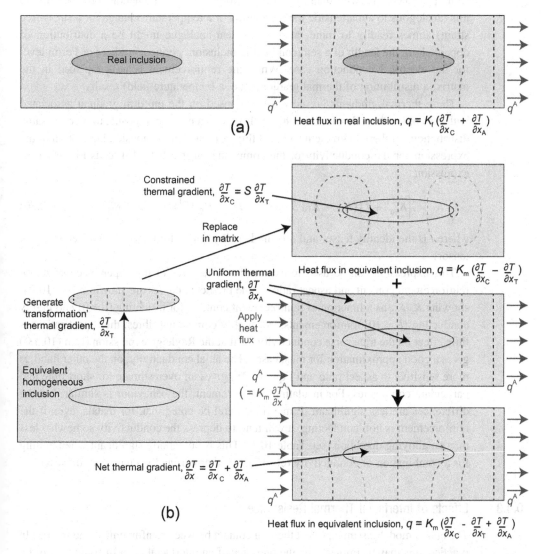

Fig. 10.23 Schematic illustration [3] of the Eshelby method for heat conduction, showing thermal fields for: (a) the real composite ($K_f > K_m$) and (b) the equivalent homogeneous inclusion, with and without an applied heat flux. Thermal gradients can be deduced from the spacings of the isotherms. The thermal gradient is uniform in the inclusion, but variable in the matrix.

the (a) real and (b) equivalent homogeneous systems, for a case in which the conductivity of the inclusion is greater than that of the matrix. Attention is concentrated on thermal gradients (i.e. the spacings between the isotherms).

The equivalent of the 'stress-free strain' introduced for the mechanical Eshelby treatment is a uniform thermal gradient generated within the reinforcement. This is termed a 'transformation thermal gradient', $(\partial T/\partial x)_T$. It can be imagined as being due to a set of distributed heat sources and sinks within the reinforcement, but it is important to recognise that there is no heat flow down this thermal gradient. (This is analogous to a strain produced otherwise than by a stress.) No physical phenomenon corresponding to this (analogous to a martensitic transformation or a temperature change for a stress-free strain) comes readily to mind, although a near-analogue might be a distribution of dopant along the length of a semiconducting inclusion, giving a gradient of Fermi level for the electrical conduction case. When the reinforcement is now replaced in the matrix, a distribution of thermal gradients (i.e. a temperature field) results.

These thermal gradients are now superimposed on the uniform gradient associated with the applied heat flux for this (thermally) homogeneous composite, to give the same distributions of thermal gradient and heat flux as in the real composite. Derivation of the expression for the conductivity of the composite on this basis [3] leads to the tensor expression

$$K_c = \left[K_m^{-1} + f\{(K_m - K_f)[S - f(S - I)] - K_m\}^{-1}(K_f - K_m)K_m^{-1}\right]^{-1} \quad (10.36)$$

where I is the identity tensor and S is the Eshelby tensor for conduction (a second-rank tensor).

This expression can be used to explore basic features of the dependence of K_c on reinforcement content and aspect ratio. Some predicted curves are shown in Fig. 10.24, showing K_c/K_m as a function of reinforcement content, for insulating ($\beta = 0$) and highly conducting ($\beta = 10$) reinforcements. For highly conducting fibres, the aspect ratio has little effect on the transverse conductivity, so that the Rayleigh expression (Eqn (10.33)) gives a good approximation for most cases. The axial conductivity, on the other hand, is more sensitive to aspect ratio and Eqn (10.28) gives an overestimate for short fibre and particulate composites. For insulating reinforcement, the behaviour is similar, but the differences are less significant. It may in general be noted that, for metals, even if the reinforcement is non-conducting, it will tend to depress the conductivity somewhat less than an alloying operation (see Table 10.1). This is potentially significant when aiming for a combination of conductivity (electrical or thermal) and strength (or stiffness).

10.4.3   Effects of Interfacial Thermal Resistance

The above models assume perfect thermal contact between reinforcement and matrix. In practice, this may be impaired by the presence of an interfacial layer of some sort, or by voids or cracks in the vicinity of the interface. Furthermore, even in the absence of such barriers to heat flow, there may be some loss of heat transfer efficiency across the interface if the carriers are different in the two constituents, as with metal–ceramic

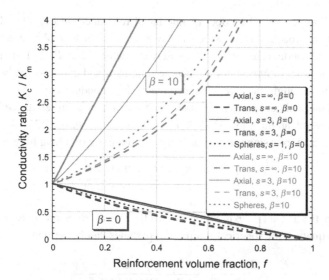

Fig. 10.24 Eshelby (Eqn (10.36)) plots of axial and transverse composite conductivity ratios, as a function of reinforcement content, for insulating ($\beta = 0$) and highly conducting ($\beta = 10$) reinforcements with different aspect ratios.

systems. Such a thermal resistance is characterised by an *interfacial heat transfer coefficient* or *thermal conductance*, h (W m^{-2} K^{-1}), defined as the proportionality constant between the heat flux through the interface and the temperature drop across it

$$q_i = h\Delta T_i \qquad (10.37)$$

An analogous expression applies to flow of electrical current across an interface, with the temperature drop being replaced by a voltage drop and the units of the interfacial electrical conductance being S m^{-2} (Ω^{-1} m^{-2}). If the resistance of the interface can be considered to arise from the presence of an interfacial layer of resistivity ρ and thickness w, then its conductance is given by $1/(\rho w)$.

Interfacial resistance has no effect on the axial conductivity of long fibre or infinite plate composites (except where it arises from interfacial porosity or reaction layers which change the effective cross-sectional area of matrix or reinforcement). There will also be no effect when the reinforcement is effectively an insulator ($\beta \approx 0$). For other cases, a finite value for h leads to a reduction in the conductivity of the composite, since the flow of heat (or electrical current) meets extra increments of resistance each time it crosses an interface. There is a *scale effect* here, since interfaces will be crossed more frequently with finer reinforcements. The dimensionless constant that determines this frequency is the ratio of the conductance of the characteristic heat flow distance within the reinforcement (= conductivity/distance) to the conductance of the interface (= h). This ratio is often termed the *Biot number, Bi*. For transverse heat flow with long fibres, it is given by

$$Bi = \frac{K_f}{r h} \qquad (10.38)$$

where r is the fibre radius. The effect of interfacial resistance on heat flow becomes negligible as Bi approaches zero, which is favoured by: (a) low fibre conductivity, (b) high interfacial conductance and (c) large fibre diameter. When $Bi \gg 1$, on the other hand, the interfacial resistance starts to become dominant and can effectively exclude heat from flowing through the fibres, irrespective of their conductivity.

Modelling of thermal (and electrical) conductivity with interfacial resistance has been addressed by several authors [36,47–49]. Hasselman and Johnson [36] derived the following analytical expression for the transverse conductivity of a long fibre composite

$$\frac{K_{2c}}{K_m} = \frac{(1-f) + \beta(1+f) + Bi(1-f)}{(1+f) + \beta(1-f) + Bi(1+f)} \tag{10.39}$$

This reduces to the Rayleigh expression (Eqn (10.33)) as Bi approaches zero. The corresponding equation for spherical reinforcing particles can be written as

$$\frac{K_c}{K_m} = \frac{2(1-f) + \beta(1+2f) + 2Bi(1-f)}{(2+f) + \beta(1-f) + Bi(2+f)} \tag{10.40}$$

This reduces to the Maxwell expression (Eqn (10.35)) as Bi becomes small.

Predictions from Eqns (10.39) and (10.40) are shown in Fig. 10.25. Fig. 10.25(a) shows how the transverse conductivity of a long fibre composite is reduced as Bi is increased (h decreases), until, for $Bi \sim 100$, the conductivity of the fibres is unimportant, since they are then insulated from the matrix by the high interfacial resistance. The effect of reinforcement shape is illustrated by the plots in Fig. 10.25(b), which compares transverse conductivities of long fibre composites with values for particles (spheres). Particulate composites have slightly higher conductivities, as for the case of perfect interfacial contact (Fig. 10.24). It should, however, be noted that Eqns (10.39) and (10.40) are not reliable at high reinforcement contents, and do not correctly predict that $K_c/K_m \rightarrow \beta$ as $f \rightarrow 1$ for all Bi (except in the case of $\beta = 0$). This effectively arises from the implicit assumption that the significance of the interface will increase continuously as the reinforcement content is raised, whereas in practice the interfacial area must start to decrease again as f approaches 100%.

In fact, all of the approaches outlined above tend to become somewhat unreliable at very high values of f, even if they do correctly predict that $K_c/K_m \rightarrow \beta$ as $f \rightarrow 1$. For example, if the reinforcement is effectively insulating (which is quite common with MMCs, particularly for electrical conduction), then it is clear that the composite conductivity will be zero, not only when $f = 1$, but also when f is sufficiently close to unity for the metallic 'matrix' to be present as isolated islands which do not interconnect with each other.

The theoretical approaches outlined above, which all effectively depend on some sort of averaging being applied to the composite, are unsuited to treating this regime and models that focus on the *connectivity* of the conducting phase are required. This is the domain of *percolation treatments* [50–52], the details of which are beyond the scope of this book. They are commonly applied to problems such as fluid flow through porous media, as well as to heat and electric charge transport through an insulating constituent

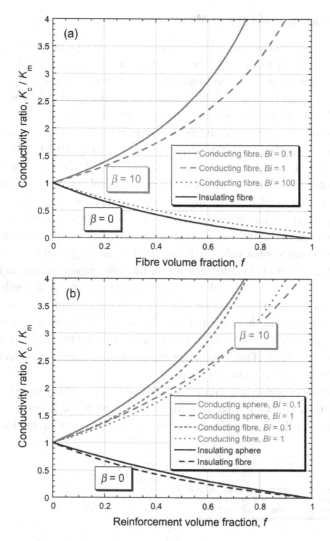

Fig. 10.25 Plots illustrating the effect of interfacial resistance on the transverse conductivity of long fibre composites, showing (a) how it approaches that for insulating fibres as Bi becomes large (poor interfacial contact), even with highly conducting fibres ($\beta = 10$), and (b) a comparison with the case of spherical particles. These curves were obtained using Eqns (10.39) and (10.40).

containing interconnected conductors. The threshold level exhibited by a particular system is expected to be sensitive to its geometrical characteristics. For example, an insulating matrix containing parallel, uniformly dispersed metallic wires will have zero transverse conductivity up to a very high metal content (~90%, depending on the packing arrangement), whereas metallic foams usually remain conducting even when the metal volume fraction is less than 5%. High aspect ratio fibres of small diameter, which tend to be highly flexible, can provide connectivity at very low contents – values as low as 0.05% have been reported [53] for carbon nanotubes in polymeric matrices. The *percolation threshold* can therefore vary over a very wide range.

10.4.4   Measured Conductivities of Polymer Matrix Composites

A classical set of measurements on the conductivity of epoxy–carbon fibre laminae and laminates was carried out some time ago by Pilling et al. [54]. A few general points should be noted about such systems. First, the fibres have much higher (axial) conductivities (K_f ~ 10–100 W m^{-1} K^{-1}) than the matrix (K_m ~ 0.2 W m^{-1} K^{-1}) – see Table 10.1. Also, as for stiffness, these fibres are anisotropic, often with substantially lower transverse conductivities. Furthermore, there are variants of carbon fibres with significantly different properties – see Fig. 10.20 – which is not really true for glass fibres. A comparison is shown in Fig. 10.26 between measured transverse conductivities of unidirectional laminae and predictions from Eqn (10.39), for three values of Bi. This suggests that the interfacial thermal resistance is not very significant – i.e. that Bi is relatively small. This is probably correct, although the uncertainty about the transverse conductivity of the fibre should be noted – this plot is based on a value of 10 W m^{-1} K^{-1}, as suggested by Pilling et al. [54]. Even with this (low) value, the presence of the fibres raises the conductivity by a factor of ~2–5, relative to that of the matrix. However, this is a much smaller increase than that in the axial direction, for which (using a simple rule of mixtures and a fibre conductivity in this case [54] of ~20 W m^{-1} K^{-1}), the factor would be ~40–60 (and larger still for HM-type carbon fibres, for which the axial conductivity [54] is ~80 W m^{-1} K^{-1}). This strong anisotropy in the thermal conductivity of carbon fibre composites is worthy of note.

Polymer composites reinforced with other types of fibre are likely to exhibit much lower conductivities, and also lower anisotropy, with most of the fibres in use being much less conductive than carbon fibres. (Of course, polymers can be reinforced with

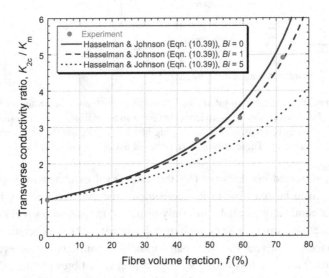

Fig. 10.26 Comparison between experimental data [54] for the transverse conductivity of epoxy–carbon (HS) laminae (at ambient temperature) and predictions obtained using Eqn (10.39). The matrix conductivity is 0.23 W m^{-1} K^{-1}.

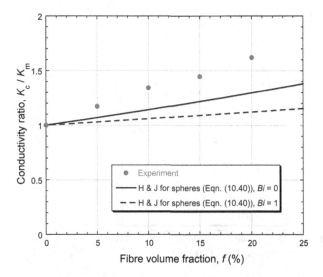

Fig. 10.27 Comparison between experimental data [55] for the conductivity of LDPE–glass chopped fibre composites (at ambient temperature) and predictions obtained using Eqn (10.40). The matrix conductivity is 0.29 W m^{-1} K^{-1}.

metallic fibres, but in practice such combinations are not common.) Fig. 10.27 compares measured conductivities [55] of composites made by randomly dispersing chopped glass fibres into a matrix of LDPE (low-density polyethylene) with predictions from Eqn (10.39), for two values of *Bi*. This equation actually relates to spherical reinforcing particles, but such composites often have similar properties to those of composites containing the same volume fraction of randomly oriented short fibres – the stiffness, for example, is usually very close in the two cases. The conductivity of these glass fibres (~1 W m^{-1} K^{-1}) is about four times that of the matrix. It can be seen that, with these relatively low fibre contents, the measured conductivities are above those predicted, even for perfect interfacial contact (*Bi* = 0). As suggested by the authors of the paper [55], this is probably a consequence of 'connectivity' effects (see Section 10.4.3), with networks of directly contacting fibres being established throughout the sample. These fibres had high aspect ratios (~300–400), favouring the formation of such networks at low volume fractions. Nevertheless, it is clear that all such composite systems are likely to be relatively poor thermal conductors.

10.4.5 Measured Conductivities of Metal Matrix Composites

Conductivities of MMCs are of particular interest in view of the impairment of conductivity (electrical and thermal) that usually arises on addition of solute to a pure metal (with the intention of strengthening it). Adding ceramic reinforcement introduces potential for strengthening while the matrix retains a high conductivity. As for PMCs, the axial conductivity of unidirectional long fibre composites should be reliably predicted by the rule of mixtures (Eqn (10.28)), with the interfacial conductance playing no

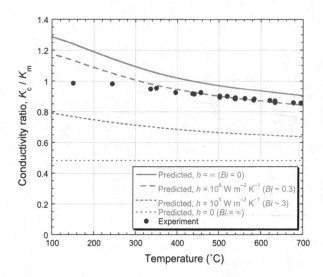

Fig. 10.28 Experimental data [56] for the thermal conductivity ratio, as a function of temperature, for Ti–6Al–4V reinforced with 35 vol% of continuous SiC monofilaments, for heat flow transverse to the fibre axis. Predictions from Eqn (10.39) are also shown, for different values of h. As temperature is raised over this range, fibre conductivity falls from about 16 to 12 W m^{-1} K^{-1}, while K_m rises from about 8 to 15 W m^{-1} K^{-1}. (This rise is associated with changes in phonon scattering in this alloy [27].)

role. If the reinforcement has a much lower conductivity than the matrix, then this conductance is also irrelevant for transverse conduction.

However, there are some MMC systems for which the conductivities of the two constituents are at least comparable and in these cases the interfacial conductance is relevant when considering transverse properties. Eqn (10.29) can be used in conjunction with experimental conductivity data to estimate the interfacial conductance. For example, Fig. 10.28 shows measured conductivities [56] over a range of temperature for Ti-6Al-4V reinforced with SiC monofilaments, plotted as a ratio to that of the matrix. Also shown are predictions from Eqn (10.29) corresponding to several h values, using the fibre radius (~50 μm). These data are consistent with a value for h of approximately 10^6 W m^{-2} K^{-1}. This represents relatively poor thermal contact for a fibre/matrix interface, but it can be seen that in this system the overall conductivity would not have been much greater had the interface been perfect ($h = \infty$). This illustrates the size effect, since the large diameter of the SiC monofilaments means that the interfacial conductance is not of much significance (unless it becomes very low indeed). It may also be noted that, although data in Table 10.1 suggest that there is scope for raising the conductivity of Ti-6Al-4V by addition of SiC, the monofilament in this composite has a relatively low conductivity (~15 W m^{-1} K^{-1}), primarily as a consequence of the phonon scattering at grain boundaries, which is favoured by its very fine grain size [57]. The presence of impurities may also be partly responsible for this value being so low. Measured axial and transverse conductivities [56] are thus very similar to that of the unreinforced matrix in this system.

Further points can be illustrated by thermal conductivity data for particulate-reinforced Ti. In Fig. 10.29(a), experimental data [56] for a Ti-10%SiC$_p$ composite, produced by a powder blending and extrusion route, have been plotted for a range of measurement temperatures. Also shown are predictions from Eqn (10.40) corresponding to several h values, using experimental conductivity values for matrix (~18 W m^{-1} K^{-1} over the range of temperature) and reinforcement (~100 W m^{-1} K^{-1} at room temperature, falling to 65 W m^{-1} K^{-1} at 700°C), and a particle radius of 10 μm. Note that the SiC conductivity, which was measured on sintered material with a grain size comparable to the particle diameter of the powder used in making the composite, is appreciably higher than that of the fine-grained SiC monofilament referred to above.

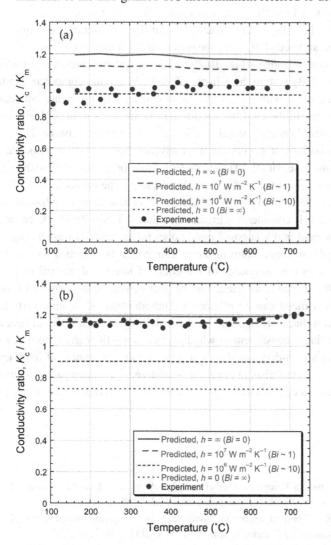

Fig. 10.29 Plots [56] of thermal conductivity ratio, as a function of temperature, for two particulate MMCs, showing experimental data and predictions from Eqn (10.40), based on different values for h. Plots are shown for (a) Ti–10%SiC and (b) Ti–10%TiB$_2$.

As with the monofilament-reinforced Ti-6Al-4V/SiC system, these data are consistent with an h value of $\sim 10^6$ W m^{-2} K^{-1}. In this case, however, a consequence of the smaller size of the reinforcement, compared with the monofilament in the system referred to in Fig. 10.28, is that the overall conductivity is considerably lower than would be the case with perfect interfacial contact. Indeed, the composite conductivity is quite close to that expected with an insulating interface ($h = 0$). It is clear that relatively poor interfacial contact (which is in this case associated with a strong tendency for interfacial chemical reaction to occur during processing of Ti–SiC composites [58,59]), can lead to significant degradation of composite conductivity when the reinforcement is relatively fine.

From the data in Fig. 10.29(b), it is evident that the interfacial characteristics are different for the Ti–TiB$_2$ system from those for Ti–SiC particulate composites. (The conductivity of sintered TiB$_2$ was measured to be about 90 W m^{-1} K^{-1} over the complete temperature range.) Fig. 10.29(b) shows that the interfacial conductance is higher in the Ti–TiB$_2$ system, by a factor of about 10, giving composite conductivities close to those expected with perfect interfacial contact. The difference between the two systems has been attributed [56] to differences in the extent and significance of interfacial reaction. While chemical reaction does take place between Ti and TiB$_2$ (to form the monoboride), it occurs more slowly and has a smaller associated volume change than is the case with the Ti–SiC system. This apparently leads to the retention of higher interfacial conductance after typical processing operations.

It is clear from the data in Fig. 10.28 and 10.29 that, although the handbook thermal conductivity of SiC is considerably greater than that of Ti and its alloys, the introduction of SiC reinforcement into a Ti-based matrix does not necessarily raise its conductivity – either because the microstructure of the reinforcement is such that its conductivity is relatively low or as a consequence of the effect of interfacial thermal resistance. The data for Ti–TiB$_2$, on the other hand, confirm that genuine enhancement of the thermal conductivity of a metal can be effected by introduction of ceramic particles. These results do suggest that there is no inherent inefficiency in the transfer of heat across an interface separating a constituent in which electrons are the carrier from one in which heat transfer is phonon-dominated – or at least that any increment of interfacial resistance associated with this change in carrier is relatively small compared to other sources of interfacial resistance, such as the presence of reaction layers or fine cracks/pores, etc.

References

1. Bowles, DE and SS Tompkins, Predictions of coefficients of thermal expansion for unidirectional composites. *Journal of Composite Materials* 1989; **23**(4): 370–388.
2. Schapery, RA, Thermal expansion coefficients of composite materials based on energy principles. *Journal of Composite Materials* 1968; **2**(3): 380–404.
3. Clyne, TW and PJ Withers, *An Introduction to Metal Matrix Composites*. Cambridge University Press, 1993.
4. Lamé, G, *Leçons sur la Théorie de l'Élasticité*. Gauthier-Villars, 1852.

5. Lee, JK, YY Earmme, HI Aaronson and KC Russel, Plastic relaxation of the transformation strain energy of a misfitting spherical precipitate: ideal plastic behaviour. *Metallurgical Transactions A* 1980; **11**: 1837–1847.

6. Warwick, CM and TW Clyne, Development of a composite coaxial cylinder stress analysis model and its application to SiC monofilament systems. *Journal of Materials Science.* 1991; **26**: 3817–3827.

7. Kural, MH and BK Min, The effects of matrix plasticity on the thermal deformation of continuous fibre graphite/metal composites. *Journal of Composite Materials* 1984; **18**: 519–535.

8. Wolff, EG, BK Min and MH Kural, Thermal cycling of a unidirectional graphite–magnesium composite. *Journal of Materials Science.* 1985; **20**: 1141–1149.

9. Masutti, D, JP Lentz and F Delannay, Measurement of internal stresses and of the temperature dependence of the matrix yield stress in metal matrix composites from thermal expansion curves. *Journal of Materials Science. Letts.* 1990; **9**: 340–342.

10. Withers, PJ, Mapping residual and internal stress in materials by neutron diffraction. *Comptes Rendus Physique* 2007; **8**(7–8): 806–820.

11. Kobayashi, S, K Terada and N Takeda, Evaluation of long-term durability in high temperature resistant CFRP laminates under thermal fatigue loading. *Composites Part B: Engineering* 2003; **34**(8): 753–759.

12. Schwenker, SW and D Eylon, Creep deformation and damage in a continuous fiber-reinforced Ti-6Al-4V composite. *Metallurgical and Materials Transactions A: Physical Metallurgy and Materials Science* 1996; **27**(12): 4193–4204.

13. McLean, M, Creep deformation of metal-matrix composites. *Composites Science and Technology* 1985; **23**(1): 37–52.

14. Nieh, TG, Creep rupture of a silicon carbide reinforced aluminium composite. *Metallurgical Transactions A* 1984; **15**: 139–146.

15. Nardone, VC and JK Tien, On the creep rate dependence of particle strengthened alloys. *Scripta Metallurgica* 1986; **20**: 797–802.

16. Jarry, P, W Loué and J Bouvaist, Rheological behaviour of SiC/Al composites, in *Proceedings of the Sixth International Conference on Composite Materials combined with the Second European Conference on Composite Materials*, 20–24 July 1987, Imperial College of Science and Technology, London. Elsevier, 1987, pp.2.350–2.361.

17. Furness, JAG and TW Clyne, The application of scanning laser extensometry to explore thermal cycling creep of metal matrix composites. *Materials Science and Engineering A* 1991; **141**: 199–207.

18. Furness, JAG, *Thermal Cycling Creep of Aluminium-Based Composites*, PhD thesis, University of Cambridge, 1991.

19. Wu, MY and OD Sherby, Superplasticity in a silicon carbide whisker reinforced aluminium alloy. *Scripta Metallurgica* 1984; **18**: 773–776.

20. Jansson, S, DJ Dal Bello and FA Leckie, Transverse and cyclic thermal loading of the fiber reinforced metal-matrix composite SCS6/Ti-15-3. *Acta Metallurgica Materialia* 1994; **42**(12): 4015–4024.

21. Newaz, GM and BS Majumdar, A comparison of mechanical response of MMC at room and elevated temperatures. *Composites Science and Technology* 1994; **50**: 85–90.

22. Kalton, AF, P Feillard and TW Clyne, The effect of interfacial structure on the failure of long fibre reinforced titanium under transverse loading, in *High Performance Materials in Engine Technology*, Vincenzini, P, editor. Techna Srl., 1995, pp. 285–292.

23. Kowalski, IM, New high performance domestically-produced carbon fibres. *SAMPE Journal* 1987; **32**: 953–963.

24. Ting, JM and ML Lake, Vapor grown carbon fiber reinforced aluminum composites with very high thermal conductivity. *Journal of Materials Research* 1995; **10**: 247–250.

25. Mukai, K, K Asaka, XL Wu, T Morimoto, T Okazaki, T Saito and M Yumura, Wet spinning of continuous polymer-free carbon-nanotube fibers with high electrical conductivity and strength. *Applied Physics Express* 2016; **9**(5).

26. Morelos-Gomez, A, M Fujishige, SM Vega-Diaz, I Ito, T Fukuyo, R Cruz-Silva, F Tristan-Lopez, K Fujisawa, T Fujimori, R Futamura, K Kaneko, K Takeuchi, T Hayashi, YA Kim, M Terrones, M Endo and MS Dresselhaus, High electrical conductivity of double-walled carbon nanotube fibers by hydrogen peroxide treatments. *Journal of Materials Chemistry A* 2016; **4** (1): 74–82.

27. Klemens, PG and RK Williams, Thermal conductivity of metals and alloys. *International Metals Reviews* 1986; **31**: 197–215.

28. Petry, W, A Heiming, J Trampenau, M Alba, C Herzig, HR Schober and G Vogl, Phonon dispersion of the bcc phase of group-IV metals. 1: bcc titanium. *Physical Review B* 1991; **43** (13): 10933–10947.

29. Goldsmith, A, TE Waterman and HJ Hirschhorn, *Handbook of Thermophysical Properties of Solid Materials*. Macmillan, 1961.

30. Geiger, GH and DR Poirier, *Transport Phenomena in Metallurgy*. Addison-Wesley Publishing Company, 1973.

31. Lovell, MC, AJ Avery and MW Vernon, *Physical Properties of Materials*. Van Nostrand Rheinhold, 1976.

32. Pollock, DD, *Physical Properties of Materials for Engineers*. CRC Press, 1982.

33. West, EG, *Copper and its Alloys*. Ellis Horwood, 1982.

34. Nye, JF, *Physical Properties of Crystals: Their Representation by Tensors and Matrices*. Clarendon, 1985.

35. Benveniste, Y, Effective thermal conductivity of composites with a thermal contact resistance between the constituents: non-dilute case. *Journal of Applied Physics* 1987; **61**: 2840–2843.

36. Hasselman, DPH and LF Johnson, Effective thermal conductivity of composites with interfacial thermal barrier resistance. *Journal of Composite Materials* 1987; **21**: 508–515.

37. Hatta, H, Thermal diffusivities of composites with various types of filler. *Journal of Composite Materials* 1992; **26**(5): 612.

38. Qu, XH, L Zhang, M Wu and SB Ren, Review of metal matrix composites with high thermal conductivity for thermal management applications. *Progress in Natural Science: Materials International* 2011; **21**(3): 189–197.

39. Pietrak, K and TS Wisniewski, A review of models for effective thermal conductivity of composite materials. *Journal of Power Technologies* 2015; **95**(1): 14–24.

40. Rayleigh, JW, On the influence of obstacles arranged in rectangular order upon the properties of a medium. *Philosophical Magazine* 1892; **34**: 481–507.

41. Davies, WEA, The dielectric constant of fibre composites. *Journal of Physics D* 1974; **7**: 120–130.

42. Maxwell, JC, *A Treatise on Electricity and Magnetism* Clarendon Press, 1873.

43. Hatta, H and M Taya, Equivalent inclusion method for steady state heat conduction in composites. *International Journal of Engineering and Science* 1986; **24**: 1159–1172.

44. Hatta, H and M Taya, Thermal conductivity of coated filler composites. *Journal of Applied Physics* 1986; **59**: 1851–1860.

45. Taya, M, Modelling of physical properties of metallic and ceramic composites: generalized Eshelby model, in *Mechanical and Physical Behaviour of Metallic and Ceramic Composites, 9th Risø International Symposium on Metals & Materials Science*, Roskilde, Denmark. Riso National Laboratory, 1988, pp. 201–231.

46. Gordon, FH and TW Clyne, Transport properties of short fibre SiC-reinforced titanium, in *Metal Matrix Composites – Processing, Microstructure and Properties, 12th Risø International Symposium on Metals & Materials Science*, Roskilde, Denmark. Riso National Laboratory, 1991, pp. 361–366.

47. Fadale, TD and M Taya, Effective thermal conductivity of composites with fibre–matrix debonding. *Journal of Materials Science Letters*. 1991; **10**: 682–684.

48. Hasselman, DPH, KY Donaldson and JR Thomas, Effective thermal conductivity of uniaxial composite with cylindrically orthotropic carbon fibres and interfacial barrier resistance. *Journal of Composite Materials* 1993; **27**(6): 637–644.

49. Markworth, AJ, The transverse thermal conductivity of a unidirectional fibre composite with fibre-matrix debonding. *Journal of Materials Science Letters* 1993; **12**(19): 1487–1489.

50. Last, BJ and DJ Thouless, Percolation and electrical conductivity. *Physical Review Letters* 1971; **27**: 1719–1721.

51. Stauffer, D and A Aharony, *Introduction to Percolation Theory*, 2nd edition. Taylor & Francis, 1992.

52. Sun, K, ZD Zhang, L Qian, F Dang, XH Zhang and RH Fan, Dual percolation behaviors of electrical and thermal conductivity in metal–ceramic composites. *Applied Physics Letters* 2016; **108**(6): 061903.

53. Lisunova, MO, YP Mamunya, NI Lebovka and AV Melezhyk, Percolation behaviour of ultrahigh molecular weight polyethylene/multi-walled carbon nanotubes composites. *European Polymer Journal* 2007; **43**(3): 949–958.

54. Pilling, MW, B Yates, MA Black and P Tattersall, Thermal-conductivity of carbon fiber-reinforced composites. *Journal of Materials Science* 1979; **14**(6): 1326–1338.

55. Kalaprasad, G, P Pradeep, G Mathew, C Pavithran and S Thomas, Thermal conductivity and thermal diffusivity analyses of low-density polyethylene composites reinforced with sisal, glass and intimately mixed sisal/glass fibres. *Composites Science and Technology* 2000; **60**(16): 2967–2977.

56. Gordon, FH, SP Turner, R Taylor and TW Clyne, The effect of the interface on the thermal conductivity of Ti-based composites. *Composites* 1994; **25**: 583–592.

57. Lepetitcorps, Y, M Lahaye, R Pailler and R Naslain, Modern boron and SiC CVD filaments: a comparative study. *Composites Science and Technology* 1988; **32**(1): 31–55.

58. Choi, SK, M Chandrasekaran and MJ Brabers, Interaction between titanium and SiC. *Journal of Materials Science* 1990; **25**: 1957–1964.

59. Reeves, AJ, R Taylor and TW Clyne, The effect of interfacial reaction on thermal properties of titanium reinforced with particulate SiC. *Materials Science and Engineering* 1991; **141**: 129–138.

11 Surface Coatings as Composite Systems

Certain types of multi-component configuration, while not constituting conventional composite materials, can nevertheless be treated using the approaches and methodologies of composite theory. These include layered systems, particularly the simplest one of a (planar) substrate with a coating ('deposit') on one face. This is effectively a real version of the 'slab model' that is commonly used to predict composite properties such as stiffness, conductivity, etc. Such property prediction is, of course, valid for a substrate/coating system, although often of limited interest. However, the slab model can be extended in directions that are potentially more useful for coatings. In particular, the creation of curvature, which has only been touched on so far, in the context of (asymmetrical) laminates, can be predicted and utilised (for example to measure residual stresses in coatings). A central concept here concerns the misfit strain, which is used in earlier parts of the book (particularly as it relates to the Eshelby method).

11.1 Curvature in Substrate–Coating Systems

11.1.1 A Substrate–Deposit System with a Uniform Misfit Strain

An important concept in layered (and other) systems is that of a misfit strain – i.e. a difference between the stress-free dimensions of two or more constituents that are bonded together. It is relevant to composite materials and also to macroscopic systems such as two or more components that are bolted or welded together in some way. In general, this strain is a tensor, with principal axes and three principal values. For a layered system, however, the focus is on a single (in-plane) direction, so that the strain can be treated as a scalar. Fig. 11.1 shows how the creation of such a misfit strain can be represented in a substrate–deposit system.

Misfit strains can arise in a number of ways. One of the simplest is differential thermal contraction, but anything that creates a difference between the (in-plane) stress-free dimensions of a substrate and a coating (or a surface layer of a substrate) has a similar effect. These include phase changes, setting of a resin, plastic deformation, creep, etc., as well as phenomena such as atomic bombardment, which can create stresses during formation of a coating. It is also possible for such processes to modify the value of an existing $\Delta\varepsilon$. For example, creep could occur due to the presence of residual stresses, in such a way as to reduce them.

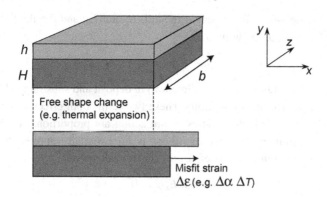

Fig. 11.1 Creation of a uniform misfit strain, $\Delta\varepsilon$, in a bilayer system.

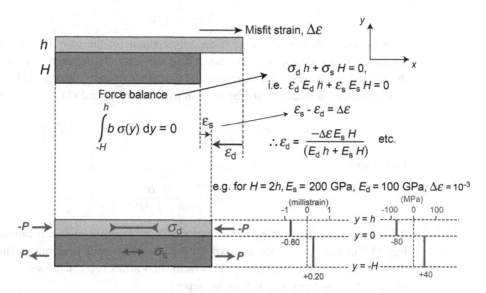

Fig. 11.2 Use of force and strain balances to find the distributions of stress and strain resulting from the imposition of a misfit strain between deposit and substrate.

11.1.2 Force and Strain Balances

The concept of imposing force and strain balances so as find the (in-plane) stresses and strains that arise when the misfit strain is accommodated (i.e. when the two constituents are constrained to have the same in-plane dimensions) is used in Section 10.1.2 (to predict composite expansivity values). This is illustrated for this more general case of any misfit strain in Fig. 11.2. (The subscripts d and s represent 'deposit' and 'substrate' – i.e. the upper and lower layers.)

The equations involved, which correspond to Eqns (10.1) and (10.2), are shown in the figure, which also illustrates the outcome for a specific case (a misfit strain of

1 millistrain and some values for Young's moduli, E_s and E_d, and the thickness ratio, h/H). The force balance can be written as

$$\sigma_d h + \sigma_s H = 0 \tag{11.1}$$

where σ_d and σ_s are the stresses (in the x direction) in deposit and substrate, while h and H are their thicknesses (in the y direction). These two terms both effectively represent forces (since the areas on which the stresses are acting are proportional to the thicknesses). They must sum to zero because there is no externally applied load. This equation can also be written in terms of strains (in the x direction)

$$\varepsilon_d E_d h + \varepsilon_s E_s H = 0 \tag{11.2}$$

Furthermore, the strain balance, which is simply the requirement for the overall misfit strain to be partitioned in some way between the two constituents, can be expressed as

$$\Delta\varepsilon = \varepsilon_s - \varepsilon_d \tag{11.3}$$

Care is needed here with signs. If the misfit strain for the case shown (stress-free dimension of the deposit greater than that of the substrate) is taken as positive, then the above form is correct (recognising that ε_s is positive and ε_d is negative).

The solution is readily found by combining Eqns (11.2) and (11.3)

$$\varepsilon_s = \Delta\varepsilon + \varepsilon_d = \frac{-\varepsilon_d E_d h}{E_s H} \rightarrow \varepsilon_d E_s H + \varepsilon_d E_d h = -\Delta\varepsilon E_s H$$

from which

$$\varepsilon_d = \frac{-\Delta\varepsilon E_s H}{(E_d h + E_s H)} \tag{11.4}$$

It is clear that ε_s, and also σ_d and σ_s, can readily be found from this. It can be seen in Fig. 11.2 that, for the example shown, 80% of the misfit strain is accommodated in the deposit and 20% in the substrate. (This factor of 4 arises because the substrate is both twice as thick and twice as stiff as the deposit.)

11.1.3   Moment Balance

This system differs from that used to derive the expansivity of a fibre composite, since the two forces (P and $-P$), being displaced laterally, now create a *bending moment*. This will tend to cause *curvature* of the bilayer. (This is analogous to the 'out-of-plane' distortions that can arise in (asymmetric) laminates.) The bending moment must be balanced by an internal moment set up by the stresses created within the layers – this is what happens when any beam is subjected to such a moment. Using the moment balance to find the curvature (and the associated stress and strain distributions) is slightly more complex than applying the force balance. Details are available elsewhere [1]. Referring to Fig. 11.3, the forces P and $-P$ generate the moment, given by

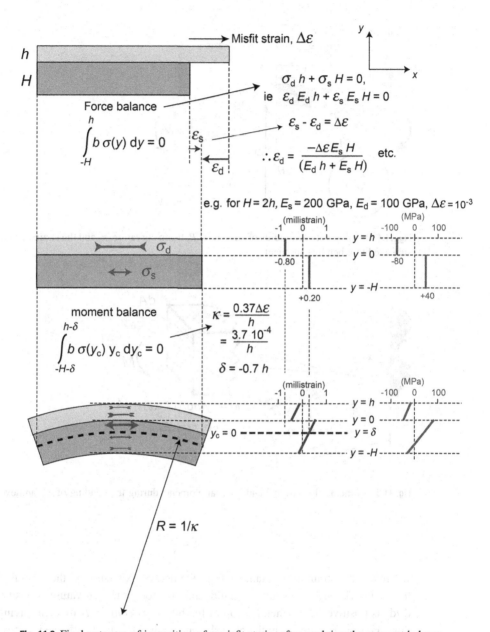

Fig. 11.3 Final outcome of imposition of a misfit strain, after applying the moment balance.

$$M = P\left(\frac{h+H}{2}\right) \tag{11.5}$$

This will create a curvature, κ. Often the most helpful interpretation of the curvature of a beam is to consider it as the through-thickness gradient of the strain (along the length of the beam). This is illustrated by Fig. 11.4, which shows that the curvature can be expressed as

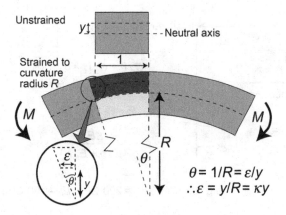

Fig. 11.4 Relation between radius of curvature, R, beam curvature, κ, and the strains within a beam subjected to a bending moment.

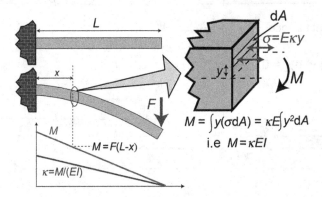

Fig. 11.5 Balancing the external and internal moments during the bending of a (cantilever) beam.

$$\kappa = \frac{\varepsilon}{y} \tag{11.6}$$

where ε is the strain at a distance y from the neutral axis (strictly, the neutral plane) of the beam. The sign convention should again be noted. The curvature is positive when $d\varepsilon/dy$ is positive – i.e. when the top of the bilayer is *convex*. A moment giving rise to this sign of curvature, as in the case shown in Fig. 11.3, is also defined as positive.

The relationship between an applied moment and the curvature induced by it is obtained by recognising that the external moment must be balanced by an internal one, which is given by the sum of the incremental moments generated by the stresses within the beam – see Fig. 11.5 (showing the example of an end-loaded cantilever). Since κ is the strain gradient, the stress at a distance y from the neutral axis is given by

$$\sigma = E\kappa y \tag{11.7}$$

The moment due to this stress, acting on the incremental area dA, is

$$dM = (\sigma dA)y$$

so the total moment is therefore

$$M = \int \sigma y dA = \kappa E \int y^2 dA = \kappa E l \tag{11.8}$$

where I is the *second moment of area* (sometimes, rather misleadingly, termed the *moment of inertia*). It depends only on the sectional shape of the beam. For simple shapes, such as uniform circular or rectangular section beams, there are well-known expressions for I. The product of E and I, often termed the *beam stiffness*, Σ, can be regarded as the beam bending analogue of the Young's modulus for simple axial loading (with M being analogous to the stress and κ analogous to the strain).

It can be seen that, for the situation shown in Fig. 11.2, the force P can be expressed as

$$P = \frac{2\kappa\Sigma}{h+H} \tag{11.9}$$

with the beam stiffness given in this case by

$$\Sigma = b \int_{-H-\delta}^{h-\delta} E(y_c)y_c^2 dy_c = bE_d h\left(\frac{h^2}{3} - h\delta + \delta^2\right) + bE_s H\left(\frac{H^2}{3} + H\delta + \delta^2\right) \tag{11.10}$$

where b is the width (Fig. 11.1), and δ is the distance between the neutral axis ($y_c = 0$) and the interface ($y = 0$)).

The expression for the value of δ is straightforward to obtain. Consider the system shown in Fig. 11.6, representing a two-layer system (with no misfit strain) being subjected to an external bending moment. The force balance

$$b \int_{-H}^{h} \sigma(y)dy = 0 \tag{11.11}$$

can be divided into contributions from the two constituents and expressed in terms of the strain

$$b \int_{0}^{h} E_d \varepsilon(y)dy + b \int_{-H}^{0} E_s \varepsilon(y)dy = 0 \tag{11.12}$$

which can then be written in terms of the curvature (through-thickness strain gradient) and the distance from the neutral axis

$$b \int_{0}^{h} E_d \kappa(y - \delta)dy + b \int_{-H}^{0} E_s \kappa(y - \delta)dy = 0 \tag{11.13}$$

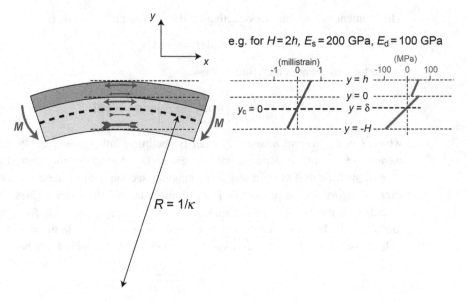

Fig. 11.6 Location of the neutral axis of a bimaterial beam.

Removing the width, b, and curvature, κ, which are constant, gives

$$E_d\left[\frac{y^2}{2} - \delta y\right]_0^h + E_s\left[\frac{y^2}{2} - \delta y\right]_{-H}^0 = 0$$

$$\therefore E_d\left(\frac{h^2}{2} - \delta h\right) + E_s\left(\frac{-H^2}{2} - \delta H\right) = 0$$

This can be rearranged to give an expression for δ

$$\delta(E_d h + E_s H) = \frac{1}{2}(E_d h^2 - E_s H^2)$$

$$\therefore \delta = \frac{h^2 E_d - H^2 E_s}{2(h E_d + H E_s)} \tag{11.14}$$

The magnitude of P is found by expressing the misfit strain as the difference between the strains resulting from application of the P forces (i.e. by writing a strain balance):

$$\Delta\varepsilon = \varepsilon_s - \varepsilon_d = \frac{P}{H b E_s} + \frac{P}{h b E_d}, \qquad \therefore \frac{P}{b} = \Delta\varepsilon\left(\frac{h E_d H E_s}{h E_d + H E_s}\right) \tag{11.15}$$

For the case shown, $\Delta\varepsilon$ is positive (with ε_s positive and ε_d negative). Combination of this with Eqns (11.9), (11.10) and (11.14) gives a general expression for the curvature, κ, arising from imposition of a uniform misfit strain, $\Delta\varepsilon$

$$\kappa = \frac{6 E_d E_s (h + H) h H \Delta\varepsilon}{E_d^2 h^4 + 4 E_d E_s h^3 H + 6 E_d E_s h^2 H^2 + 4 E_d E_s h H^3 + E_s^2 H^4} \tag{11.16}$$

The stress at the interface ($y = 0$), and at the free surfaces ($y = h$ or $-H$), can be written in terms of the base level in each constituent (arising from the force balance) and the change due to the stress gradient (= curvature × stiffness)

$$\sigma_d\big|_{y=h} = \frac{-P}{bh} + E_d\kappa(h - \delta) \quad \text{and} \quad \sigma_d\big|_{y=0} = \frac{-P}{bh} - E_d\kappa\delta \qquad (11.17)$$

$$\sigma_s\big|_{y=-H} = \frac{P}{bH} - E_s\kappa(H + \delta) \quad \text{and} \quad \sigma_s\big|_{y=0} = \frac{P}{bH} - E_s\kappa\delta \qquad (11.18)$$

Since the gradient is linear in each constituent, this gives the complete stress profile.

11.1.4 Stresses, Strains and Curvatures in Substrate–Deposit Systems

An example of an outcome of applying the above equations is shown in Fig. 11.3, for particular values of $\Delta\varepsilon$, h/H, E_s and E_d. A point worth noting here is that the imposition of a misfit strain creates distributions of stress and strain that do *not* have values of zero at the neutral axis. This is in contrast to the outcome when a bending moment is externally applied, as depicted in Fig. 11.6. This is because a misfit strain leads to a requirement for both force and moment balances to be satisfied. What does in fact happen is that, after the force balance has been satisfied (middle of Fig. 11.3), the adoption of curvature (lower part of Fig. 11.3) creates *no change* in the stresses and strains at the neutral axis. It also creates gradients of strain (in both constituents) equal to the curvature. Of course, the stresses are related to the strains via the Young's modulus – but see Section 11.2.1.

Inspection of Eqn (11.16) shows that, for a given thickness ratio, h/H, the curvature is inversely proportional to substrate thickness, H. This *scale effect* is important in practice, since it means that it is only with relatively thin substrates that curvatures are likely to become large. Some predicted curvatures (arising from misfit strains due to differential thermal expansion) are shown in Fig. 11.7. Curvatures below ~0.1 m^{-1} (radius of curvature, $R > 10$ m) are in general not very noticeable (and can be difficult to measure accurately).

11.2 Curvatures and Their Measurement in Real Systems

11.2.1 Poisson Effects and the Biaxial Modulus

Several points should be noted regarding the behaviour of real systems. One of these is that, while it is possible for a misfit strain (and resultant stress and strain distributions) to be generated in just one in-plane direction, this is in fact unusual. Much more common is for the misfit strain, and hence the stress and strain distributions, to be *biaxial*, and in many cases *equal biaxial*. For example, provided the expansivities and stiffnesses are isotropic in the plane of the coating, the same misfit strain will arise in all in-plane directions (which can be represented by two such strains, in directions that are normal to each other, with it being immaterial which directions are chosen). When one or both

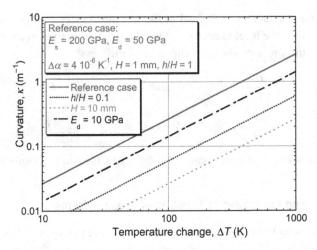

Fig. 11.7 Predicted curvature (obtained using Eqn (11.16)), as a function of the change in temperature, for four different substrate–deposit combinations.

materials are anisotropic, then the stress and strain state is likely to be *unequal biaxial*, although this also is relatively rare. Similar arguments apply when misfit strains are being generated in other ways, although it is certainly true that not all cases are equal biaxial.

The main modification to the treatment that is required by the presence of more than one in-plane stress (and strain) relates to Poisson effects. For an equal biaxial stress state, there is a stress (with the same magnitude as σ_x) in the z direction, which induces a Poisson strain in the x direction. (It also induces a Poisson strain in the y direction, but this is not relevant for present purposes.) For isotropic stiffness and no through-thickness stress ($\sigma_y = 0$), the strain in the x direction is given by

$$\varepsilon_x E = \sigma_x - \nu(\sigma_y + \sigma_z) = \sigma_x(1 - \nu) \tag{11.19}$$

so the ratio of stress to strain in the x direction can be expressed as

$$\frac{\sigma_x}{\varepsilon_x} = \frac{E}{(1 - \nu)} = E' \tag{11.20}$$

Thus, when there is an equal biaxial stress state – the most common situation when a misfit strain is imposed – this modified form of the Young's modulus, E' (often called the *biaxial modulus*) must be used instead of E in all of these preceding expressions.

11.2.2   The Stoney Equation: The Thin Coating Limit

The Stoney equation [2] originated in 1909 – long before the above relationships were developed – but is still in widespread use. There are reviews [3–5] that cover its origins and subsequent developments in the area. It relates the curvature of a substrate with a thin coating to the stress level within the coating (for an equal biaxial case). It is easy to show that the Stoney equation can be derived from Eqn (11.16) for the case of a coating that is much thinner than the substrate ($h \ll H$). All of the denominator terms except the last one can then be discarded and $(h + H) = H$ can be assumed, so that

$$\kappa = \frac{6E'_d h}{E'_s H^2} \Delta \varepsilon \qquad (11.21)$$

with the biaxial moduli now being used. Furthermore, the $h \ll H$ condition allows the assumption to be made that all of the misfit strain is accommodated in the coating (deposit), so only the coating is under stress. The misfit strain can thus be expressed as

$$\Delta \varepsilon = \frac{-\sigma_d}{E'_d} \qquad (11.22)$$

recognising that, with the convention we are using for $\Delta \varepsilon$, a positive value will generate a negative value for σ_d (i.e. a compressive stress). Substitution of this then leads to the Stoney equation

$$\kappa = \frac{-6E'_d h \sigma_d}{E'_s H^2 E'_d} = \frac{-6h\sigma_d}{E'_s H^2}$$

$$\therefore \sigma_d = \frac{-E'_s H^2}{6h(1 - \nu_s)} \kappa \qquad (11.23)$$

The minus sign is not always included, but the curvature should have a sign and, using the convention that a convex upper surface corresponds to positive curvature, this implies a negative deposit stress. This equation allows a coating stress to be obtained from a (measured) curvature. Only E and ν values for the substrate are needed – this is convenient, since they are often known (whereas those of the coating may not be). A single stress value is obtained – if the coating is thin, then any through-thickness variations will be small. It is really the misfit strain that is the more fundamental measure of the characteristics of the system, but a stress value is often regarded as more easily interpreted.

The Stoney equation is easy to use and, indeed, is still widely used. Cases in which it applies, at least approximately, are sometimes referred to as Stoney conditions. However, it does have the limitation of being accurate only in a regime in which the curvatures tend to be relatively small. In some applications – such as with semiconductor wafers – surfaces are very smooth, so that highly accurate optical methods of curvature measurement are feasible and this is not such a problem. However, when curvatures are relatively high (or need to be high for reliable measurement), the Stoney equation may be inaccurate.

When the coating is not very thin compared with the substrate, the adoption of curvature can effect relatively large changes in stress levels and high through-thickness gradients can result. It can be seen from Eqns (11.17) and (11.18) that (for a given value of h/H), since P is proportional to H and κ is inversely proportional to H, the stresses (at $y = -H$, 0 and h) do not depend on H, i.e. the *stress distribution is independent of scale*. However, the *curvature is not*. Substrates must be fairly *thin* if *measurable curvatures* are required, although the maximum thickness could be as small as 50 μm, or as large as 50 mm, depending on various factors. Some indications of the regimes of h/H within which the Stoney equation is likely to be fairly accurate are given in Fig. 11.8.

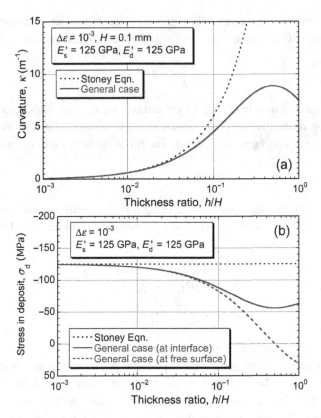

Fig. 11.8 Predicted dependence on thickness ratio of (a) curvature and (b) stress in deposit (coating), obtained using Eqns (11.16)–(11.18) and the Stoney equation (Eqn (11.23)).

11.2.3   An Educational 'Coating Mechanics' Package

Manipulation of the above set of equations is straightforward, but can be facilitated by use of resources within the *Coating Mechanics* package, one of the Teaching and Learning Packages (TLPs) within the DoITPoMS set of educational resources. The URL is www.doitpoms.ac.uk/tlplib/coating_mechanics/index.php. A screenshot is shown in Fig. 11.9 from the *Force and Moment Balances* page, which allows prediction of the curvature, and plotting of associated stress and strain distributions, for user-specified cases. There are also pages covering other aspects, such as the regimes of reliability of the Stoney equation.

11.2.4   Experimental Measurement of Curvature

There are several types of approach to measurement of curvature, ranging from optical (including interferometric) sensing (which often requires smooth surfaces) to mechanical actuation via a transducer. Also, changing curvatures can be monitored (e.g. during temperature change or during deposition of a coating). Such procedures can be very

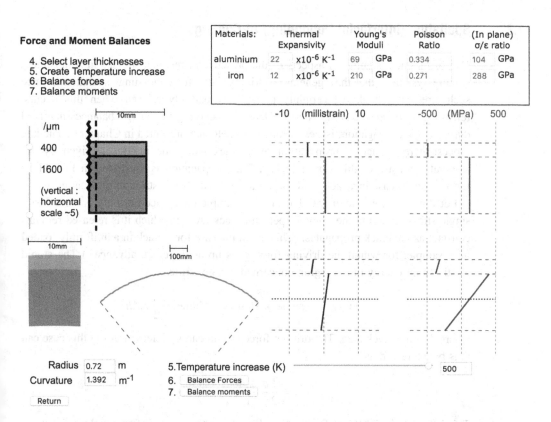

Fig. 11.9 Screenshot from the 'Force and Moment Balances' page in the DoITPoMS Teaching and Learning Package on Coating Mechanics, which is accessible at www.doitpoms.ac.uk/tlplib/coating_mechanics/misfit_force_balance.php.

informative regarding the development of stresses and strains in substrate–coating systems. Of course, it is also possible to measure strains directly, using techniques such as strain gauges, X-ray diffraction, photo-elasticity (with polarised light), digital image correlation and moiré interferometry. However, these all suffer from certain limitations, notably that in general they relate only to strains in a thin surface layer.

There are several alternative approaches to the measurement of curvature. Among the most accurate are optical (reflection and interferometric) techniques [6–9], although there are often requirements regarding the reflectivity of the surface. Other techniques include (contacting and non-contacting) profilometry [10,11]. Various approaches are possible for the deduction of stress profiles from curvature measurements, some based on progressive removal of material [12] and others relying on a model for stress generation during coating formation [13]. There have been investigations [13–15] based on carrying out continuous *in situ* monitoring of curvature changes. Correlation with direct measurements of strain [15,16] or with other information about stress generation during the process [17] has also been common. In general, curvature measurement is a powerful tool for investigation of residual stresses (in multilayer systems) and their origins.

11.3   Spallation (Interfacial Debonding) of Coatings

One of the most significant aspects of the presence of residual stresses in a substrate–coating system is that they generate a driving force for debonding (spallation), since such stresses will almost certainly be at least partially relaxed when this occurs, releasing stored elastic strain energy. Details concerning the energy balances involved during crack propagation, issues of mode mix, etc. are presented in Chapter 9. The key process here is that of propagation of a crack along the interface, driven by the associated release of this stored energy. This propagation is illustrated in Fig. 11.10 for a (Stoney) case in which there is just a single (uniform) stress in the coating.

For this situation, the energy balance is a simple one, since not only is there just a single stress level, but the volume (per unit crack area) in which it is released remains constant as the crack propagates. (This is not the case for a crack in a uniformly loaded bulk sample, for which the driving force goes up as the crack advances.) The stored elastic strain energy in the region concerned can be written

$$\text{energy} = \frac{1}{2}\,\text{stress} \times \text{strain} \times \text{volume} = \frac{1}{2}\sigma\varepsilon Ah$$

where A is the crack area. The driving force (strain energy release rate) in this case can thus be expressed as

$$G_\text{i} = \frac{1}{2}\frac{\sigma_\text{d}\varepsilon_\text{d}Ah}{A} = \frac{E_\text{d}\varepsilon_\text{d}^2 h}{2}$$

This is for a uniaxial stress. For the (much more common) case of an equal biaxial stress state, there are two stresses and the ratio of in-plane stress to in-plane strain is the biaxial

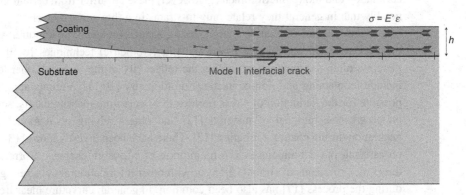

Stored elastic strain energy / unit area =

$$\frac{1}{2}\frac{\text{stress} \times \text{strain} \times \text{volume}}{\text{area}} = \frac{1}{2}\frac{\sigma\,\varepsilon\,(Ah)}{A} = \frac{E\,\varepsilon^2\,h}{2}$$

for biaxial case, this = $E'\varepsilon^2 h = G_\text{i}$ (strain energy release rate)

Coating

Substrate   Mode II interfacial crack

$\sigma = E'\varepsilon$

h

$\sigma = E'\varepsilon$

Fig. 11.10 Schematic depiction of the propagation of an interfacial crack (coating spallation) for a Stoney ($h \ll H$) case, driven by the release of elastic strain energy stored as a uniform residual stress in the coating.

modulus, E'_d. Assuming that both of these stresses can be fully relaxed while crack advance takes place, this equation has the form

$$G_i = 2\frac{E'_d\varepsilon_d^2 h}{2} = E'_d\varepsilon_d^2 h \tag{11.24}$$

Propagation will be energetically favoured if this driving force is equal to or greater than the (mode II – i.e. shearing mode) fracture energy of the interface, G_{ic}

$$E'_d\varepsilon_d^2 h \left(= \frac{\sigma_d^2 h}{E'_d} \right) \geq G_{ic} \tag{11.25}$$

This takes no account of any barrier to initiation of the crack. In many cases, however, there are likely to be relatively large defects present in the interface, so the above condition may well lead to spallation. It certainly means that the coating is (thermodynamically) unstable. Immediate implications are (unsurprisingly) that high stresses and brittle interfaces (low G_{ic}) make debonding more likely. Also clear (and widely observed) is that thicker coatings are more likely to debond than thinner ones.

These effects are of considerable practical significance. Spallation of coatings can be highly deleterious and this may be promoted by changes that can occur under service conditions. A good example is provided by thermal barrier coatings (TBCs), which are of central importance [18,19] in raising the maximum permissible temperature for the gas entering a turbine (and hence in improving the efficiency of aero-engines and power generation plants). These coatings, which are commonly made of yttria-stabilised zirconia (YSZ) and deposited either by plasma spraying or physical vapour deposition (PVD), need to be relatively thick in order to provide thermal protection for the substrate (usually a turbine blade or nozzle guide vane made of an Ni-based superalloy). Residual stresses are readily generated in these coatings, notably as a result of differential thermal contraction with the substrate. (Substrate expansivities are $\sim$15 microstrain K^{-1} and, while that of YSZ is relatively high for a ceramic at $\sim$10 microstrain K^{-1}, this still means that a misfit strain (equal biaxial) of $\sim$5 millistrain is created on cooling through a temperature drop of 1000 K – such drops are common, since the temperature of gas entering the turbine may well be around 1500°C or more.)

Of course, the magnitude of the associated strain energy release rate rises, not only with the thickness of the coating, but also with its stiffness, as indicated by Eqn (11.24). In the as-produced state, both PVD and plasma-sprayed TBCs have relatively low stiffness ($\sim$20–40 GPa) as a consequence of the presence of porosity and micro-cracks. This gives them good *strain tolerance*. However, after prolonged exposure to high temperatures under service conditions, they can undergo (diffusional) sintering that raises their stiffness considerably. Typical associated microstructural changes [20] are shown in Fig. 11.11 for a plasma-sprayed TBC, where it can be seen that sintering has promoted improved bonding between individual splats, raising both the conductivity and the stiffness. Such changes are reflected in the measured stiffness values [21] for these coatings, as a function of time and temperature, shown in Fig. 11.12. While these values are still well below the nominal stiffness of (fully dense) YSZ (over 200 GPa), it can be seen that substantial increases can occur within short periods, particularly for temperatures above about 1200°C.

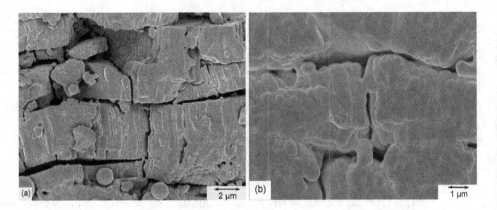

Fig. 11.11 SEM fracture surfaces [20], showing typical splat structures in plasma-sprayed YSZ coatings (a) as-sprayed and (b) after ten hours at 1400°C.
Reprinted from Surface and Coatings Technology, 203, S. Paul, A. Cipitria, S.A. Tsipas and T.W. Clyne, Sintering characteristics of plasma sprayed zirconia coatings containing different stabilisers, 1069–1074, © 2009, with permission from Elsevier.

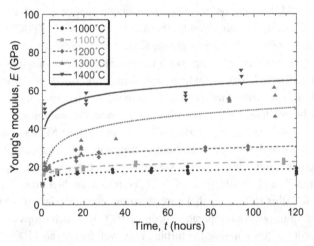

Fig. 11.12 Young's modulus data [21] obtained by cantilever bend testing of detached YSZ TBCs coats that had been subjected to various prior heat treatments.

Such changes often correlate quite closely with the likelihood of spallation [22,23]. An example outcome [23] is shown in Fig. 11.13, which indicates that, using the simple approach embodied in Eqn (11.25), observed conditions under which spallation was observed (in a range of cases) were consistent with a fairly constant value for the interfacial fracture energy. Those results relate to a case in which no changes were occurring in the microstructure of the interfacial region during the heat treatments (since the substrate was a ceramic), but it should be recognised that, for a superalloy substrate (often with a so-called 'bond coat') such changes [24,25] may occur (particularly

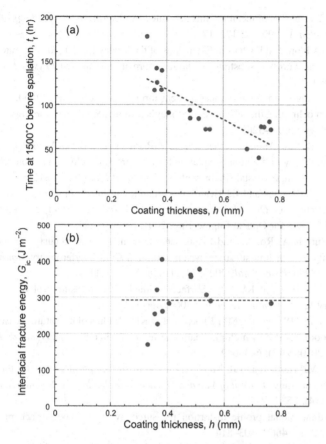

Fig. 11.13 Data [23] for spallation of YSZ coatings on alumina substrates, showing (a) lifetimes at 1500°C (with quenching to ambient temperature once per hour), plotted against coating thickness; and (b) inferred interfacial fracture energy for a subset of these, obtained using information about the changing stiffness of the coatings.

Reprinted from Acta Materialia, 61, M. Shinozaki and T.W. Clyne, A methodology, based on sintering-induced stiffening, for prediction of the spallation lifetime of plasma-sprayed coatings, 579–588, © 2013, with permission from Elsevier.

concerning the so-called thermally grown oxide (TGO) there), which could cause changes in the interfacial fracture energy. It is also worth noting that the mode mix (Section 9.2.1) may vary under different conditions. These are complex phenomena, but, as with most situations in which crack propagation needs to be analysed and understood, a fracture mechanics (energy-based) approach is usually the most fruitful and reliable one.

References

1. Clyne, TW, Residual stresses in surface coatings and their effects on interfacial debonding. *Key Engineering Materials* 1996; **116/117**: 307–330.

2. Stoney, GG, The tension of metallic films deposited by electrolysis. *Proceedings of the Royal Society of London A* 1909; **82**: 172–175.

3. Freund, LB, JA Floro and E Chason, Extensions of the Stoney formula for substrate curvature to configurations with thin substrates or large deformations. *Applied Physics Letters* 1987; **74**: 1987–1989.

4. Mezin, A, Coating internal stress measurement through the curvature method: a geometry-based criterion delimiting the relevance of Stoney's formula. *Surface & Coatings Technology* 2006; **200**(18–19): 5259–5267.

5. Janssen, GCAM, MM Abdalla, F van Keulen, BR Pujada and B van Venrooy, Celebrating the 100th anniversary of the Stoney equation for film stress: developments from polycrystalline steel strips to single crystal silicon wafers. *Thin Solid Films* 2009; **517**: 1858–1867.

6. Schmitz, TL, AD Davies and CJ Evans, Uncertainties in interferometric measurements of radius of curvature, in *Optical Manufacturing and Testing IV*, Stahl, HP, editor. SPIE: International Society for Optical Engineering, 2001, pp. 432–447.

7. Park, TS, S Suresh, AJ Rosakis and J Ryu, Measurement of full-field curvature and geometrical instability of thin film-substrate systems through CGS interferometry. *Journal of the Mechanics and Physics of Solids* 2003; **51**(11–12): 2191–2211.

8. Wang, J, P Shrotriya and KS Kim, Surface residual stress measurement using curvature interferometry. *Experimental Mechanics* 2006; **46**(1): 39–46.

9. Abdelsalam, DG, MS Shaalan, MM Eloker and D Kim, Radius of curvature measurement of spherical smooth surfaces by multiple-beam interferometry in reflection. *Optics and Lasers in Engineering* 2010; **48**(6): 643–649.

10. Thomas, ME, MP Hartnett and JE McKay, The use of surface profilometers for the measurement of wafer curvature. *Journal of Vacuum Science & Technology A: Vacuum Surfaces and Films* 1988; **6**(4): 2570–2571.

11. Lee, DH, 3-dimensional profile distortion measured by stylus type surface profilometer. *Measurement* 2013; **46**(1): 803–814.

12. Sglavo, VM, M Bonafini and A Prezzi, Procedure for residual stress profile determination by curvature measurements. *Mechanics of Materials* 2005; **37**(8): 887–898.

13. Gill, SC and TW Clyne, Investigation of residual stress generation during thermal spraying by continuous curvature measurement. *Thin Solid Films* 1994; **250**: 172–180.

14. Leusink, GJ, TGM Oosterlaken, G Janssen and S Radelaar, The evolution of growth stresses in chemical-vapor-deposited tungsten films studied by in-situ wafer curvature measurements. *Journal of Applied Physics* 1993; **74**(6): 3899–3910.

15. Gergaud, P, O Thomas and B Chenevier, Stresses arising from a solid state reaction between palladium films and Si(001) investigated by in situ combined X-ray diffraction and curvature measurements. *Journal of Applied Physics* 2003; **94**(3): 1584–1591.

16. Totemeier, TC and JK Wright, Residual stress determination in thermally sprayed coatings: a comparison of curvature models and X-ray techniques. *Surface and Coatings Technology* 2006; **200**(12–13): 3955–3962.

17. Dean, J, T Gu and TW Clyne, Evaluation of residual stress levels in plasma electrolytic oxidation coatings using a curvature method. *Surface & Coatings Technology* 2015; **269**: 47–53.

18. Padture, NP, M Gell and EH Jordan, Thermal barrier coatings for gas-turbine engine applications. *Science* 2002; **296**(5566): 280–284.

19. Clarke, DR, M Oechsner and NP Padture, Thermal-barrier coatings for more efficient gas-turbine engines. *MRS Bulletin* 2012; **37**(10): 891–898.

20. Paul, S, A Cipitria, SA Tsipas and TW Clyne, Sintering characteristics of plasma sprayed zirconia coatings containing different stabilisers. *Surface and Coatings Technology* 2009; **203**: 1069–1074.

21. Thompson, JA and TW Clyne, The effect of heat treatment on the stiffness of zirconia top coats in plasma-sprayed TBCs. *Acta Materialia* 2001; **49**(9): 1565–1575.

22. Théry, PY, M Poulain, M Dupeux and M Braccini, Spallation of two thermal barrier coating systems: experimental study of adhesion and energetic approach to lifetime during cyclic oxidation. *Journal of Materials Science* 2009; **44**(7): 1726–1733.

23. Shinozaki, M and TW Clyne, A methodology, based on sintering-induced stiffening, for prediction of the spallation lifetime of plasma sprayed coatings. *Acta Materialia* 2013; **61**(2): 579–588.

24. Nusier, SQ and GM Newaz, Growth of interfacial cracks in a TBC/superalloy system due to oxide volume induced internal pressure and thermal loading. *International Journal of Solids and Structures* 2000; **37**(15): 2151–2166.

25. Naumenko, D, V Shemet, L Singheiser and WJ Quadakkers, Failure mechanisms of thermal barrier coatings on MCrAlY-type bondcoats associated with the formation of the thermally grown oxide. *Journal of Materials Science* 2009; **44**: 1687–1703.

12 Highly Porous Materials as Composite Systems

Composites are essentially materials comprising two or more distinct constituents, integrated into a single entity. An important aspect of composite theory concerns the properties that the material exhibits, expressed in terms of those of the constituents and the architecture of the integration. A case of interest is that of a two-constituent system in which one of them is just a void – possibly a vacuum, although more commonly a gas phase. Of course, voids have properties that are substantially different from those of constituents in conventional composites. For example, the stiffness will be effectively zero and the conductivity will tend to be very low. In practice, many materials contain at least some porosity, with the potential to affect certain properties, but in most cases it would not be considered appropriate to classify them as composites. However, very high porosity levels (say, >~30–40%) can justify treatment as a separate type of (composite) material. Sometimes the term 'foam' is used in such cases, although the word does carry connotations that would not necessarily apply to all highly porous materials. In this chapter, some composite theory approaches are applied to such materials and information is provided about their 'microstructure' (pore architecture), production and potential benefits.

12.1 Types of Highly Porous Material and Their Production

Materials that might be described as foams are in widespread use, for applications ranging from domestic sponges to cores of sandwich panels. Many are based on polymeric materials, particularly rubbers, but there is also interest in metallic and ceramic foamed materials. There are also many natural materials that could be regarded as foams, including cork, coral, bone and sandstone. Other types of highly porous material have fibrous structures, and would in most cases not be described as foams. There are several authoritative texts [1–5] covering the basics of foams and 'lattice' or 'fibre network' structures. In this section, a brief overview is given of the main types of highly porous material and how they are produced.

12.1.1 Gas Evolution to Produce Foams

Many foams are produced via the evolution of gas within the material while the 'matrix' is in an easily deformable state. The gas is created by what are often termed 'blowing agents'. Depending on exactly how this gas evolution takes place, this tends to produce an assembly of 'closed cells' – i.e. the voids are occluded (isolated) – although this is far from universal. A lack of connectivity between neighbouring cells does have certain consequences for the properties of the foam – for example, they tend to be more or less impermeable to fluids. In fact, in terms of property optimisation, it is common to aim for a high pore content in combination with a relatively fine structure of closed cells, although this does depend on the properties being sought.

In the case of polymeric foams, blowing agents are classified as either chemical or physical, in terms of the gas evolution mechanism. Chemical agents [5] include isocyanate and water (for polyurethane rubber foams), azo-hydrazine (for thermoplastic and elastomeric foams) and sodium bicarbonate (for thermoplastic foams). The gases evolved are commonly nitrogen, carbon dioxide, ammonia, etc. Physical blowing agents are gases that do not take part in chemical reactions during the process. These include volatile hydrocarbons, such as propane and butane, chlorofluorocarbons (CFCs) and hydrochlorofluorocarbons (HCFCs), as well as simple gases such as nitrogen and carbon dioxide. Originally, vaporisation of organic chemicals that readily dissolved in the polymer, such as HCFCs, was commonly used. As the environmental problems that can arise with such gases became clearer, there was a move towards gases such as carbon dioxide. Of course, while these also have associated environmental issues, they are less severe than those that can arise with various organic compounds. The procedure employed with a physical blowing agent typically involves saturating the polymer with the (inert) gas concerned, under high pressure, heating it to around its glass transition temperature, reducing the external pressure so that the material becomes supersaturated (and cells start to nucleate and grow), and finally dropping the temperature sharply, so that cell growth is inhibited and a relatively fine-scale structure is produced. Further control over the structure is sometimes achieved via the addition of particulate fillers to the polymer.

As might be expected, both the porosity level and the cell size and uniformity can vary over a wide range, depending on various details of the processing conditions. There is often perceived to be an incentive to refine the cell size. What might be termed a micro-cellular structure would typically contain cells below 100 μm in size, whereas a nano-cellular foam would have a cell size around 1 μm or below. This is illustrated by the micrographs [6] in Fig. 12.1. These were both produced by the same type of route, but with different conditions, additives, etc. In both cases, the porosity level is of the order of 90%, which is a typical value for many such foams.

It is also possible to produce metallic foams via the use of blowing agents, and in fact this has been a very active area over recent decades [7–10]. For metals, these agents are all chemical, with various gas-generating reactions being exploited. Metals present a challenge in terms of the matrix being much more fluid in liquid form than a polymeric melt – the viscosity of most metals is similar to that of water. This creates problems, since it promotes rapid cell growth, leading to ruptured cell walls and coarse structures. In practice, it is common to introduce significant levels of ceramic particulate to stabilise the cell wall structure, although this does tend to cause embrittlement – see Section 12.2.2. Even when such measures are employed, metallic foams made via this route do tend to be coarse, although, as can be seen in Fig. 12.2, relatively fine

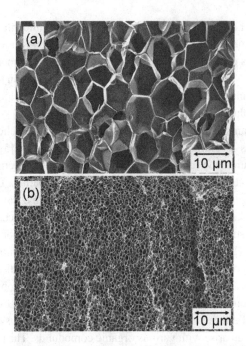

Fig. 12.1 SEM micrographs [6] of (a) micro-cellular and (b) nano-cellular polymer foams, produced using a CO_2 physical blowing agent.

Reprinted from European Polymer Journal, 73, Chimezie Okolieocha, Daniel Raps, Kalaivani Subramaniam and Volker Altstädt, Microcellular to nanocellular polymer foams: Progress (2004–2015) and future directions – A review, 500–519, © 2015, with permission from Elsevier.

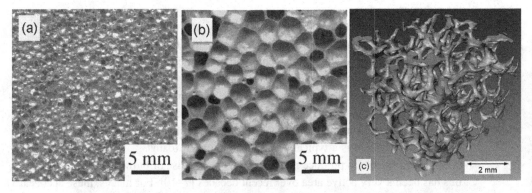

Fig. 12.2 SEM micrographs [8] of Al alloy foams produced using blowing agents of (a) $CaCO_3$ and (b) TiH_2, and (c) tomographic reconstruction [11] of the Duocel material (produced by ERG). For materials (a) and (b), the porosity level is about 80%, while for (c) it is about 90%.

(a,b) Reprinted from Composites Science and Technology, 63, V. Gergely, D.C. Curran and T.W. Clyne, The FOAMCARP process: foaming of aluminium MMCs by the chalkaluminium reaction in precursors, 2301–2310, © 2003, with permission from Elsevier. (c) Reprinted from Composites Science and Technology, 71, C.S. Dunleavy, J.A. Curran and T.W. Clyne, Plasma electrolytic oxidation of aluminium networks to form a metal-cored ceramic composite hybrid material, 908–915, © 2011, with permission from Elsevier.

structures (sub-millimetre cell size) can be produced using this route [8], if suitable procedures are employed. Also shown in Fig. 12.2 is a 3D tomographic reconstruction [11] of 'Duocel', a material produced via a casting (replication) route, which is often regarded as a foam (although no blowing agents are used and its structure is in some ways closer to that of a fibre network material than to a blown foam).

12.1.2   Syntactic Foams

This term is applied to composite materials produced by introducing a dispersion of hollow spheres (sometimes called 'micro-balloons' or 'cenospheres') into a matrix of metal, polymer or ceramic. (In this context, the word 'syntactic', which commonly refers to grammatical syntax, means 'put together'.) Commonly, the matrix, often in powder form initially, is blended with the spheres before being put into a mould and consolidated in some way. Alternatively, the matrix may be injected under pressure into the spaces between an assembly of spheres. Such procedures can allow good control over the size, content and distribution of the voids (and normally ensures that they are occluded), but is rather cumbersome and imposes limits on the porosity level (and cannot normally be used to produce very fine voids). An example microstructure is shown in Fig. 12.3, where it can be seen that cell sizes down to a few tens of microns can be produced. A variety of spheres are available for this purpose, including a number based on glass or on various types of polymer.

12.1.3   Fibre Network Materials

While syntactic foams, and those produced by gas evolution, usually contain cells that are at least predominantly closed, there are also many highly porous materials in which the cells are all open (interconnected), if indeed individual cells can be identified at all. For example, an important class of material comprises a bound assembly of slender

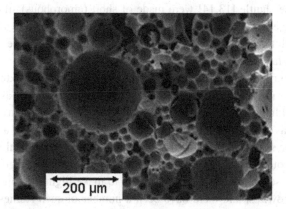

Fig. 12.3 SEM micrograph [12] of a syntactic foam comprising hollow thermoplastic spheres (micro-balloons) in an epoxy resin matrix. The porosity level is about 60%.

Fig. 12.4 SEM micrographs [15] of a space shuttle tile, showing (a) a near-surface region and (b) a higher magnification view in the interior. The porosity level is about 90%. Reprinted from Clyne, T. W. et al., 2006, Porous materials for thermal management under extreme conditions, with permission from The Royal Society.

members, such as fibres, wires, rods, ribbons, etc. Such materials are in some respects very well established and understood. They have been used since antiquity in various guises, including paper (papyrus), woven textiles and chain-mail. Although, like many 'foams', they commonly have porosity levels in the vicinity of 80–95%, the term would not normally be applied to them. (Probably by association with familiar sights, such as the froth on beer, emissions from fire extinguishers, etc., 'foams' are often taken to have closed cells.) However, in terms of engineering usage, fibre network materials are probably of even wider utility than foams, being of interest for applications as diverse as heat exchangers, filters, catalyst supports, biomedical scaffolds, ropes, etc.

Such materials can be produced in a variety of ways, depending partly on whether the members are metallic, ceramic or polymeric. Bonding is often provided in some way at fibre contact points, although in some cases natural entanglements might be sufficient to supply the necessary integrity. Fig. 12.4 shows that the thermal protection tiles used on the outside of the space shuttle [13,14] were made of short (amorphous) silica fibres, assembled randomly together with some inter-fibre bonding produced by the addition of colloidal silica, although there was also a concentration of a high-temperature cement in the regions near the surface.

A range of manufacturing techniques can be used, depending to some extent on the diameter of the fibres or rods. At the upper end of the range, (individually) assembling and bonding (e.g. welding) large-diameter (>~1 mm) metal wires or rods would normally be regarded as manufacturing a structure, rather than producing a material, although there are some scenarios, such as the production of cores for metal sandwich panels [2,15–17], that could be classified as falling into an intermediate category. At the other end of the length scale, it is, of course, possible to make highly porous material from nanofibres or nanowires, and there have been a number of such studies [18,19], although there are various practical difficulties and the mechanical properties of such materials, where reported, have often been poor – see also Chapter 14, which concerns

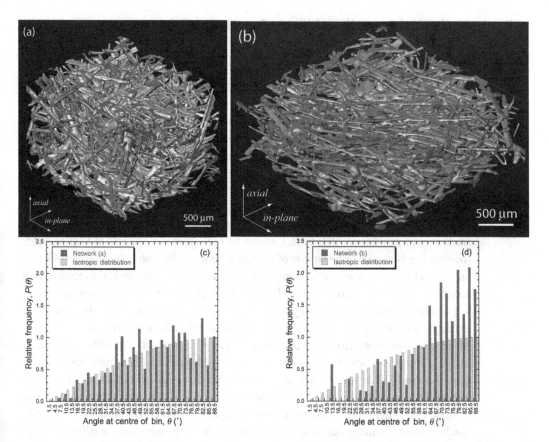

Fig. 12.5 (a) and (b) are tomographic reconstructions [20] from two fibre network materials (brazed assemblies of melt-spun stainless steel fibres), while (c) and (d) are corresponding fibre segment orientation distributions (in the form of frequency plots for the inclination to the axial direction). Reprinted from Clyne, T. W. et al., 2018. Copyright © 2006 WILEY-VCH Verlag GmbH & Co. KGaA, Weinheim.

'nano-composites' in general. In fact, most of the interest is focused on an intermediate to coarse range of fibre diameter, from a few microns up to about 1 mm.

A notable feature of fibre network materials is that it is often possible (and desirable) to introduce (controlled) anisotropy – i.e. a non-random distribution of the fibre orientations. Information about this type of characteristic can readily be obtained using X-ray computed tomography (CT). An illustration of this is provided in Fig. 12.5, which shows structural visualisations in the case of two metallic fibre network materials (produced with and without directional pressure while consolidation took place). Such pressure, or similar configurations such as the rolling processes used in making paper, tend to create a planar random orientation distribution, while in other cases there may be a tendency towards uniaxial alignment – for example, via drawing or extrusion processing. It may also be noted that, for some materials, such as textiles, a well-defined and regular fibre architecture is created. Something similar could be said about the materials that make up string, ropes, etc.

12.2   Mechanical Properties of Highly Porous Materials

12.2.1   Stiffness

It is helpful to treat foams and fibre network materials separately, in terms of both stiffness and other mechanical properties. In fact, the stiffness of a 'composite' comprising a dispersed set of spherical (or elongated) voids in a homogeneous matrix is readily predicted, using the treatments described in Chapter 6 (and employing a value of zero for the Young's modulus of the 'reinforcement'). While the shear lag model can in principle be applied to a 'short fibre' with an aspect ratio of unity (approximating to equiaxed particles), it is explained there that this is likely to be rather inaccurate, even using the modified shear lag approach – see Figs 6.6 and 6.7. The Eshelby method, on the other hand, reliably captures the effect of a dispersion of spheres, at least up to volume fractions of the order of 70%. In fact, as can be seen from the contents of Section 6.2.5, the outcome is close to a linear rule of mixtures, between the matrix stiffness at 0% voids and zero at 100%. For syntactic foams, which do have the structure of a dispersed set of spherical voids, this is expected to be fairly accurate (in the range 40–80% porosity level).

For foams with higher porosity levels than this – and the range 60–95% might be expected for those made using blowing agents – the cells are often some way from being spherical, being closer to cubes in some cases, or in reality probably approximating to a more isotropic shape, such as a tetrakaidecahedron (identified originally by Lord Kelvin [21] as the (tessellating) shape with the smallest surface area) or close alternatives, such as a pentagonal dodecahedron [22–26]. In fact, while the detailed geometry of such representations is important for prediction of some characteristics (such as liquid drainage during foam production), it is not so critical for stiffness. It has also been common to treat 'honeycombs' – i.e. elongated cells in which the cell structure lies only in one plane and a simple hexagonal shape tessellates (and hence can be used as the building block). As might be expected, there are some differences between open cell and closed cell cases. It should be recognised that, in the regime of very high porosity levels (when the cell walls have high length–thickness ratios), elastic deformation increasingly occurs by bending of them (rather than stretching or compression). In such cases, the exact cell shape – for example whether the walls exhibit curvature or have corrugations, may have an effect on the stiffness [27,28].

In fact, (high porosity level) open cell foams are sometimes treated as if they were composed of a set of struts, the bending of which dominates the elastic response (stiffness). In this regime, there is a close parallel with the behaviour of fibre network materials (in which case the slender structural elements would be fibre segments between bonds). This can allow large elastic deflections in response to low applied loads – i.e. give a low stiffness. It should be highlighted at this point that this may be a key reason why the material is suitable for the application concerned, since it facilitates shape change (including those from differential thermal contraction and other constraint-related effects) without large stresses being created.

Of course, another key attribute of highly porous materials is that they are often very light (have a low density). This alone can make them attractive for structural purposes.

An illustration of this can be seen in Fig. 1.4, which is a map of stiffness against density, with an area representing (polymer-based) foams being included. The significance of the low density does vary with the application (relevant merit index). An example for which the merit index is very sensitive to density ($M = E/\rho^3$) is provided by the objective of minimising the mass of a beam required to have a certain stiffness. However, examples of highly porous materials finding useful applications because they offer a combination of a certain stiffness with a low density are probably less common than those in which a low stiffness (making the material more strain tolerant) is important (provided it is combined with reasonably high strength – there are very few applications in which a low strength is at all attractive!). Of course, the exact stiffness characteristics will depend on the fibre architecture, particularly the fibre orientation distribution. Ropes, for example, do often need to have fairly high axial stiffness (and certainly a high axial strength), but it is the fact that the fine individual strands can bend without generating high stress that confers the flexibility.

Many fibre network materials are, of course, much more isotropic than ropes. A model for the stiffness of this type of material (referred to by them as an open cell foam) was developed by Gibson and Ashby [24]. This is based on three-point bending of beams under a normal load, assuming an orthogonal set of fibres – i.e. one-third of them being oriented along each of the three cube directions. They are taken to be simply supported cylindrical beams lying parallel or normal to the applied load. The Young's modulus predicted using this model can be expressed in terms of the length of fibre segments, L, and the fibre diameter, D:

$$E_n = \frac{3\pi E_f}{4\left(\frac{L}{D}\right)^4} \tag{12.1}$$

For this model, the fibre volume fraction, f, is fixed by the fibre segment aspect ratio (L/D), and so does not appear explicitly in Eqn (12.1). In fact, since f (and hence the relative density of the foam) is inversely proportional to $(L/D)^2$, it follows that the (relative) stiffness is proportional to the square of the (relative) density. This relationship is commonly quoted (for a range of types of foam), although it is not expected to apply to those with a structure that approximates to a set of dispersed spherical voids (such as a syntactic foam), for which the dependence is expected to be more or less linear.

A slightly more complex analytical model was developed by Markaki and Clyne [29,30], again based on bending of individual fibre segments (sections between joints). As with the Gibson and Ashby model, it is assumed that the deflections and strains exhibited by the network arise predominantly from this type of deformation. (This is expected to be acceptable provided the segments are relatively slender – i.e. $L/D > \sim 3$–4.) In the basic form of the model, an isotropic fibre orientation distribution is assumed: while this will not always be the case, it is often expected to be more realistic than the assumption of a cubic array. The loading situation is depicted schematically in Fig. 12.6, focusing on the elastic deformation exhibited by a single fibre segment, lying with its axis at an angle θ to the loading direction. When a uniaxial stress is applied, a force W acts on it in the loading direction, generating a bending moment. Application

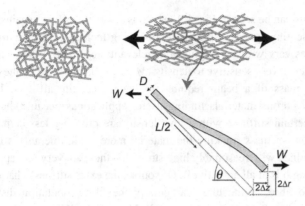

Fig. 12.6 Geometry of a (beam bending) model [31] for prediction of fibre network stiffness. Reprinted from Composites Science and Technology, 65, T.W. Clyne, A.E. Markaki and J.C. Tan, Mechanical and magnetic properties of metal fibre networks, with and without a polymeric matrix, 2492–2499, © 2005, with permission from Elsevier.

of standard cantilever bending mechanics [30,31] leads to expressions for the axial and transverse deflections, as a function of the applied stress σ (assuming uniform partitioning of the load to all fibres)

$$\Delta z = \frac{4\sigma L^3 \sin^2\theta}{3E_{tf}D^2} \tag{12.2}$$

$$\Delta r = \frac{4\sigma L^3 \sin\theta \cos\theta}{3E_{tf}D^2} \tag{12.3}$$

The macroscopic deflection in the loading direction, and hence the overall strain, can be obtained by summing the contributions from the deflections of individual fibres. Doing this in a rigorous manner is clearly complex, since the deflections exhibited by individual fibre segments will be influenced by the configuration of neighbouring segments. However, if these interactions are neglected, then the net strain can be obtained by simple integration. If the fibre orientation distribution is isotropic, so that it exhibits a $\sin\theta$ probability about any given axis, then the axial strain is given by

$$\varepsilon_z = \frac{\Delta Z}{Z} = \frac{\int_0^{\pi/2} \Delta z \sin\theta d\theta}{\int_0^{\pi/2} z \sin\theta d\theta} = \frac{\int_0^{\pi/2} \frac{4\sigma L^3 \sin^3\theta}{3E_{tf}D^2} d\theta}{\int_0^{\pi/2} \left(\frac{L}{2}\cos\theta\right)\sin\theta d\theta} \tag{12.4}$$

$$\therefore \varepsilon_z = \left(\frac{8\sigma}{3E_{tf}}\right)\left(\frac{L}{D}\right)^2 \frac{\int_0^{\pi/2} \sin^3\theta d\theta}{\int_0^{\pi/2} \cos\theta \sin\theta d\theta} = \left(\frac{8\sigma}{3E_{tf}}\right)\left(\frac{L}{D}\right)^2 \frac{2/3}{1/2} = \left(\frac{32\sigma}{9E_{tf}}\right)\left(\frac{L}{D}\right)^2$$

$$\tag{12.5}$$

so the Young's modulus of the network is given by

$$E_n = \frac{9E_f f}{32\left(\frac{L}{D}\right)^2} \tag{12.6}$$

It can be seen that this formulation resembles the Gibson and Ashby outcome (Eqn (12.1)), recognising that in their case f is inversely proportional to $(L/D)^2$.

Similarly, the transverse (radial) strain is given by

$$\varepsilon_r = \frac{\Delta R}{R} = \frac{\int_0^{\pi/2} \Delta r \sin^3\theta \, d\theta}{\int_0^{\pi/2} r \sin\theta \, d\theta} = \left(\frac{8\sigma}{3E_f f}\right)\left(\frac{L}{D}\right)^2 \frac{\int_0^{\pi/2} \sin^2\theta \cos\theta \, d\theta}{\int_0^{\pi/2} \sin^2\theta \, d\theta} \tag{12.7}$$

$$\therefore \varepsilon_r = \left(\frac{8\sigma}{3E_f f}\right)\left(\frac{L}{D}\right)^2 \frac{1/3}{\pi/4} = \left(\frac{32\sigma}{9\pi E_f f}\right)\left(\frac{L}{D}\right)^2 \tag{12.8}$$

The Poisson ratio, ν ($=\varepsilon_r/\varepsilon_z$) is thus predicted to have a constant value of $1/\pi$ (~ 0.32), independent of the fibre volume fraction and fibre segment aspect ratio. Since the material is isotropic, E_n and ν together fully define the elastic behaviour.

Predictions from these two models are shown in Fig. 12.7, together with experimental data [30] for materials made by sintering of stainless steel fibres, an example of which is shown in Fig. 12.5(a). The segment aspect ratio values for the experimental data were

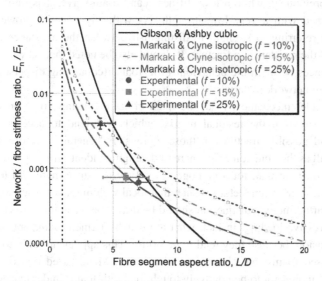

Fig. 12.7 Comparison [30–32] between theory and experiment for the relative stiffness of fibre networks, with different fibre volume fractions. For the Gibson and Ashby model (Eqn (12.1)), the fibre volume fraction is fixed by the segment aspect ratio, while the two are independent in the Markaki and Clyne model (Eqn (12.6)).

obtained from images of that type. The Gibson and Ashby model predicts a sharper fall in stiffness with increasing L/D than the Markaki and Clyne model. However, increases in L/D would often be accompanied by reductions in fibre content, f, making the effective plot for the Markaki and Clyne model somewhat steeper than the individual curves. In any event, it can be seen that both of these models are fairly reliable for network materials that are approximately isotropic. The Markaki and Clyne model can also be applied for known (anisotropic) fibre orientation distributions (with Eqns (12.5) and (12.7) being numerically integrated) and good agreement with experiment has also been observed in such cases [32].

It is clear from Fig. 12.7 that the stiffness of this type of material is sensitive to the fibre architecture. For a given porosity level, such as the 90% curve ($f = 10\%$), the stiffness drops by about two orders of magnitude on moving from a structure composed of short struts ($L/D \sim 3$) to one made up of long, slender elements ($L/D \sim 20$). This is very different from the situation with spherical voids, where the stiffness essentially depends only on the porosity level (although values above about 60–70% cannot readily be obtained with such a geometry). It is also worth noting, however, that, while the stiffness of a fibre network material is sensitive to the architecture (fibre volume fraction and aspect ratio of fibre segments, plus fibre orientation distribution, if it is not isotropic), it is independent of scale, so there is no (stiffness-related) advantage to be gained by refining the structure. This independence does not, however, extend to all mechanical properties of fibre network materials – see below.

12.2.2 Deformation and Fracture of Foams

As mentioned previously, while a low stiffness can be attractive, it is unlikely to be useful if combined with a very low strength. As with any material, however, it is important when referring to strength to distinguish between the resistance to plastic deformation and the resistance to fracture. It should also be noted that, as with stiffness, there are differences between the behaviour of a 'foam' (containing dispersed cells or voids) and a fibre network material.

For a foam under mechanical load, the main effect of having voids present is to concentrate the stress in the residual matrix, which will tend to accelerate plastic deformation (and possibly fracture) in those regions. In general, therefore, the yield stress tends to fall as the void content is increased, although identifying a yield stress for a foam is more difficult than for a homogeneous material, since there will tend to be regions where the (pore) architecture is such that local yielding takes place well before general yield. Furthermore, a distinction should be drawn between the behaviour under tensile or compressive loading. In fact, what are usually termed foams are commonly very weak in tension, because they tend to be brittle. This applies not only to ceramic- and polymer-based foams, but also to metallic foams. Metal-based materials are, of course, expected in general to be relatively tough, but with foams under tensile loading there is often a tendency for cracks to propagate through a neighbouring series of cell walls, with little energy being absorbed and what constitutes flaws of significant size being present from the start. (An effect of cell size is to be expected here, with a finer structure being beneficial.) In addition to the early onset of cracking, it is common for metallic foams to contain dispersed ceramic particulate, introduced in order to stabilise the cells during production (using blowing agents). This almost inevitably causes embrittlement of the walls [33,34], promoting their fracture. The upshot is that virtually all foams tend to be very weak (and brittle) in tension (although, in some loading situations, their low stiffness can allow them to 'escape' the load).

In practice, this has led to a tendency for foams to be used mainly under compressive loading. In this case, while individual cell walls may buckle (elastically or plastically) and/or fracture, the deformation tends to be spread over a large volume and there is at least the potential for substantial absorption of energy (particularly when considered per unit mass, rather than per unit of original volume). There have in fact been many experimental studies of the deformation and fracture characteristics of various kinds of foam, under both tensile and compressive loading. Example plots are shown in Fig. 12.8, which relates to several different types of Al-based foam (all with porosities in the range 80–95%). There are some differences between the plots for the different foams – for example, the variations seen in the compressive curve for the Alcan foam arise from the very coarse cell structure in that material – but in general they are all similar in character. In tension, they fracture at low levels of both stress and strain.

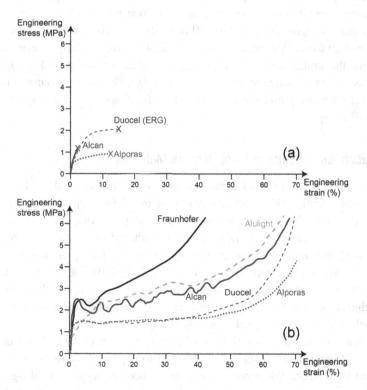

Fig. 12.8 Representative experimental stress–strain plots [35] for (a) tensile and (b) compressive loading of a variety of Al-based foams.

Genuine fracture toughness values are not really available, but these are certainly very brittle materials: if the area under a stress–strain plot is taken as an approximate indication of toughness, then, in comparison with conventional engineering materials, these would be worse than almost any other type of material. It may, however, be noted that the Duocel material (Fig. 12.2(c)) fares significantly better in tension than the other two foams: as observed in Section 12.1.1, it has a structure that is closer to that of a fibre network material than a 'blown' foam – see Section 12.2.3.

In compression, of course, the situation is different, since mode I fracture cannot occur. Failure in shear (mode II) would be possible, and indeed shear bands can form in this type of material, but this is likely to require higher stress (and strain) levels than are commonly imposed in such tests. What can be seen is that the yield stress is approximately the same in tension and compression (as expected), but that this is followed in compression by a regime approximating to a plateau, during which plasticity spreads throughout the sample and the cells progressively collapse. After this, the stress rises as the material starts to behave more like a conventional metal. This plateau regime offers potential for relatively large amounts of energy absorption, without the applied stress level becoming very high, and this has been the focus of extensive research [36–38], some of it oriented towards enhancement of the crashworthiness of vehicles. However, the scope for energy absorption (per unit weight) has in general proved to be relatively small (in comparison, for example, with that of polymer-based composites). In general, the industrial exploitation of metallic foams for such purposes has remained limited, although there is ongoing interest [39–42] in various types of hybrid structure incorporating (metallic) foam. Polymeric foams in general behave similarly to metallic foams, except that the strains tend to be even larger (and the stresses even lower), although there is less scope for permanent absorption of energy. They are, of course, widely used for general cushioning and redistribution of loads, if not for the absorption of large amounts of energy.

12.2.3 Deformation and Fracture of Fibre Network Materials

Fibre network materials do, of course, at least offer the potential for high strength (failure stress) levels. This is clearly exhibited by various types of rope and cable, although of course these are highly anisotropic materials, with negligible stiffness or strength if loaded transversely. In fact, the strength and toughness of more isotropic fibre network materials, with higher porosity levels than ropes, can also be relatively high (particularly when compared with foams). This is partly due to the fact that fibres are often themselves quite strong, but is also related to a capacity for the fibre architecture to change in response to an applied load, often such that individual fibres become reoriented so as to carry the load more effectively.

The approach illustrated in Fig. 12.6 – i.e. with the focus on beam bending within a random (isotropic) fibre network – can be used to explore the onset of yielding. The maximum local stress will occur at the fibre surface, at the position along the beam where the bending moment reaches a peak (i.e. adjacent to the joints) and is given [30,31] by

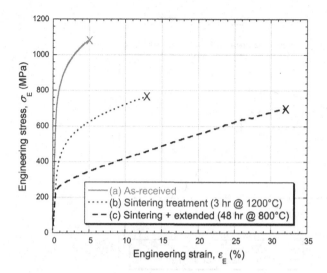

Fig. 12.9 Representative tensile stress–strain plots [43] for single fibres of austenitic stainless steel (a) as-received (melt-spun), (b) after three hours at 1200°C and (c) after treatment (b) followed by 48 hours at 800°C.

$$\sigma_{fm} = \frac{D}{2}\frac{M}{I} = \frac{D}{2}\frac{W\sin\theta(L/2)}{\left(\pi D^4/64\right)} \tag{12.9}$$

On setting this stress equal to the fibre yield stress, σ_{fY}, and substituting for W, an estimate can be obtained for the applied stress at the onset of yielding.[1]

$$\sigma_{nY} = \sigma_{fY}\frac{f}{8(L/D)} \tag{12.10}$$

In order to compare predictions such as that of Eqn (12.10) with actual behaviour, experimental stress–strain plots are needed both for individual fibres and for network materials. The single-fibre data should correspond to their condition in the network – for example, after the heat treatment involved in sintering, and any other treatment to which it has been exposed. The effect of such treatments will clearly depend on the fibres concerned, and their production method, but Fig. 12.9 shows that, for melt-spun AISI304 (austenitic) stainless steel fibres, such heat treatments can have a significant effect, reducing the yield stress and increasing the ductility. Of course, these fibres are relatively strong materials.

Stress–strain curves for fibre network materials tend to exhibit fairly complex characteristics. This is illustrated by Fig. 12.10, which shows tensile plots for two different porosity levels (85% and 90%), made from the same fibres as in Fig. 12.9, with

---

[1] The distinction between engineering and true stress and strain levels is not relevant here, since they are effectively the same at the onset of yield; however, the difference is potentially significant when considering the plastic work involved in failure of fibre network materials – i.e. their toughness – see Eqn (12.12).

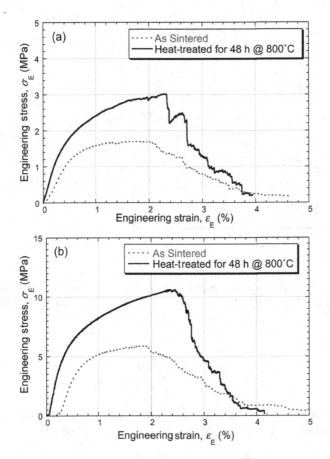

Fig. 12.10 Representative tensile stress–strain plots [43] for fibre network materials, with and without post-sintering heat treatments, having relative densities of (a) 10% and (b) 15%.

and without the same post-sintering heat treatment. The elastic strains exhibited by these materials (which are predicted to have a stiffness of ~200 MPa, i.e. ~E_f/1000 – see Fig. 12.7) are relatively small (~0.3–0.4%). This is not unexpected, since low overall strains can generate relatively high stresses in the vicinity of joints, where a small amount of plastic deformation is expected to facilitate reorientation of fibres, leading to relatively large macroscopic strains. However, this is a progressive process, which is unlikely to occur at all joints simultaneously, so the yield stress (~5–10 MPa) is not well defined – of course, it is often not very well defined for single fibres (Fig. 12.9) either.

The accumulation of damage at fibre joints and segments eventually leads to failure, with individual (joint or fibre segment) fracture events sometimes detectable as drops in stress (when viewed at high resolution). For both network densities, the heat treatment leads to an enhancement of strength (in the form of higher flow stress and peak stress, although the macroscopic ductility is not affected). The work of fracture (approximately the area under the plot) is also enhanced – see below. This is consistent with the associated increase in the ductility of the individual fibres (Fig. 12.9), which allows the overall stress to rise higher before joint damage (and rupture of individual fibres) becomes excessive. It is also noticeable that the increase in network density from 10% to 15% has a pronounced effect on the flow stress and peak stress. It may also be noted at this point that the 'toughness', taken in an indicative way to be proportional to the areas under these curves, is significantly higher than those of the foams in Fig. 12.8 – particularly when it is recognised that Duocel (Fig. 12.2(c)) has a structure that is closer to a fibre network than to the set of cells typical of a 'blown' foam.

The fibre deformation that occurs during tensile and compressive deformation of network materials has been investigated quite extensively [43–48]. In fact, during (tensile) fracture of network materials composed of (ductile and strong) metal fibres, considerable plastic work often accompanies fracture, although it is rare for extensive plastic deformation to occur homogeneously throughout the material and it tends to become concentrated in regions where rupture of fibres and joints initiate. Fig. 12.11 depicts the geometrical basis of the model of Tan and Clyne [43] for prediction of the fracture energy of a fibre network material that fractures so that there is a 'process zone', of width z_p, within which individual fibres plastically deform and rupture in a similar way to that in a single fibre tensile test.

A similar approach is adopted to that employed in treating a metal fibre-reinforced ceramic composite (Section 9.2.7). The work of fracture is obtained by summing the

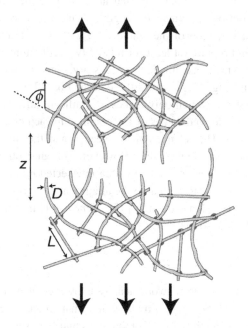

Fig. 12.11 Geometrical basis of a (fibre plastic work) model [43] for prediction of the toughness (fracture energy) of bonded (metallic) fibre network materials.
Reprinted from Tan, J. C. and Clyne, T. W., 2008. Copyright © 2008 WILEY-VCH Verlag GmbH & Co. KGaA, Weinheim.

energy needed to plastically deform and rupture all of the fibres in the section. The fracture energy can thus be expressed as

$$G_{net} = nU_s z_p \qquad (12.11)$$

where n is the number of fibres per unit sectional area and U_s is the single fibre work of deformation to fracture, in units of J m^{-1}. This is given by the area under the load–displacement plot, divided by the sample length, x, although the latter is changing during the test and so this should be written as

$$U_s = \int_0^{\delta_*} \frac{F d\delta}{x}$$

where F is the load, δ is the extension and δ_* is its value at failure. Conserving volume, x can be written as $(A_0 x_0/A)$, where A_0 is the original sectional area and A is its current value, so that U_s can be expressed as

$$U_s = \int_0^{\delta_*} \left(\frac{F}{A_0}\right)\frac{A}{x_0} d\delta = \int_0^{\delta_*} \sigma_E \frac{AA_0}{Ax} d\delta = A_0 \int_0^{\delta_*} \sigma_E \left(\frac{d\delta}{x}\right) = A_0 \int_0^{\varepsilon_{T*}} \sigma_E d\varepsilon_T \qquad (12.12)$$

The value of U_s is therefore given by the area under a plot of engineering stress against true strain, multiplied by the original sectional area. Ideally, the single-fibre experiments would be carried out on specimens of length z_p, but in practice the work done during uniform plastic deformation of the whole fibre will often dominate that associated with final necking and rupture (at least for relatively ductile fibres and high aspect ratio fibre segments), in which case this is not necessary, since U_s will then be independent of specimen length. Fig. 12.12 shows the data of Fig. 12.9, replotted as engineering stress against true strain (which only makes a significant difference at the higher strain levels), together with derived values of U_s (obtained in each case using an original fibre diameter of 70 μm). The relationship between n and the fibre volume fraction, f, depends on the network architecture. However, for an isotropic fibre orientation distribution, as noted in Section 9.2.7, the area intersected by any plane is twice the area intersected by a plane lying normal to the alignment direction of a set of parallel fibres occupying the same volume fraction [49]. Hence, n is half that for an aligned set of cylinders ($=f/(\pi D^2/4)$), so that

$$n = \frac{2f}{\pi D^2} \qquad (12.13)$$

The value of z_p is likely to depend on network density. However, it is difficult to predict this dependence, since it will be affected by how deformation and fracture take place at the fibre joints, and on the ductility of both joint areas and fibre segments. It is likely to be bounded by the segment length, L, at the lower end, and the fibre length at the upper end. Maximising its value will be helpful in optimising the toughness: using relatively long fibres and taking steps likely to improve the (fibre and joint) ductility are

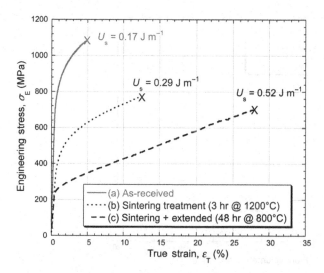

Fig. 12.12 Tensile test data of Fig. 12.9, replotted as engineering stress against true strain, with the derived values of the single-fibre work of deformation to fracture, U_s, indicated for each case.

expected to be beneficial. In practice, however, it is probably best to regard it essentially as a parameter to be estimated experimentally.

Uncertainty about the appropriate value of z_p makes examination of the validity of Eqn (12.11) a little difficult. A limited exploration has been undertaken [43] by estimating z_p for a few cases by visual inspection of samples after tensile failure. Using such values of z_p, a comparison between macroscopically measured fracture energy values (based on the area under the load–displacement plot) and predictions from Eqn (12.11) is shown in Fig. 12.13. Obviously this comparison is limited, but there is a measure of agreement. The extended heat treatment (raising the ductility) has a more beneficial effect than predicted, probably because it postpones failure at the joints. The main overall outcome to note is simply that these fracture energy values (up to a few tens of kJ m^{-2}) represent very respectable toughness levels – much higher than values typically obtained (under tensile loading) for non-fibrous metallic foams [7,33,50–53].

12.3   Permeation of Fluids through Highly Porous Materials

12.3.1   Permeability and Darcy's Law

The pressure drop (Δp) across a porous medium of thickness Δx, needed to generate a fluid flux through it of Q (m^3 m^{-2} s^{-1}), is dictated by Darcy's law

$$Q = \frac{\kappa \, \Delta p}{\eta \, \Delta x} \tag{12.14}$$

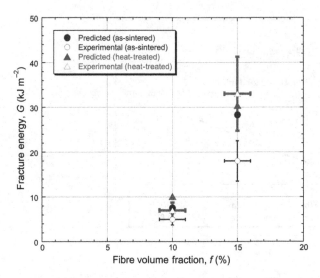

Fig. 12.13 Comparison between experimentally measured fracture energy values for network materials with two different fibre contents, with and without a post-sintering heat treatment, and predictions from Eqn (12.11), using fibre work of deformation to fracture values, U_s, obtained from single-fibre tests, and visually estimated values of the process zone width, w_p.

where η (Pa s) is the viscosity of the fluid and κ (m^2) is the specific permeability of the medium. Media with finer structures (pores) naturally tend to have lower permeabilities, leading to larger pressure drops and/or lower flow rates.

The permeability can be estimated in many cases using the (empirical) Carman–Kozeny equation [47,54]

$$\kappa = \frac{P^3}{\lambda(1-P)^2 S^2} \tag{12.15}$$

where P is the porosity level, S is the specific surface area (m^2 m^{-3}) and λ is a dimensionless constant (~5). Fine structures tend to have large values of S, leading to low permeabilities. The equation takes no account of the pore architecture (beyond its influence on S). The connectivity of the pores is in practice a key issue. A material with a high volume fraction of coarse, isolated pores would have a relatively large value of κ according to Eqn (12.15), but would in fact be impermeable – i.e. $\kappa = 0$. The equation is therefore useful for materials, such as those with a fibre network structure, having good inter-cell connectivity, but not for those with closed cells, or with limited connection between cells, which includes most materials classed as foams. Fig. 12.14 is a plot [55] of the dimensionless permeability for a number of fibre membrane materials, covering a very wide range of diameters. (These data are normalised for scale, with this set of results covering about 6 orders of magnitude in fibre radius, and hence about 12 orders of magnitude in permeability values.) It can be seen that the Carman–Kozeny formulation works well across this range, for materials of this type. Of course, many filters, as well as various types of scaffold, heat exchanger, etc., are based on fibrous structures –

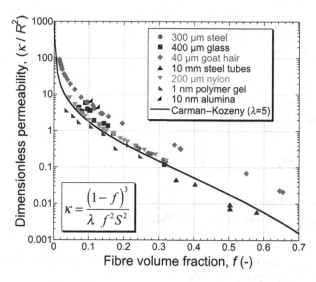

Fig. 12.14 Dimensionless permeabilities [55] of fibre-based membranes, together with a Carman–Kozeny master curve.

see below. For such applications, the fluid permeation characteristics are of central importance.

12.3.2   Filters and Scaffolds

There is interest in filtration of many species from fluids, ranging from coarse inorganic suspensions to small dissolved ions. This is illustrated in Fig. 12.15, which shows some of the terms commonly used to denote different types of filtration and corresponding length scales. Small molecules and ions cannot be removed via mechanical entrapment and require precipitation or osmotic separation. However, provided a suitably fine permeable medium is available, mechanical filtration can be effective for very small species (down to ~1 nm), although this may be at the cost of very low fluid flow rates. The majority of filters (and scaffolds for other purposes, such as support of catalysts or bioactive agents) are based on fibre assemblies of some sort.

An example of an important and demanding filtration requirement is that for automotive diesel engine exhaust, from which very fine (10–100 nm) carbon particles [57] – see Fig. 12.16 – must be removed. Such fine particles, which are likely to reach the lungs (and promote long-term pulmonary damage), are present in many combustion exhausts – candle flames are particularly rich in them – and the level in diesel exhaust is relatively high.

A diesel particulate filter (DPF) must therefore remove very fine particles from a high flux stream of hot gas, without significantly impeding the flow. A typical gas flow rate for a diesel car is ~250 kg h^{-1}, its exhaust pipe diameter is ~150 mm and a back-pressure above ~100 mbar across the filter would impair operation of the engine. The filter

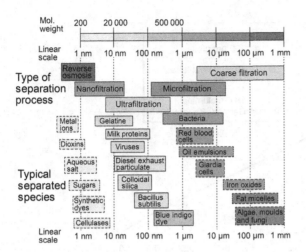

Fig. 12.15 Schematic depiction of the scales involved in various types of filtration process, and some of the corresponding target species [56].
Reprinted from Su, V., Terehov, M. and Clyne, T. W., 2012. Copyright © 2012 WILEY-VCH Verlag GmbH & Co. KGaA, Weinheim.

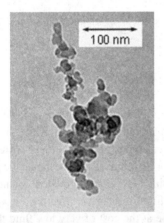

Fig. 12.16 TEM micrograph [57] showing diesel exhaust soot particles.
Reprinted from Ristovski, Z. D. et al., 2012. © 2011 The Authors. Respirology © 2011 Asian Pacific Society of Respirology.

surface area is raised by creating a honeycomb structure – see Fig. 12.17(a). This has a high surface area, so that the flux through the filter wall, Q, is kept relatively low ($\sim$0.1 m^3 m^{-2} s^{-1} – i.e. a gas velocity of $\sim$0.1 m s^{-1}). The wall thickness, Δx, is also kept low ($\sim$1 mm). However, this still presents a challenge in terms of the performance of the filter material. The Carman–Kozeny equation was used to create Fig. 12.17(b), which gives the pressure drop as a function of pore size, based on the filter material having the structure of a set of parallel cylinders, with a porosity level of 50%. (The exact pore architecture is not important for the purposes of this estimate.)

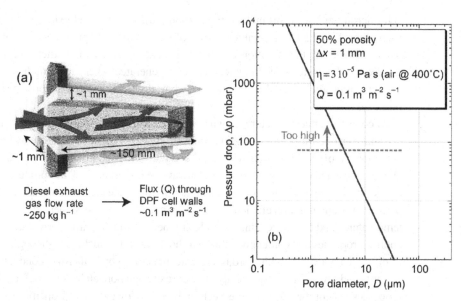

Fig. 12.17 (a) Schematic [58] of gas flow through a DPF and (b) predicted pressure drop across it, as a function of pore size.
Image © Johnson Matthey PLC.

Fig. 12.18 SEM micrograph [59] showing the pore architecture of a SiC-based DPF.
Reprinted from Applied Thermal Engineering, 62, Chahid Benaqqa, Moussa Gomina, Arnaud Beurotte, Michel Boussuge, Benoît Delattre, Karine Pajot, Edouard Pawlak and Fabiano Rodrigues, Morphology, physical, thermal and mechanical properties of the constitutive materials of diesel particulate filters, 599–606, © 2014, with permission from Elsevier.

This plot suggests that pore dimensions finer that a few microns will lead to unacceptably high pressure drops and in fact DPFs are normally made by lightly sintering together relatively coarse particulate [59], creating pores with dimensions of tens of microns – see Fig. 12.18. This appears problematic for filtration of substantially sub-micron particles. Fortunately, the carbon particles in diesel exhaust adhere well to each other (Fig. 12.16), so that a network of them builds up quickly in the pores. When 'clean', however, the filtration efficiency is relatively low, and filters are repeatedly returned to this state, since they need to be 'regenerated' every few hundred kilometres (by injecting fuel into the exhaust, so that the accumulated carbon particles are burnt off).

Of course, the microstructure shown in Fig. 12.18 does not incorporate fibres, although it does have an open structure with good connectivity between 'cells'. There may be scope for development of DPF materials with improved performance, possibly incorporating fibres, although cost and ease of processing will always be important for mass-produced components such as these. It should also be noted that performance requirements are not limited to permeability and filtration efficiency. Diesel particulate filters are subjected not only to high temperatures and thermal cycling, but also to thermal shock (particularly during 'regeneration'). Properties that confer good thermal shock resistance include high toughness, which is difficult to achieve with porous ceramic materials, high thermal conductivity, low stiffness and low thermal expansivity. The porosity and pore architecture affect all of these properties except the last one (see Section 10.1.2). However, the properties of the (fully dense) base materials are also relevant and these do influence the choice of material. In fact, most DPFs are currently made of either cordierite (created by mixing and sintering talc, kaolin and alumina powders) or SiC, which is more expensive, but has a higher thermal conductivity and better thermal stability.

12.4   Thermal Properties of Highly Porous Materials

12.4.1   Heat Transfer Mechanisms in Highly Porous Materials

A schematic representation is shown in Fig. 12.19 of the mechanisms by which heat transfer can occur in porous materials. Characteristics of conduction in solids, with electrons and/or phonons as carriers, and of radiative transmission, are described in standard sources [60–62], and also outlined in Section 10.4.2. Mechanisms of conduction in gases are also well established, with conductivity being dependent on the mean free path, λ, which in turn is a function of temperature and pressure [63] – see Fig. 12.19. At ambient temperature and pressure, λ has a value of about 60 nm, falling to 2 nm at 30 atmospheres and rising to 0.4 μm at 2000 K.

The gas conductivity within a pore is close to that in the free gas (e.g. $K_{air} \sim 0.025$ W m^{-1} K^{-1}), provided the dimensions of the pore are much larger than the mean free path ($L > \sim 10\lambda$). However, it falls below the free gas value if the pore structure is finer than this and can approach that due solely to gas molecules colliding with walls (Knudsen conduction) if L is less than λ. This would require an exceptionally fine pore structure (unless the gas pressure were low and the temperature high), but even moderately fine structures ($L < \sim 1$ μm) can lead to conductivities significantly below that of the free gas. Of course, all gas conductivities are normally much lower than those of solids, which usually fall between ~ 1 and ~ 300 W m^{-1} K^{-1}.

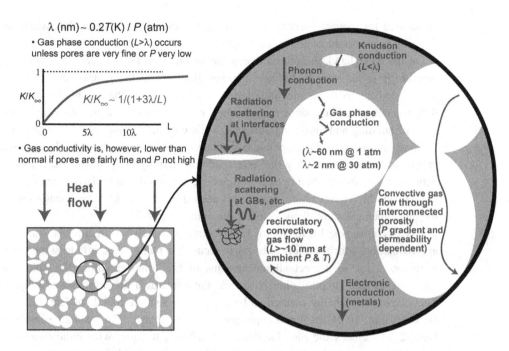

Fig. 12.19 Mechanisms of heat transfer in porous materials [15].

The scale of the pore structure (and the grain size) can also affect radiative heat transfer, since radiation is scattered by interfaces (and by grain boundaries). A fine-scale structure thus tends to result in increased scattering and reduced transmission. Another effect of scale relates to the possibility of convection within or between pores [64–66]. Convection through the inter-connected space in a fibre network material or a foam with high cell connectivity can be highly significant. In fact, such materials can be attractive as heat exchangers [67], using either gaseous or liquid fluids. It is also possible for heat transfer to occur via recirculatory convection within closed pores, although this tends to be significant only if they are very large [15].

12.4.2 Thermal Applications of Highly Porous Materials

Highly porous materials are often potentially attractive in terms of offering good thermal insulation. This applies to all types of foam and fibre network material, although of course polymers can be used only over a relatively narrow temperature range and even metals may be unsuitable at very high temperatures, due to incipient melting and/ or excessive oxidation. Furthermore, while radiative transport of heat within cells (open or closed) can in principle make a contribution to the overall heat transfer, in practice it is often negligible, at least over the range of temperature for which polymers and light metals, such as Al or Mg, can be used. At sufficiently high temperature, radiative exchange across pores can become significant, particularly in ceramic materials such as those used for thermal barrier coatings [68], but even in such cases this usually only happens at temperatures above about 1000°C.

On the other hand, the convective contribution often tends to increase as the temperature is raised, mainly because thermal gradients in the sample then tend to be higher, and this is sometimes interpreted as an increase in the radiative contribution. This is a complex issue, since radiative heat transfer depends on the absolute temperatures of emitting and absorbing surfaces, while conductive and convective heat fluxes are proportional only to temperature differences, at least in principle. In any event, experimental data concerning the 'effective' conductivity of open cell foams and fibre network materials are often open to various interpretations, particularly when they refer to relatively high temperature, and care is needed in relating the experimental set-up to the phenomena that are likely to affect the outcome [69–71].

It is, of course possible to employ highly simplified analytical treatments. An obvious approximation, which is likely to be fairly accurate in many cases, is to treat the pores as insulators – i.e. to neglect all mechanisms of heat transfer within the void space. Another possible simplification is to treat the voids as a dispersed set of ellipsoids. This allows the Eshelby method to be used – see Section 10.4.2. In fact, the curves shown in Fig. 10.24 for $\beta = 0$ correspond to such a case, with insulating inclusions. Fig. 12.20 includes the plot for dispersed spheres, together with indications of the approximate ranges observed experimentally for several different types of highly porous material [15]. Of course, close agreement is not expected here, partly because the Eshelby model is not really appropriate at very high inclusion contents. (While the equations concerned can be employed for contents above that corresponding to the maximum packing density of spheres, the correlation with the physical situation then becomes poor.) Also, various effects cannot be captured in this way, such as the

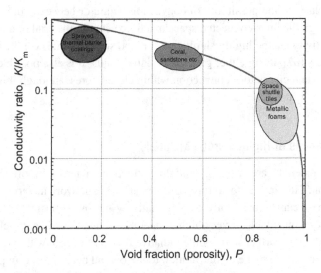

Fig. 12.20 Approximate observed ranges of thermal conductivity and porosity for a few different types of highly porous material, plotted together with predictions from the Eshelby model for dispersed insulating spheres [15].

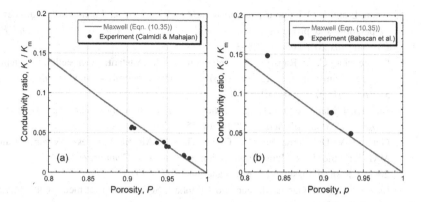

Fig. 12.21 Comparisons between experimental conductivity data for Al-based foams and predictions from the Maxwell equation (Eqn (10.35)), with the conductivity of the pores set to zero, for (a) Duocel [72] and (b) Alporas [73] foams. In both cases, the conductivity of the alloy was taken to be 120 W m^{-1} K^{-1}.

(ceramic) thermal barrier coatings having multiple interfaces normal to the heat flow direction. Nevertheless, it can be seen that, as a crude guideline, an approach like this could be helpful.

Expanding slightly on this, and noting that, for the case of spherical inclusions, the Eshelby treatment gives an outcome that is very close to that of the Maxwell equation (Eqn (10.35)), comparisons are presented in Fig. 12.21 between predictions from that equation and experimental data for two types of Al foam. The first of these comparisons [72], shown in Fig. 12.21(a), relates to Duocel (Fig. 12.2(c)), which is effectively a fibre network material, while the second (Fig. 12.21(b)) concerns a closed cell foam (Alporas). It can be seen that, in both cases, the agreement is quite good. Of course, metals are good thermal conductors and neglect of any heat flow through the pores is likely to be less reliable for ceramic and polymeric foams. If more accurate modelling is required, the best approach is probably to capture the pore architecture tomographically and then use numerical modelling to predict the heat flow – this could include convective and/or radiative contributions if necessary.

References

1. Ashby, MF, AG Evans, NA Fleck, LJ Gibson, JW Hutchinson and HNG Wadley, *Metal Foams: A Design Guide.* Butterworth-Heinemann, 2000.
2. Deshpande, VS, MF Ashby and NA Fleck, Foam topology bending versus stretching dominated architectures. *Acta Materialia.* 2001; **49**: 1035–1040.
3. Ashby, MF, Hybrids to fill holes in material property space. *Philosophical Magazine* 2005; **85**(26–27): 3235–3257.
4. Ashby, MF, The properties of foams and lattices. *Philosophical Transactions of the Royal Society A: Mathematical Physical and Engineering Sciences* 2006; **364**(1838): 15–30.

5. Aram, E and S Mehdipour-Ataei, A review on the micro- and nanoporous polymeric foams: preparation and properties. *International Journal of Polymeric Materials and Polymeric Biomaterials* 2016; **65**(7): 358–375.

6. Okolieocha, C, D Raps, K Subramaniam and V Altstadt, Microcellular to nanocellular polymer foams: progress (2004–2015) and future directions – a review. *European Polymer Journal* 2015; **73**: 500–519.

7. Banhart, J, Manufacture, characterisation and application of cellular metals and metal foams. *Progress in Materials Science* 2001; **46**: 559–632.

8. Gergely, V, DC Curran and TW Clyne, The FOAMCARP process: foaming of aluminium MMCS by the chalk–aluminium reaction in precursors. *Composites Science and Technology,* 2003; **63** (special issue on porous materials): 2301–2310.

9. Mukherjee, M, F Garcia-Moreno and J Banhart, Solidification of metal foams. *Acta Materialia* 2010; **58**(19): 6358–6370.

10. Heim, K, GS Vinod-Kumar, F Garcia-Moreno and J Banhart, Stability of various particle-stabilised aluminium alloys foams made by gas injection. *Journal of Materials Science* 2017; **52**(11): 6401–6414.

11. Dunleavy, CS, JA Curran and TW Clyne, Plasma electrolytic oxidation of aluminium networks to form a metal-cored ceramic composite hybrid Material. *Composites Science and Technology* 2011; **71**(6): 908–915.

12. Dando, KR, WM Cross, MJ Robinson and DR Salem, Production and characterization of epoxy syntactic foams highly loaded with thermoplastic microballoons. *Journal of Cellular Plastics* 2018; **54**(3): 499–514.

13. Daryabeigi, K, Thermal analysis and design optimization of multilayer insulation for reentry aerodynamic heating. *Journal of Spacecraft and Rockets* 2002; **39**(4): 509–514.

14. Nakamura, T and T Kai, Combined radiation-conduction analysis and experiment of ceramic insulation for reentry vehicles. *Journal of Thermophysics and Heat Transfer,* 2004; **18**(1): 24–29.

15. Clyne, TW, IO Golosnoy, JC Tan and AE Markaki, Porous materials for thermal management under extreme conditions. *Philosophical Transactions of the Royal Society A: Maths, Physics & Engineering Science* 2006; **364**(1838): 125–146.

16. Sypeck, DJ and HNG Wadley, Cellular metal truss core sandwich structures. *Advanced Engineering Materials* 2002; **4**(10): 759–764.

17. Lee, SC, F Barthelat, JW Hutchinson and HD Espinosa, Dynamic failure of metallic pyramidal truss core materials: experiments and modelling. *International Journal of Plasticity* 2006; **22**: 2118–2145.

18. Jung, SM, DJ Preston, HY Jung, ZT Deng, EN Wang and J Kong, Porous Cu nanowire aerosponges from one-step assembly and their applications in heat dissipation. *Advanced Materials* 2016; **28**(7): 1413–1419.

19. Li, H, YJ Li, LL Sun and XL Zhao, One-step, template-free electrochemical preparation of three-dimensional porous Au nanowire network and its enhanced activity toward methanol electrooxidation. *Electrochimica Acta* 2013; **108**: 74–78.

20. Tan, JC, JA Elliott and TW Clyne, Analysis of tomography images of bonded fibre networks to measure distributions of fibre segment length and fibre orientation. *Advanced Engineering Materials* 2006; **8**(6): 495–500.

21. Thomson, W, On the division of space with minimum partitional area. *Philosophical Magazine and Journal of Science* 1887; **24**: 503–514.

22. White, PL and LH Van Vlack, A comparison of two- and three-dimensional size distributions in a cellular material. *Metallography* 1970; **3**: 241–258.

23. Weaire, D and R Phelan, A counter-example to Kelvin's conjecture on minimal surfaces. *Philosophical Magazine Letters* 1994; **69**: 107–110.

24. Gibson, LJ and MF Ashby, *Cellular Solids: Structure and Properties*, 2nd edition. Cambridge University Press, 1997.

25. Daxner, T, RD Bitsche and HJ Bohm, Space-filling polyhedra as mechanical models for solidified dry foams. *Materials Transactions* 2006; **47**(9): 2213–2218.

26. Hedayati, R, M Sadighi, M Mohammadi-Aghdam and H Hosseini-Toudeshky, Comparison of elastic properties of open-cell metallic biomaterials with different unit cell types. *Journal of Biomedical Materials Research Part B: Applied Biomaterials* 2018; **106**(1): 386–398.

27. Andrews, E, W Sanders and LJ Gibson, Compressive and tensile behaviour of aluminum foams. *Materials Science and Engineering A: Structural Materials, Properties, Microstructure and Processing* 1999; **270**(2): 113–124.

28. Pardo-Alonso, S, E Solorzano, J Vicente, L Brabant, ML Dierick, I Manke, A Hilger, E Laguna and MA Rodriguez-Perez, mu CT-based analysis of the solid phase in foams: cell wall corrugation and other microscopic features. *Microscopy and Microanalysis* 2015; **21**(5): 1361–1371.

29. Markaki, AE and TW Clyne, Magneto-mechanical stimulation of bone growth in a bonded array of ferromagnetic fibres. *Biomaterials* 2004; **25**(19): 4805–4815.

30. Markaki, AE and TW Clyne, Magneto-mechanical actuation of bonded ferromagnetic fibre arrays. *Acta Materialia* 2005; **53**(3): 877–889.

31. Clyne, TW, AE Markaki and JC Tan, Mechanical and magnetic properties of metal fibre networks, with and without a polymeric matrix. *Composites Science and Technology* 2005; **65**(15–16): 2492–2499.

32. Clyne, TW, AE Markaki and J Dean, Mechanical properties of metallic fiber network materials, in *Comprehensive Composite Materials II.* Clyne TW, editor. Elsevier, 2018, pp. 173–187.

33. Markaki, AE and TW Clyne, The effect of cell wall microstructure on the deformation and fracture of aluminium-based foams. *Acta Materialia* 2001; **49**(9): 1677–1686.

34. Islam, MA, MA Kader, PJ Hazell, AD Brown, M Saadatfar, MZ Quadir and JP Escobedo, Investigation of microstructural and mechanical properties of cell walls of closed-cell aluminium alloy foams. *Materials Science and Engineering A: Structural Materials, Properties, Microstructure and Processing* 2016; **666**: 245–256.

35. Andrews, E, W Sanders and LJ Gibson, Compressive and tensile behaviour of aluminium foams. *Materials Science and Engineering A: Structural Materials, Properties, Microstructure and Processing* 1999; **270**: 113–124.

36. Mukai, T, H Kanahashi, T Miyoshi, M Mabuchi, TG Nieh and K Higashi, Experimental study of energy absorption in a close-celled aluminum foam under dynamic loading. *Scripta Materialia* 1999; **40**(8): 921–927.

37. Kanahashi, H, T Mukai, Y Yamada, K Shimojima, M Mabuchi, T Aizawa and K Higashi, Improvement of crashworthiness in ultra lightweight metallic foam by heat-treatment for microstructural modification of base material. *Materials Transactions* 2001; **42**(10): 2087–2092.

38. Zhang, X and H Zhang, Optimal design of functionally graded foam material under impact loading. *International Journal of Mechanical Sciences* 2013; **68**: 199–211.

39. Babbage, JM and PK Mallick, Static axial crush performance of unfilled and foam-filled aluminum-composite hybrid tubes. *Composite Structures* 2005; **70**(2): 177–184.

40. Guden, M, S Yuksel, A Tasdemirci and M Tanoglu, Effect of aluminum closed-cell foam filling on the quasi-static axial crush performance of glass fiber reinforced polyester composite and aluminum/composite hybrid tubes. *Composite Structures* 2007; **81**(4): 480–490.

41. Sun, GY, GY Li, SJ Hou, SW Zhou, W Li and Q Li, Crashworthiness design for functionally graded foam-filled thin-walled structures. *Materials Science and Engineering A: Structural Materials, Properties, Microstructure and Processing* 2010; **527**(7–8): 1911–1919.

42. Duarte, I, M Vesenjak, L Krstulovic-Opara and ZR Ren, Crush performance of multifunctional hybrid foams based on an aluminium alloy open-cell foam skeleton. *Polymer Testing* 2018; **67**: 246–256.

43. Tan, JC and TW Clyne, Ferrous fibre network materials for noise reduction in gas turbine aeroengines, part II: thermomechanical stability. *Advanced Engineering Materials* 2008; **10**: 201–209.

44. Zhou, W, Y Tang, MQ Pan, XL Wei and JH Xiang, Experimental investigation on uniaxial tensile properties of high-porosity metal fiber sintered sheet. *Materials Science and Engineering A: Structural Materials, Properties, Microstructure and Processing* 2009; **525**(1–2): 133–137.

45. Zhou, W, Y Tang, B Liu, R Song, LL Jiang, KS Hui, KN Hui and HM Yao, Compressive properties of porous metal fiber sintered sheet produced by solid-state sintering process. *Materials & Design* 2012; **35**: 414–418.

46. Huang, X, QH Wang, W Zhou and JR Li, A simple fracture energy prediction method for fiber network based on its morphological features extracted by X-ray tomography. *Materials Science and Engineering A: Structural Materials, Properties, Microstructure and Processing* 2013; **585**: 297–303.

47. Varley, MC, S Neelakantan, TW Clyne, J Dean, RA Brooks and AE Markaki, Cell structure, stiffness and permeability of freeze-dried collagen scaffolds in dry and hydrated states. *Acta Biomaterialia* 2016; **33**: 166–175.

48. Zhao, TF, CQ Chen and ZC Deng, Elastoplastic properties of transversely isotropic sintered metal fiber sheets. *Materials Science and Engineering A: Structural Materials, Properties, Microstructure, and Processing* 2016; **662**: 308–319.

49. Underwood, EE, *Quantitative Stereology*. Addison-Wesley Publishing Company, 1970.

50. Gibson, LJ, Mechanical behaviour of metallic foams. *Annual Reviews in Materials Science* 2000; **30**: 191–227.

51. Olurin, OB, NA Fleck and MF Ashby, Deformation and fracture of aluminium foams. *Materials Science and Engineering A: Structural Materials, Properties, Microstructure, and Processing* 2000; **291**(1–2): 136–146.

52. Motz, C and R Pippan, Fracture behaviour and fracture toughness of ductile closed-cell metallic foams. *Acta Materialia* 2002; **50**(8): 2013–2033.

53. Sugimura, Y, J Meyer, MY He, H BartSmith, J Grenstedt and AG Evans, On the mechanical performance of closed cell Al alloy foams. *Acta Materialia* 1997; **45**(12): 5245–5259.

54. Valdes-Parada, FJ, JA Ochoa-Tapia and J Alvarez-Ramirez, Validity of the permeability Carman–Kozeny equation: a volume averaging approach. *Physica A: Statistical Mechanics and its Applications* 2009; **388**: 789–798.

55. Jackson, GW and DF James, The permeability of fibrous-porous media. *Canadian Journal of Chemical Engineering* 1986; **64**: 364–374.

56. Su, V, M Terehov and TW Clyne, Filtration performance of membranes produced using nanoscale alumina fibers (NAF). *Advanced Engineering Materials* 2012; **14**(12): 1088–1096.

57. Ristovski, ZD, B Miljevic, NC Surawski, L Morawska, KM Fong, F Goh and IA Yang, Respiratory health effects of diesel particulate matter. *Respirology* 2012; **17**: 201–212.

58. Twigg, MV and PR Phillips, Cleaning the air we breathe: controlling diesel particulate emissions from passenger cars. *Platinum Metals Review* 2009; **53**(1): 27–34.

59. Benaqqa, C, M Gomina, A Beurotte, M Boussuge, B Delattre, K Pajot, E Pawlak and F Rodrigues, Morphology, physical, thermal and mechanical properties of the constitutive materials of diesel particulate filters. *Applied Thermal Engineering* 2014; **62**(2): 599–606.

60. Siegel, R and JR Howell, *Thermal Radiation Heat Transfer.* McGraw-Hill, 1972.

61. Berman, R, *Thermal Conduction in Solids.* Clarendon, 1976.

62. Klemens, PG and RK Williams, Thermal conductivity of metals and alloys. *International Metals Reviews* 1986; **31**: 197–215.

63. Loeb, LB, *The Kinetic Theory of Gases.* McGraw-Hill, 1934.

64. Lu, TJ, HA Stone and MF Ashby, Heat transfer in open-cell metal foams. *Acta Materialia* 1998; **46**(10): 3619–3635.

65. Lu, TJ, Heat transfer efficiency of metal honeycombs. *International Journal of Heat and Mass Transfer* 1999; **42**: 2031–2040.

66. Zhao, CY, T Kim, TJ Lu and HP Hodson, Thermal transport in high porosity cellular metal foams. *Journal of Thermophysics and Heat Transfer* 2004; **18**(3): 309–317.

67. Golosnoy, IO, A Cockburn and TW Clyne, Optimisation of metallic fiber network materials for compact heat exchangers. *Advanced Engineering Materials* 2008; **10**(3): 210–218.

68. Golosnoy, IO, A Cipitria and TW Clyne, Heat transfer through plasma-sprayed thermal barrier coatings in gas turbines: a review of recent work. *Journal of Thermal Spray Technology* 2009; **18**: 809–821.

69. Ranut, P, On the effective thermal conductivity of aluminum metal foams: review and improvement of the available empirical and analytical models. *Applied Thermal Engineering* 2016; **101**: 496–524.

70. Zhao, CY, Review on thermal transport in high porosity cellular metal foams with open cells. *International Journal of Heat and Mass Transfer* 2012; **55**(13–14): 3618–3632.

71. Veyhl, C, T Fiedler, O Andersen, J Meinert, T Bernthaler, IV Belova and GE Murch, On the thermal conductivity of sintered metallic fibre structures. *International Journal of Heat and Mass Transfer* 2012; **55**(9–10): 2440–2448.

72. Calmidi, VV and RL Mahajan, Effective thermal conductivity of high porosity fibrous metal foams. *Journal of Heat Transfer ASME* 1999; **121**(2): 466–471.

73. Babcsan, N, I Meszaros and N Hegman, Thermal and electrical conductivity measurements on aluminum foams. *Materialwissenschaft und Werkstofftechnik* 2003; **34**(4): 391–394.

13   Bio-Composites and Recycling

Composite materials, or at least materials that could be regarded as composites, are widespread in nature. This is, of course, a reflection of the many gains in 'efficiency' that can be made by integration of two or more constituents. Moreover, the development of artificial composite materials, for mechanical and/or other purposes, has benefited considerably from insights gained by examining bio-composites, and by their direct utilisation. The kingdoms of both plants (wood, grasses, straw, etc.) and animals (bone, skin, teeth, marine shells, corals, etc.) offer many examples of highly successful materials that are essentially composites. Their importance relates, not only to lessons about structure–property relationships, but also to the issue of degradation and recycling. While the 'rotting' of wood is often regarded as its Achilles' heel, viable recycling strategies are increasingly required for all materials (and manufactured composites are often perceived as being unsatisfactory in this respect). It is clearly not appropriate in a book of this type to provide great detail about natural materials, or indeed about recycling, but a few of the main principles and issues involved are briefly summarised here.

13.1   Biomaterials as Composite Systems

In order to survive and grow, living organisms need to produce matter, using whatever raw materials (organic and inorganic) are accessible to them. They do this in a vast and complex variety of ways, although, apart from very simple species such as viruses, it always involves the (re)production of cells of various types. Different types of cell have different functions, but many of them must meet engineering challenges of a similar type to those faced when making artificial materials, with common requirements including high stiffness, strength and toughness, in combination with low density, good environmental stability, etc. Of course, the 'design' procedures involved in development of these 'materials' are (as for those controlling evolution of the complete organism) those of natural selection. This is in many ways a very inefficient process, but, given sufficient time, it can certainly lead to highly optimised 'products', although it should always be borne in mind that they are optimal for the particular requirements and environment of the organism concerned.

Certain common features can be identified for at least most natural composite materials. One of the most striking is that they often exhibit hierarchical structures over length scales that cover several orders of magnitude. Of course, manufactured materials can exhibit structures that could be regarded in this way. For example, metals have an atomic-scale (crystallographic) structure, but are also organised into assemblies of grains, perhaps with precipitates or other second phases present. However, natural materials commonly exhibit much more complex hierarchical structures than this, with organisational features on perhaps four or five different length scales. Such structures are, of course, related to how the material is produced, although they are also linked to the complexity of the basic molecular building blocks, which tends to be high for many natural materials.

An example is shown in Fig. 13.1, which is a compilation of images [1] relating to the structure of spruce. The macroscopic building blocks are hollow tubes ('wood cells'). These are efficient in terms of being lightweight and having good bending stiffness, but also serve as conduits for the transport of sap (water containing nutrients, etc.). In fact, even this picture is something of an over-simplification, since the walls of wood cells actually have complex multilayer structures and individual polymeric cellulose molecules are assembled into both amorphous and crystalline arrangements. Some of this complexity is represented [2] in Fig. 13.2. Nevertheless, wood is basically a uniaxial (long) fibre composite, essentially made up of cellulose fibrils in a lignin matrix. Its most striking characteristics, such as its anisotropy and its relatively high toughness, can be directly related to this basic structure, using the approaches outlined in this book.

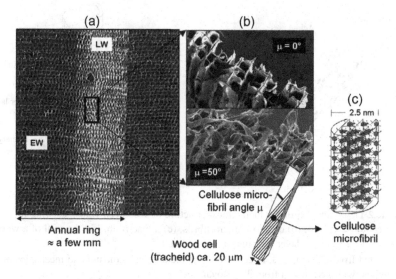

Fig. 13.1 Images [1] illustrating the hierarchical structure of spruce wood, comprising (a) an optical cross-section showing earlywood (EW) and latewood (LW) in an annual ring; (b) SEM micrographs of fracture surfaces of cells with two different micro-fibril angles; and (c) perspective view of the crystal structure of the cellulose molecule.

Reprinted from Progress in Materials Science, 52, Peter Fratzl and Richard Weinkamer, Nature's hierarchical materials, 1263–1334, © 2007, with permission from Elsevier.

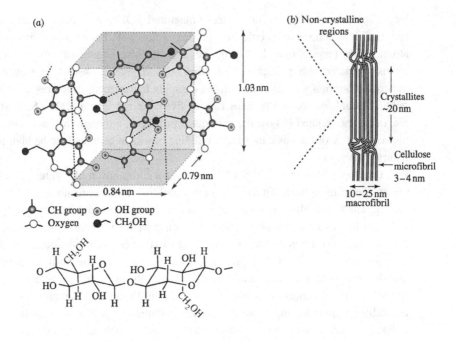

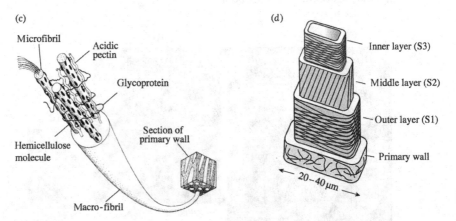

Fig. 13.2 Depictions [2] of the hierarchical structure of wood cell walls, showing (a) a unit cell of a cellulose crystal, (b) cellulose micro-fibrils, (c) a macro-fibril in the wall of a wood cell and (d) a typical multilayer structure of a wood cell.

Reprinted from Gibson, L. J. et al., 2012, The hierarchical structure and mechanics of plant materials, with permission from The Royal Society.

13.2   Cellulose-Based Fibres and Composites

13.2.1   Overview of the Growth and Structure of Wood

While all types of wood have certain features in common, their detailed structures cover an enormous range, with over 25 000 different species of trees being known [3]. A distinction is commonly drawn between *softwoods* and *hardwoods*, although the basis for this does not actually relate to hardness, or indeed to any mechanical characteristics. It concerns the type of seed they produce. In general, hardwoods come from deciduous trees ('angiosperms') that lose their leaves annually, while softwoods are from conifers ('gymnosperms') that remain evergreen. Hardwoods usually grow more slowly and tend to be more dense (and therefore 'hard'), although this is not an entirely reliable guide – for example, balsa, which is one of the lightest and softest woods, is actually classed as a hardwood. Nevertheless, there are some clear tendencies in the characteristics of hardwoods and softwoods. Hardwoods tend not only to be denser, but also to be darker in colour, to contain less sap, to grow more slowly (have finer growth rings), to be more resistant to rotting (and other types of biodegradation) and to be more expensive. Detailed information about the classification and characterisation of softwoods and hardwoods is available in the literature [4]. Issues such as the micro-fibril angle (Fig. 13.1) and its influence on properties are often of interest [5].

It may also be noted that the cell structure of hardwoods tends to be more complex than that of softwoods. The latter usually comprise axial cells (tracheids), together with a small proportion of radially oriented cells (parenchyma). Hardwoods often have more complex arrangements, containing constituents that are often termed 'vessels' and others called 'fibres' – it should be recognised that the terminology associated with wood can sometimes be variable and confusing.

Freshly felled wood contains a lot of water, since most of the cells, particularly in the outer layers (sapwood), are full of it. For most purposes, the wood is dried – often termed 'seasoning' – before being used. This may be allowed to occur naturally over a period of months or years, or it can be accelerated artificially. The target moisture level in the wood is usually around 10 wt%. The main factors relevant to this can be seen in Fig. 13.3, which shows, for a typical wood, how the equilibrium moisture content tends to vary with the relative humidity of the environment. The 'fibre saturation point' (~30%) corresponds to the cellulose in the cell walls being fully hydrated (but with no free water in the cells). Higher levels, which can arise if the wood is actually in contact with water (as opposed to humid air), such that there is free water in the cells, renders the wood vulnerable to fungal attack. With levels in the approximate range 15–30%, mould attack can occur. Furthermore, even if there is no biological degradation, changes in moisture content can led to dimensional changes – i.e. distortions and perhaps residual stresses. These arise because (anisotropic) length increases take place when cellulose micro-fibrils absorb water. The moisture content can also affect mechanical properties – see below.

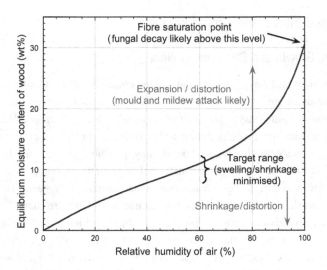

Fig. 13.3 Approximate dependence of the equilibrium moisture content of a typical (soft)wood on the relative humidity of surrounding air, with indications of how this relates to various types of degradation.

13.2.2   Elastic Constants of Wood

Both the elastic constants and the strength properties of wood tend to be quite highly anisotropic. The main effects are a consequence of wood being essentially a uniaxial fibre composite, with the cellulose micro-fibrils in the cell walls largely aligned along the axis of the trunk (i.e. parallel to the 'grain'). However, as outlined above, the structure is more complex than this and, in particular, it is not a transversely isotropic material (as a uniaxial fibre composite would be). The trunk as a whole exhibits cylindrical polar symmetry (leaving aside the inhomogeneities associated with the growth rings). It is thus an *orthotropic* material. However, the anisotropy of a particular piece of wood (a 'plank') depends on how it is cut from the trunk and may be rather complex. Nevertheless, it is common to identify axial (z), radial (r) and hoop (θ) directions and attempt to measure properties relative to them.

Some typical outcomes are presented in Table 13.1, which lists measured [3] elastic constants[1] for typical softwoods and hardwoods. In general, hardwoods are appreciably denser than softwoods, largely because they tend to contain less air (or sap prior to felling and 'seasoning'). However, in hardwoods, most of what is space in softwoods is effectively filled with 'matrix' (lignin), rather than cellulose, so the axial stiffness (E_z) of hardwoods is often fairly similar to that of softwoods (of the order of 10–15 GPa). Also, for both types of wood, the $\nu_{\theta z}$ and ν_{rz} Poisson ratios (i.e. the strain ratios when loaded in the z direction – see Section 3.3.2) are much larger than $\nu_{z\theta}$ and ν_{zr}. This is because

---

[1] There are 12 listed in the table. However, orthotropic materials have only nine independent elastic constants – see Section 4.3.2 and Fig. 4.8. There are inter-relationships between those in the table that would allow them all to be obtained if the values of nine were provided.

Table 13.1 Generic values [3] for the elastic constants of 'standard' wood in 'air dry' conditions.

Wood type	Density (g cm^{-3})	Young's moduli (GPa)			Shear moduli (GPa)			Poisson ratios (-)					
		E_z	E_r	E_θ	$G_{\theta z}$	G_{zr}	$G_{r\theta}$	$\nu_{z\theta}$	ν_{zr}	$\nu_{\theta z}$	ν_{rz}	$\nu_{\theta r}$	$\nu_{r\theta}$
Hardwood	0.65	14	1.8	1.0	0.97	1.3	0.37	0.03	0.05	0.46	0.39	0.68	0.38
Softwood	0.45	13	1.0	0.64	0.75	0.86	0.08	0.02	0.03	0.42	0.38	0.49	0.31

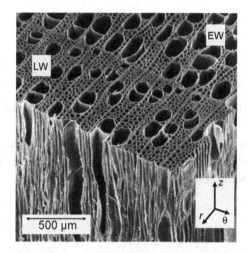

Fig. 13.4 SEM micrograph [3] of the cell structure of a typical wood, showing the nomenclature of the coordinate system used here.
Reprinted from Comptes Rendus de l'Academie des Sciences. Serie 2b. Mecanique, 329, Bernard Thibaut, Joseph Gril and Mériem Fournier, Mechanics of wood and trees: some new highlights for an old story, 701–716, © 2001, with permission from Elsevier.

the fibrils (in the cell walls) strongly inhibit contraction in the z direction under transverse loading. The differences between the two transverse directions (θ and r) arise from the cell structure – a typical example [3] of which is shown in Fig. 13.4. It can be seen from Table 13.1 that $\nu_{\theta r}$ is greater than $\nu_{r\theta}$. This is consistent with the cell structure, since the cellular configuration is such that contraction in the hoop (θ) direction is likely to take place more easily than in the radial direction. This is also reflected in $\nu_{\theta z}$ being larger than ν_{rz} and in E_r being larger than E_θ. Of course, these values are likely to be sensitive to the details of the cell structure, which can be complex (and can include the presence of 'ray' cells aligned in the radial direction).

It is also important to appreciate that the moisture content can have noticeable effects on the mechanical properties (both elastic constants and strength/toughness). The effect on elastic properties is illustrated by Fig. 13.5, which relates to beech (a hardwood) tested in tension [6]. It can be seen that there is a general trend, on raising the water content, for the stiffness to go down and the Poisson ratios to rise, although the effect is relatively small. (Typical moisture contents in practice, of wood being used for

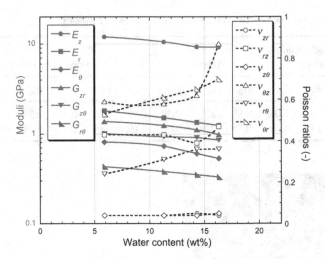

Fig. 13.5 Experimental data [6] for the elastic constants of beech, as a function of its water content.

structural purposes, are usually of the order of 10% – see below.) The reduction in stiffness is associated with hydration of the cellulose micro-fibrils in the cell walls, which tends to weaken the bonding between adjacent cellulose molecules.

13.2.3 Strength Properties of Wood

Analogous effects are commonly observed with respect to strength properties of wood. Fig. 13.6 shows how the tensile strength [6] of beech, in the three principal directions, varies with water content. The strong anisotropy – i.e. an axial strength of the order of 100 MPa, compared with transverse strengths of around 10–20 MPa – is again clear. Also, the effect of moisture on strength is similar to that on stiffness, with a relatively small, but quite noticeable, reduction being observed as the level rises to around 15%. The strain to failure, on the other hand, tends to rise slightly. These effects are also primarily due to hydration of the cellulose, reducing the strength of the micro-fibrils and allowing some sliding between adjacent molecules before failure occurs.

Figs 13.5 and 13.6 concern (hard)wood samples with well-defined water contents (in the range likely to be present during normal use). In practice, there is interest in how properties (particularly strength) are likely to be impaired by exposure to relevant environments, which might lead to water levels outside this range. It is also worth noting that softwoods are likely to be more susceptible to damage than hardwoods. Fig. 13.7 shows experimental data [7] for Scots pine (a softwood), both heartwood and sapwood, after immersion in moist, fertile soil for different periods (up to 11 months). Such treatments are likely to cause biological damage rather than just changing the moisture content. It can be seen that progressive reductions occurred in both stiffness and strength, up to about 50% for the sapwood and 30% for the heartwood. These changes are greater than those seen in Figs 13.5 and 13.6, although of course both the stiffness and strength of softwoods tend in any event to be lower than those of

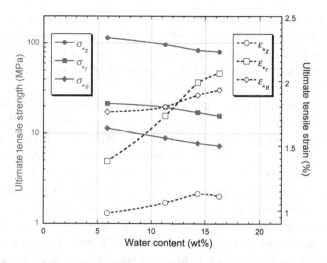

Fig. 13.6 Experimental data [6] for the tensile strength, and strain to failure, of beech, as a function of its water content, when loaded in different directions.

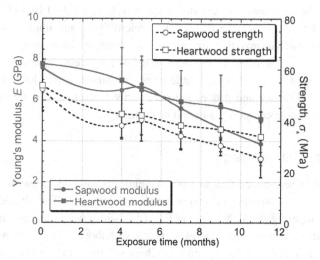

Fig. 13.7 Experimental data [7] for the stiffness and strength (obtained in bending tests) of Scots pine, tested along the grain, after exposure to moist soil for different periods.

hardwoods. It was also noted that the water content of both increased during the treatment and they also underwent some loss of (dry) mass. The latter is sometimes used to monitor rotting, but these changes were relatively small (~5–10%) and significant strength degradation can occur without much loss of mass.

Apart from weight loss measurement, there was no detailed study of the nature or extent of the rotting that occurred in the samples of Fig. 13.7. Nevertheless, it is well established that, if wet and biologically active environments are maintained for extended periods, then biological degradation (rotting) can severely damage the cellular structure and dramatically impair the mechanical properties. Needless to say, outcomes will depend in a complex manner on the type of wood and the biological environment, since the degradation is essentially one of cell wall constituents becoming damaged, and possibly consumed, by the action of biological species such as fungi [8]. (Of course, such effects are important, not only in terms of usage of wood, but also in relation to recycling and sustainability – see Section 13.4.)

While high water contents can clearly be deleterious, it should be noted that there is no incentive to reduce the water content of wood down to very low levels. Partly this is due to a tendency for very dry wood to absorb water from the air (causing dimensional changes), but it is also because such wood tends to be brittle. ('Strength' values can also be reduced by severe drying, but the effect on toughness is often more pronounced.) Reliable data for the fracture toughness of wood, as a function of water content, are in relatively short supply. Such testing can be difficult for anisotropic materials, since generating stable crack growth under well-defined (e.g. plane strain, mode I) conditions requires care with such materials – see Section 9.2.4. Moreover, energy absorption commonly occurs via processes such as fibre pull-out (see Section 9.2.3), which take place in the wake of the crack and lead to R-curve behaviour (see Section 9.2.4) – i.e. a dependence of the measured toughness on crack length [9].

Nevertheless, the brittleness of very dry wood is well established [9–11]. Complete removal of water tends to damage the cellulose micro-fibrils and impair some of the energy-absorbing mechanisms that operate during fracture. Moreover, it has been shown [12] that, once damaged in this way, the full toughness cannot be recovered by re-hydration, even if there has been little or no permanent weight loss. (Of course, this relates to felled timber and it does not necessarily mean that a living tree cannot repair structural damage caused by a period of drought.)

13.2.4   Tension–Compression Asymmetry

Finally, it should be recognised that, as with conventional uniaxial composites (Section 8.1.3), wood tends to be stronger in tension than in compression. (This should be borne in mind when comparing strength values in Fig. 13.7, which were obtained in bending, causing failure on the compression side of the beam, with those in Fig. 13.6, which were obtained from tensile testing.) As explained in Chapter 8, this weakness in compression is due to a tendency for local regions to undergo some kind of lateral collapse (forming what are often termed 'kink bands'). This is more likely in composite materials with weak shear properties parallel to the fibre axis, and also when volume changes can occur during deformation. The latter is clearly promoted by the presence of voids and the presence of the cellular spaces in wood does make it particularly susceptible to this kind of failure.

It is sometimes stated that the (axial) tensile strength of wood can be up to three times greater than the corresponding strength in compression. In fact, factors as large as 3 are unusual, but nevertheless there is a clear tendency of this type across a range of woods, with ratios of around two being quite common. Some representative data [13,14] are shown in Fig. 13.8, where it can be seen that, as expected, the moisture content is relevant. Nevertheless, for the three woods concerned (all hardwoods), the ratio of

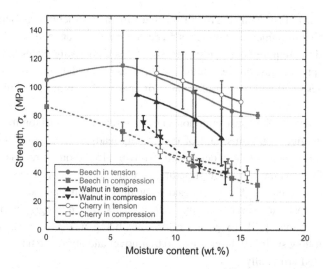

Fig. 13.8 Experimental data for the strength (failure stress) parallel to the grain (z direction), under tensile and compressive loading, as a function of water content, for beech [13] and also for walnut and cherry [14].

tensile to compressive strength varies between about 1.5 and 2. This issue should certainly be borne in mind when assessing how components made of wood are likely to fail. Another feature evident in Fig. 13.8 is that there is in general more scatter in the tensile strengths than for those in compression. Of course, wood does tend to contain defects of various types and these are likely to affect failure in tension rather more strongly than in compression. It may also be noted that some tensile–compressive asymmetry is commonly observed for the elastic properties of wood, although the differences are usually much less than for strength and in some cases it might be questioned whether genuinely elastic behaviour is being monitored.

13.3   Ceramic-Based Bio-Composites

13.3.1   Biomineralisation

As for wood, biominerals are composite materials exhibiting complex hierarchical structures, produced by cell growth occurring at ambient temperatures. However, there are some differences between the two. While all constituents in wood could be described as 'organic', biominerals comprise inorganic ('ceramic') crystals held together by organic constituents. Furthermore, while biominerals can certainly exhibit anisotropy, they cannot all be regarded as resembling (uniaxial) fibre composites. Nevertheless, they are very different from conventional ceramics. For example, most mollusc shells are made of nacre, which is calcium carbonate containing about 1% organic matter. It essentially comprises layers of interlocking aragonite ($CaCO_3$) platelets separated by thin layers of proteins such as collagen. However, the toughness of nacre is much greater [15] than that of conventional $CaCO_3$ (in either polycrystalline or single crystal form). The detailed growth mechanisms and structures associated with organisms in which biomineralisation takes place cover a wide range, depending on the environment and the requirements. However, structures comprising plates separated by thin, tenacious protein layers are common and it is clear that this often leads to deflection of cracks and stimulation of a range of energy-absorbing mechanisms – see Section 9.2.

Calcium carbonate is the most common ceramic created during biomineralisation, particularly by marine organisms, and of course this leads to the high abundance of chalk in geological strata. Many details are known [15] about how the chemical environment influences the fine-scale structure. A wide variety is also observed in the coarser-scale architecture of biomineralised structures, which can be complex and intricate. An example can be seen in Fig. 13.9, which shows micrographs of skeletal elements of sea urchins [15]. Extensive efforts are being made to obtain improved understanding of how such structures arise, so that novel and attractive materials of this type can be created artificially.

Rates of growth (mineralisation) can be high. For example, the rate of creation of $CaCO_3$ in coral reefs [16] is ~2–6 kg m^{-2} yr^{-1}. (The corresponding figure for wood creation in a commercial forest is typically about one order of magnitude lower than this.) Rates of bone growth are usually in a considerably lower range, but of course a key issue affecting growth rates concerns the supply of raw material. For materials such as coral, Ca^{2+} ions are picked up very efficiently from sea water and it is often the availability of dissolved inorganic carbon (DIC) that is rate-determining [16]. Common sources include dissolved CO_2 molecules and ions such as HCO_3^- and CO_3^{2-}. The details can be sensitive to the composition of the sea water, including its pH. As with creation of wood-like materials, it is certainly not a simple matter to replicate natural biomineralisation processes without using living cells, although progress in this direction is being made.

13.3.2   Structure of Cortical and Cancellous Bone

There is, of course, considerable interest in biomineralisation to form bone, particularly in the human body. A depiction of a typical hierarchical structure of bone is shown [17] in Fig. 13.10. Some superficial similarities to the structure of wood (Fig. 13.2) are apparent, at least for cortical bone, which does have a fairly strong directionality (alignment of fibrils). In detail, of course, there are many differences. There are also differences between bone and biomineralisation products such as mollusc shells and coral, a primary one being that the main mineral present is hydroxy-apatite (HA) rather than calcium carbonate. It is often given the formula $Ca_{10}(PO_4)_6(OH)_2$ and may be described as predominantly calcium phosphate. It commonly forms hexagonal crystals. However, in detail the structure of bone mineral is complex, commonly incorporating reductions in the Ca content of HA, and the presence of ions such as fluoride or carbonate. The mineral content of cortical bone, at about 50–70 vol%, is considerably lower than that of materials such as coral or mollusc shells.

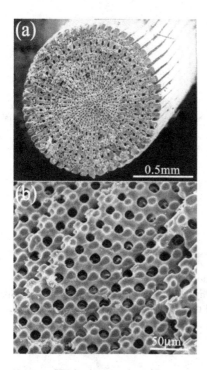

Fig. 13.9 SEM micrographs [15] of (a) a spine and (b) a skeletal plate of the sea urchin *Heliocidaris erythrograma.*
Reprinted from Calcium carbonate in biomineralisation and biomimetic chemistry, F. C. Meldrum, International Materials Reviews, 2003, Taylor & Francis, reprinted by permission of the publisher (Taylor & Francis Ltd http://www.tandfonline.com)

Most of the bone mass in a typical animal is termed cortical, although sometimes the terms 'compact' or 'lamellar' are used. It forms the thick outer layer (cortex) of long, load-bearing bones. The remainder of the bone mass, located in the central regions of long bones, and also in the end (joint) regions, is termed cancellous, although again other terms are used, such as 'trabecular' or 'spongy'. Cancellous bone consists of plates (trabeculae) and bars of bone interspersed with small, irregular cavities that contain red bone marrow. The mineral part is thus quite highly porous. A consequence of these features is that cortical bone is stiffer and stronger than cancellous bone, particularly along the bone axis (direction of alignment of the collagen fibrils) and it is also denser and harder.

13.3.3 Mechanical Properties of Bone

The mechanical properties of bone are naturally of considerable practical importance, particularly its fracture toughness. As with wood, however, meaningful fracture toughness data are not easy to acquire, and for similar reasons – i.e. anisotropy and inhomogeneity are complications, R-curve effects may be significant and the condition of the material (moisture content, porosity and general 'health') can vary substantially.

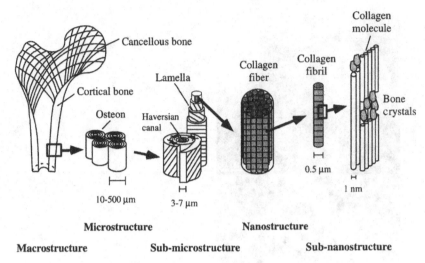

Microstructure   Nanostructure

Macrostructure   Sub-microstructure   Sub-nanostructure

Fig. 13.10 Hierarchical structural organisation of bone [17].
Reprinted from Medical Engineering & Physics, 20, Jae-Young Rho, Liisa Kuhn-Spearing and
Peter Zioupos, Mechanical properties and the hierarchical structure of bone, 92–102, © 1998,
with permission from Elsevier.

Nevertheless, there have been many investigations [17–22] and there is in general a
good level of agreement, at least concerning stiffness and strength. The (axial) stiffness
of cortical bone is usually in the range 10–20 GPa and the tensile strength around
100 MPa. (Cancellous bone, which is more porous, is weaker and less stiff.)

It is perhaps worth noting that both of these values are broadly similar to those of
wood (parallel to the grain). A noticeable difference between the two materials,
however, is that the compressive strength of (cortical) bone tends to be similar to, or
rather greater than, its tensile strength, whereas wood is generally weaker in compression. This is unsurprising, since bone is basically a ceramic and such materials are
usually stronger in compression, when mode I crack propagation cannot occur. Of
course, the situation is more complex than this, since bone is considerably tougher than
typical homogeneous ceramics – see below. Furthermore, its resemblance to a uniaxial
fibre composite is less pronounced than for wood and it is less porous, so the formation
of 'kink bands' (Section 13.2.4) is unlikely.

As outlined above, the fracture toughness of bone can be difficult to measure and
tends to be somewhat variable, partly because of its anisotropy. As outlined in Section
9.2.4, fracture energy (G_c) values for various types of (polymer) composite might range
from around 1 kJ m^{-2} up to several tens of kJ m^{-2} (for fracture paths normal to fibre
alignment directions). Values for wood tend to lie in the lower end of this range. Bone is
usually a little less tough than wood, which is unsurprising in view of issues such as the
slightly more limited scope for fibre pull-out.

Some representative experimental data [20] are presented in Fig. 13.11, which shows
the outcome of 140 compact tension tests. These cover all three modes of fracture

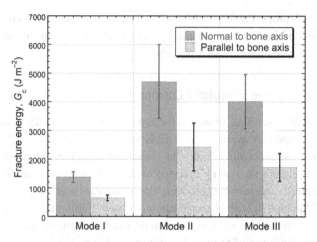

Fig. 13.11 Experimental data [20] for the fracture energy of (fresh adult bovine femoral) cortical bone, for crack paths normal and parallel to the femoral axis, loaded in the three fracture modes.

(Section 9.1.5) for crack propagation parallel and normal to the bone axis. As expected, values are higher for shearing (II) and tearing (III) modes, although such loading would in practice be fairly unusual for a bone. (The most common loading configuration for a bone is in bending, creating mode I conditions on the tension side.) Values are also higher for cracking normal to the bone axis, which again would be the most common type of fracture in practice. The most relevant value in Fig. 13.11 is thus around 1.5 kJ m^{-2}. This is certainly a respectable toughness – better than many engineering ceramics – although it is evidently not in a range such that fracture is highly unlikely. Furthermore, values appreciably lower than this may be obtained for bone that is in relatively poor condition. On the other hand, some types of bone, such as that of elk antlers [21], can be considerably tougher than this.

13.4 Recycling of Composite Materials

13.4.1 Life Cycles of Natural Composites

As with all living organisms, cycles have become established for absorption of natural composite materials back into the environment (in forms potentially suitable for incorporation into other organisms). For example, the rotting of wood (i.e. biological attack of its constituents by agents such as fungi) eventually results in its conversion to new cells (in a range of organisms), often with the release of small molecules such as CO_2, H_2O and CH_4. Natural degradation of bio-ceramic composites, such as bone or mollusc shells, is usually much slower overall, but still occurs, with the collagen breaking down fairly quickly. The mineral constituent (primarily calcium carbonate or calcium phosphate) may remain stable over long periods, but is likely to eventually re-enter life cycles (e.g. by being dissolved by slightly acidic rain and hence carried as calcium ions back to the sea). Calcium is the fifth most abundant element in the earth's crust, so these life cycles are not inhibited by large quantities being effectively excluded from them by being in geological strata for extended periods.

13.4.2 Reclamation and Recycling of Artificial Composites

Identification of viable recycling strategies for materials used in manufactured products in an increasingly important requirement. This is in general straightforward for metals, partly because smelting procedures tend to be tolerant of impurities. Moreover, reuse of scrap metal largely retains the energy invested in their original extraction from natural ores. One of the attractions of wood products (including cardboard, paper, etc.) is that natural cycles of the type described above can be utilised. Furthermore, by-products of conventional wood processing, such as wood-flour, sawdust, bark, off-cuts, etc., can be incorporated into other types of product in the form of fillers, etc.

However, the situation is different for polymers and hence for polymer-based composites. Many claims are made regarding the recycling of polymers, and indeed there is some real potential, but in general it is far from easy. Many polymers cannot readily be re-melted and there are serious problems related to contamination and tolerance of impurities. Landfill is clearly a very unsatisfactory solution, since typical polymers remain essentially unchanged in such an environment for very long periods. Incineration also presents a number of difficulties and disadvantages. For some applications, such as toys, use of wood instead of polymers has clear attractions, although the excellent mouldability and low cost of plastics are significant issues. It can undoubtedly be argued that, at least for certain applications, such as packaging, there are very strong drivers for replacing polymers with wood products such as paper and cardboard, and possibly with metals such as aluminium foil in some cases. Needless to say, however, many complex technical, economic and societal factors are relevant to such choices.

There is, of course, considerable interest [23,24] in the creation of artificial polymers that are biodegradable, particularly for applications such as food packaging [25]. The underlying challenges are clear. Wood-derived products such as paper and cardboard are already available, but their relatively rapid biodegradation, particularly when wet, renders them unsuitable for many such applications. A key requirement is for (artificial) products that can retain certain attributes of conventional polymers, such as water resistance and mouldability (and possibly other characteristics such as transparency), in combination with a tendency to undergo relatively rapid degradation (perhaps over periods of months or years) in natural environments. Of course, cost is also a factor.

Clearly, this is at least partly an issue of the thermodynamics and kinetics of chemical reactions, and a lot of chemistry-based studies have been undertaken. However, other aspects may be relevant, such as the attractions of a porous (permeable) structure in terms of creating a large surface area for such reactions. Permeability values, for transport of various species, are of relevance to degradation, as well as to usage [25]. Also, as with natural materials such as wood, the degradation could involve living cells or biological agents such as enzymes. There is also a move towards the production of polymers via natural processes such as fermentation, using a genetic engineering approach, rather than from petroleum products. The current position can probably be summarised by saying that, while successful products are available for relatively niche applications, such as sutures and similar bio-absorbed scaffolds, etc. for medical use, technical solutions and commercial strategies remain to be developed for large-scale polymer recycling, although progress is certainly being made.

Concerning polymer composites specifically, of course wood is often an obvious alternative, with scope for creating (desirable) anisotropy in a similar way. Products such as skis, racquets, surfboards, boats, planes, etc. can be, and in some cases still are, made of wood, rather than composite. However, there is no doubt that the artificial composites are often markedly superior (in terms of performance, manufacturing versatility, quality control, durability, etc.) and there is no realistic prospect of, for example, the production of commercial aircraft reverting to use of wood in place of composites. However, there is considerable interest [26–30] in the concept of making artificial composites using biodegradable products, particularly natural fibres of various types. This has the potential for improved sustainability, although the main difficulty in terms of recycling tends to lie with the (polymeric) matrix (and its separation from the fibres), rather than with the fibres themselves. Of course, the scale of the challenge (in tonnage terms) for recycling of composites is far smaller than that for polymers in general, so the scope for specialist solutions of various types is correspondingly greater.

References

1. Fratzl, P and R Weinkamer, Nature's hierarchical materials. *Progress in Materials Science* 2007; **52**(8): 1263–1334.
2. Gibson, LJ, The hierarchical structure and mechanics of plant materials. *Journal of the Royal Society Interface* 2012; **9**(76): 2749–2766.
3. Thibaut, B, J Gril and W Fournier, Mechanics of wood and trees: some new highlights for an old story. *Comptes Rendus de L'Academie Des Sciences Serie II Fascicule B: Mecanique* 2001; **329**(9): 701–716.
4. Ruffinatto, F, A Crivellaro and AC Wiedenhoeft, Review of macroscopic features for hardwood and softwood identification and a proposal for a new character list. *Iawa Journal* 2015; **36**(2): 208–241.
5. Lichtenegger, H, A Reiterer, SE Stanzl-Tschegg and P Fratzl, Variation of cellulose micro-fibril angles in softwoods and hardwoods: a possible strategy of mechanical optimization. *Journal of Structural Biology* 1999; **128**(3): 257–269.
6. Ozyhar, T, S Hering and P Niemz, Moisture-dependent elastic and strength anisotropy of European beech wood in tension. *Journal of Materials Science* 2012; **47**(16): 6141–6150.
7. Venalainen, M, H Partanen and A Harju, The strength loss of Scots pine timber in an accelerated soil contact test. *International Biodeterioration & Biodegradation* 2014; **86**: 150–152.
8. Bari, E, HR Taghiyari, B Mohebby, CA Clausen, O Schmidt, MAT Ghanbary and MJ Vaseghi, Mechanical properties and chemical composition of beech wood exposed for 30 and 120 days to white-rot fungi. *Holzforschung* 2015; **69**(5): 587–593.

9. Sinha, A, JA Nairn and R Gupta, The effect of elevated temperature exposure on the fracture toughness of solid wood and structural wood composites. *Wood Science and Technology* 2012; **46**(6): 1127–1149.

10. Stanzltschegg, SE, EK Tschegg and A Teischinger, Fracture energy of spruce wood after different drying procedures. *Wood and Fiber Science* 1994; **26**(4): 467–478.

11. Vasic, S and S Stanzi-Tschegg, Experimental and numerical investigation of wood fracture mechanisms at different humidity levels. *Holzforschung* 2007; **61**(4): 367–374.

12. Kifetew, G, F Thuvander, L Berglund and H Lindberg, The effect of drying on wood fracture surfaces from specimens loaded in wet condition. *Wood Science and Technology* 1998; **32**(2): 83–94.

13. Ozyhar, T, S Hering and P Niemz, Moisture-dependent orthotropic tension–compression asymmetry of wood. *Holzforschung* 2013; **67**(4): 395–404.

14. Bachtiar, EV, M Ruggeberg and P Niemz, Mechanical behavior of walnut (*Juglans regia* L.) and cherry (*Prunus avium* L.) wood in tension and compression in all anatomical directions: revisiting the tensile/compressive stiffness ratios of wood. *Holzforschung* 2018; **72**(1): 71–80.

15. Meldrum, FC, Calcium carbonate in biomineralisation and biomimetic chemistry. *International Materials Reviews* 2003; **48**(3): 187–224.

16. Allemand, D, C Ferrier-Pages, P Furla, F Houlbreque, S Puverel, S Reynaud, E Tambutte, S Tambutte and D Zoccola, Biomineralisation in reef-building corals: from molecular mechanisms to environmental control. *Comptes Rendus Palevol* 2004; **3**(6–7): 453–467.

17. Rho, JY, L Kuhn-Spearing and P Zioupos, Mechanical properties and the hierarchical structure of bone. *Medical Engineering and Physics* 1998; **20**(2): 92–102.

18. Yeni, YN, CU Brown, Z Wang and TL Norman, The influence of bone morphology on fracture toughness of the human femur and tibia. *Bone* 1997; **21**(5): 453–459.

19. Yeni, YN, CU Brown and TL Norman, Influence of bone composition and apparent density on fracture toughness of the human femur and tibia. *Bone* 1998; **22**(1): 79–84.

20. Feng, ZD, J Rho, S Han and I Ziv, Orientation and loading condition dependence of fracture toughness in cortical bone. *Materials Science and Engineering C: Biomimetic and Supramolecular Systems* 2000; **11**(1): 41–46.

21. Launey, ME, PY Chen, J McKittrick and RO Ritchie, Mechanistic aspects of the fracture toughness of elk antler bone. *Acta Biomaterialia* 2010; **6**(4): 1505–1514.

22. Libonati, F and L Vergani, Understanding the structure–property relationship in cortical bone to design a biomimetic composite. *Composite Structures* 2016; **139**: 188–198.

23. Amass, W, A Amass and B Tighe, A review of biodegradable polymers: uses, current developments in the synthesis and characterization of biodegradable polyesters, blends of biodegradable polymers and recent advances in biodegradation studies. *Polymer International* 1998; **47**(2): 89–144.

24. Luckachan, GE and CKS Pillai, Biodegradable polymers: a review on recent trends and emerging perspectives. *Journal of Polymers and the Environment* 2011; **19**(3): 637–676.

25. Siracusa, V, P Rocculi, S Romani and M Dalla Rosa, Biodegradable polymers for food packaging: a review. *Trends in Food Science & Technology* 2008; **19**(12): 634–643.

26. Fowler, PA, JM Hughes and RM Elias, Biocomposites: technology, environmental credentials and market forces. *Journal of the Science of Food and Agriculture* 2006; **86**(12): 1781–1789.

27. Ramamoorthy, SK, M Skrifvars and A Persson, A review of natural fibers used in biocomposites: plant, animal and regenerated cellulose fibers. *Polymer Reviews* 2015; **55**(1): 107–162.

28. Gurunathan, T, S Mohanty and SK Nayak, A review of the recent developments in biocomposites based on natural fibres and their application perspectives. *Composites Part A: Applied Science and Manufacturing* 2015; **77**: 1–25.

29. Pizzi, A, Wood products and green chemistry. *Annals of Forest Science* 2016; **73**(1): 185–203.

30. Keskisaari, A and T Karki, Raw material potential of recyclable materials for fiber composites: a review study. *Journal of Material Cycles and Waste Management* 2017; **19**(3): 1136–1143.

14　Scale Effects and Nano-Composites

There has been a prodigious level of interest in graphene over recent decades, and also in (the closely related) carbon nanotubes (CNTs). This has included quite a strong focus on their mechanical properties, with various claims in particular being made about high strength levels, which relate to their extremely fine scale. Since the scope for their usage in isolation is relatively limited, at least in terms of exploitation of high strength levels, a lot of attention has been directed towards the production and use of composite materials containing them as reinforcement. Unfortunately, most of the hopes originally expressed have not been fulfilled. In fact, the mechanical properties of such composites have in all cases been inferior to those of conventional carbon fibre composites. This is partly due to severe difficulties in manufacturing composites with relatively high levels of well-dispersed, well-aligned reinforcement. However, this is not the only problem, since many of the original expectations were based on incomplete understanding of the issues involved in defining the strength of a material, with particular reference to the role of toughness. These CNT-reinforced composites tend to have a low toughness, as a direct consequence of their very fine scale. In this chapter, the issue of scale-related effects is first addressed in general terms, followed by information about some specifics of using fine-scale reinforcement (particularly CNTs).

14.1　Scale Effects in Composite Systems

There has been an extraordinary level of interest over recent years in composite materials incorporating very fine-scale (nanometre) reinforcement. Much of this has been stimulated by the perception that exceptional (mechanical) properties can be exhibited by graphene (single layers of carbon atoms in a 2D hexagonal lattice) and by carbon nanotubes (CNTs), which are essentially graphene sheets rolled into tight cylinders. Prediction of the elastic properties of such 'materials' is relatively straightforward, but there are serious difficulties associated with both prediction and measurement of their 'strength' or 'toughness' in any meaningful way. Once a composite material has been produced, incorporating reinforcement of this type, then such measurements become practicable, although there is no escaping the fact that experimental outcomes have not fulfilled the initial hopes. However, before examining experimental data of this type, it is useful to consider theoretical expectations concerning the effects of simply refining the scale of the reinforcement in a composite material (without changing any properties, morphology, etc. of matrix, reinforcement or interface region).

14.1.1   Effect of Scale on Mechanical Properties

The elastic constants of a composite material, expressed in terms of those of the constituents and the architecture of the assembly, are independent of scale. This can be seen from the treatments presented in Chapters 3–5, where the scale of the structure does not figure in any way. Of course, it is conceivable that a particular type of reinforcement, such as a carbon fibre or a SiC particle, could have different elastic constants if it had a very small diameter, although this would only be possible if there was an associated change in the atomic-scale structure. This would not in general be expected to arise simply as a result of a reduction in dimensions, although it is certainly possible for a reinforcement that is only available in very fine form to have a different atomic structure (exhibiting a higher stiffness). This is the case with CNTs. Nevertheless, no change in stiffness is expected to arise simply from scale refinement.

The situation is different when the more complex mechanical properties of strength and toughness are considered. For the strength, the treatments presented in Chapter 8 again have no explicit scale dependence. However, a property that is relevant is the strength (tensile failure stress) of the fibre, and it is certainly plausible for this to exhibit a dependence on scale. In fact, for brittle fibres this is expected, since the size of the largest flaw tends to scale with the dimensions (diameter), leading to higher fracture stress values for finer fibres (assuming constant toughness). This is specifically exploited for glass fibres, which are relatively fine (~7 μm), with efforts routinely made to minimise the flaws introduced in them after drawing. The strengths of such fibres are usually in the range 3–4 GPa (see Section 2.1), which is certainly quite high. Furthermore, there is evidence that even finer fibres, such as SiC whiskers of sub-micron diameter, can exhibit slightly higher strengths than this. However, for extremely fine (~ a few nanometres) structures, such as CNTs, it is far from clear whether the fracture stress will continue to rise as the scale of the structure becomes even finer and approaches atomic dimensions. Indeed, the very concept of a 'flaw' becomes doubtful on this scale. Nevertheless, it should be recognised that very high strength values might be expected for ultra-fine fibres and that this could be reflected in the performance of the composite.

At this point, however, it should be recognised that, as emphasised in Chapter 9, it is often the toughness of the composite that is in practice more important than its 'strength'. Materials that are very 'strong', but brittle, are generally of little use for engineering purposes. Since the constituents of most composites (i.e. polymeric resins and carbon or glass fibres) are themselves brittle, it is the toughness that arises from synergistic effects (particularly fibre pull-out) that makes them so useful. This does have an explicit dependence on scale, but in fact it works in the opposite direction – i.e. coarser microstructures tend to give higher toughness. This can be seen from Eqn (9.22), which shows that the toughness is predicted to rise linearly with the scale of the structure (diameter of the fibre). The reason for this is that, as the structure becomes coarser, more work is done during fracture, because the 'process zone' located either side of the crack, within which the frictional sliding occurs, increases in volume.

There is actually a similar effect when energy is absorbed via plastic deformation of (metal) fibres – see Section 9.2.7 and Fig. 9.19. That plot includes experimental confirmation that coarser fibres lead to a higher toughness, which was obtainable because (metal) fibres with similar properties are available with a wide range of diameters. For carbon and glass fibres, the available range of diameters is much smaller. Moreover, in view of the way that most polymer composites are made, much coarser (and hence less flexible) fibres would be unattractive. Nevertheless, there is a clear implication here that making the scale of the (carbon fibre) reinforcement much finer is likely to reduce the size of the 'process zone' (in which energy is absorbed during crack propagation) and hence to make the composite less tough. This logic is based on the assumption that the geometry (pull-out aspect ratio, etc.) remains similar, but there is no particular reason to expect this to change significantly (even if the fibre strength is higher). This issue of a possible impairment of the toughness of such composites, and hence a reduction in what might commonly be perceived as their 'strength', has not always been fully understood.

14.1.2   Effect of Scale on Functional Properties

Various non-mechanical properties are, of course, of potential interest, ranging from conductivity (thermal and electrical) to expansivity, permeability, permittivity, etc. Many of these have no explicit dependence on scale, although it should certainly be recognised that a composite with fine-scale reinforcement will have a larger interfacial area (per unit volume) and that this could influence the microstructure (see Section 14.1.3) and certain properties. For example, an interface commonly offers a finite resistance (electrical and thermal), so that, if the flow path involves the crossing of interfaces, the overall conductivity will be lower for a finer structure (unless the reinforcement is insulating, so that no heat or current crosses the interface in any event). This effect is described in Section 10.4.3 and illustrated in Figs 10.25–10.29. (The Biot number, Bi, is inversely proportional to the fibre radius and the tendency in those figures for the composite conductivity to fall as Bi rises can clearly be seen.)

Other effects related to scale (as it affects the frequency of encountering interfaces) are also identifiable. For example, the permeability of a fibre network composite falls as the scale is reduced (and the fluid is forced to pass through narrower channels, creating greater resistance to viscous flow). There is an illustration of this in Fig. 12.14, where it can be seen that, for a given porosity level (fibre volume fraction), the permeability of a wide range of fibre network materials approximately scales with the inverse of the square of the fibre radius (in accordance with the Carman–Kozeny equation). This effect is clearly of importance in the context of applications such as filters, for which refinement of the scale is an obvious approach to filtration of finer species, but with the consequence of a lower permeability and hence a slower flow rate (for a given pressure gradient).

14.1.3   Effects of Scale on Microstructure

There are expected to be some effects of scale in composites that relate to microstructure, particularly in the vicinity of the interface. It is common for the matrix, and possibly the reinforcement, to be at least slightly different in this region. For example, there may be a reaction product present, or there may be a region in which some kind of inter-diffusion has occurred, or there has been a change in the degree of crystallinity, the grain structure, etc. Various examples of such effects are provided in Section 7.1 and Section 7.3. An obvious consequence of reducing the scale of the reinforcement is that this 'interface-affected' volume will then occupy a larger fraction of the total volume.

Taking the extreme example of CNT reinforcement, since the 'fibre' diameter is ~1 nm, the inter-fibre spacing (assuming a 'typical' fibre volume fraction of, say, 25%) should also be of this order. It would certainly be expected that a region of matrix this close to any reinforcement would differ significantly in microstructure from the 'normal' matrix, so all of the matrix will be 'interface-affected'. For example, if the matrix has the slightest propensity to form a carbide, then virtually all of it will be carbide, even if little or no heating has been involved in production of the composite. (Diffusion over distances of the order of 1 nm takes place very quickly, even with very low diffusion coefficients.) This rules out most metals as matrices for CNT composites. Even for polymers, it is clear that all of the matrix will be quite different from what would be expected of the bulk – there would, for example, be no scope for any rubbery behaviour, since a typical chain segment length (between pinning points) would be considerably greater than the inter-fibre spacing. Such changes would certainly be expected to present difficulties (quite separate from those associated with the production of such composites – see Section 14.3.1).

14.2   Fine-Scale, Carbon-Based Reinforcement

14.2.1   Graphene

Recent decades have seen the publication of a huge number of papers concerning graphene and its usage in composites. Comprehensive coverage is provided in a number of reviews [1–3]. The structure of graphene is, of course, very simple – it is just a layer of carbon atoms in a hexagonal array, with strong (sp^2 hybridised) σ-bonds between them. Various materials, ranging from graphite to conventional carbon fibres, are predominantly graphene sheets (in different configurations). It is well established that, as a consequence of the strength of the bonds within a graphene sheet, it has a high in-plane stiffness and strength. Of course, while a 'bond strength' translates in a logical way to a macroscopic stiffness, the concept of a macroscopic 'strength' is more complex, involving as it does both the way in which inter-atomic bonds can be broken and the energetics of propagation of a macroscopic crack. While diamond has a very high stiffness (~1000 GPa), also a consequence of strong C–C bonds, it is brittle and, even if it were easy (and cheap) to form it into large components, it would for this reason not be widely used as a structural engineering material. Moreover, graphene is essentially a 2D 'material', presenting both theoretical and practical difficulties when attempting to identify a stiffness and, particularly, a strength that is meaningful for engineering usage. There is considerable scope for confusion in this situation.

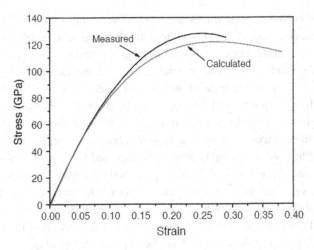

Fig. 14.1 Measured and theoretically predicted stress–strain curves [1] for in-plane elastic deformation of a graphene monolayer.
Reprinted from Composites Science and Technology, 72, Robert J. Young, Ian A. Kinloch, Lei Gong and Kostya S. Novoselov, The mechanics of graphene nanocomposites: A review, 12–18, © 2012, with permission from Elsevier.

Measurements have been made [4] of the in-plane stiffness of single graphene sheets, in bending via nanoindentation, using an AFM (atomic force microscope). This technique allows the accurate application of very small forces, and measurement of very small displacements, although there are some difficulties in the case of graphene, related to issues such as the effective thickness of a monolayer. Nevertheless, a typical [4] inferred in-plane stiffness value of around 1000 GPa is quite plausible. A comparison [1] is shown in Fig. 14.1 between measured and predicted (elastic) stress–strain curves for a graphene monolayer. It can be seen that, over this relatively large strain range, non-linear behaviour is observed, but the initial stiffness is of the order of 1000 GPa.

Strength is, of course, another matter. Results of the above type have been repeatedly used by various researchers to infer strength (ultimate tensile stress) levels that are very large – often of the order of 100 GPa. It has also been suggested that such values correspond approximately to some kind of theoretical limit (~10% of the modulus). Of course, the concept of what is effectively just an inter-atomic bond strength being used to infer a (macroscopic) strength is doubtful, to say the least. Actually, in terms of having graphene planes aligned along a loading axis, conventional carbon fibres largely provide this – see Fig. 2.1 – but their tensile strengths are typically no greater than 5 GPa. This is essentially because they are brittle materials – i.e. cracks propagate through them with little energy absorption. This would also tend to occur with an aligned assembly of graphene planes (and what happens with diamond is not so different). Of course, it might be expected that, if they could be rolled up (into nanotubes) and dispersed into a suitable matrix, then toughness might arise from similar synergistic effects to those that operate in conventional carbon fibre composites, and the expected very high strength of these nanotubes might turn out to be of benefit. This has certainly been an expectation in some circles.

14.2.2   Carbon Nanotubes and Micro-Fibrils

There have, of course, been many detailed studies [5–8] of CNTs – both single-walled and multi-walled, with much of the work oriented towards their mechanical properties. There has also been extensive investigation into what might be termed ropes, yarns or fibrils, made by spinning CNTs into something resembling a fine fibre, with good alignment of the individual tubes along the axis.

While they do require experimental ingenuity, tensile tests have been carried out on individual CNTs (single-walled and multiple-walled). Some illustrative data [6] are shown in Fig. 14.2, which refers to multiple-walled CNTs. These were attached to two opposing AFM tips, with a cantilever incorporated in the loading train to measure the applied force. It can be seen that tensile strengths were about 20–40 GPa, while the stiffness was around 250 GPa. These figures are somewhat lower than those apparent in Fig. 14.1 (obtained by bending of single graphene sheets). Broadly, it is expected that both the (axial) stiffness and strength of graphene sheets rolled into a tube would be similar to the in-plane properties of the sheets, but there must be some concerns about the interpretation of bending experiments, whereas at least the data in Fig. 14.2 were obtained in tests that involved a uniform applied tensile stress (and a fracture event).

Obvious yardsticks for all such measurements are the corresponding properties of a 'typical' conventional carbon fibre, which can be taken to have a stiffness of 300–500 GPa and a strength of 3–5 GPa – see Section 2.1.1. As might have been expected, the (axial) stiffness of CNTs is no better than that of carbon fibres – they both have good alignment of graphene planes along the axis and no scale effect is expected for stiffness – see Section 14.1.1. The strength values of the CNTs, however, do seem to offer an advantage (as might have been expected from their finer scale).

Of course, a major challenge in terms of practical usage is that of how to manufacture composite material containing (a reasonably high volume fraction of) uniformly dispersed CNTs. This is extremely difficult for polymeric matrices, which tend to have

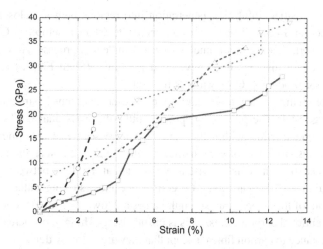

Fig. 14.2 Some representative stress–strain curves [6] for individual multi-walled CNTs.

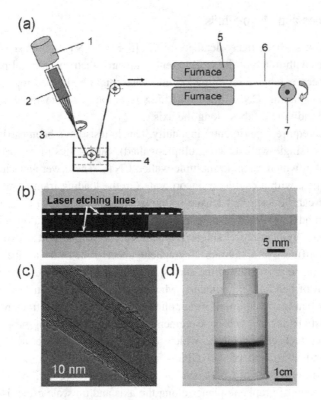

Fig. 14.3 Spinning of CNT yarns [8], showing (a) a schematic of the set-up, (b) a long strip of yarn after laser etching, (c) a TEM image of an individual multiple-walled CNT in a yarn and (d) a 40 m long yarn (of diameter 10 μm) wound onto a collecting tube.

relatively high viscosities, inhibiting penetration into regions between closely spaced CNTs – see Section 14.3.1. There has thus been interest in assembling CNTs into bundles of some sort, making them much easier to handle and process. An example of the type of procedure that has been developed in shown [8] in Fig. 14.3, where a spinning set-up for producing CNT yarns is illustrated.

Such yarns are relatively easy to test and some representative tensile strength data [8] are shown in Fig. 14.4. The 'shrinking' process pulls the individual CNTs together, raising their volume fraction within the yarn. It can be seen that this improves the strength. There is also an effect of yarn diameter, with a peak strength at about 10 μm. It may also be noted that there is a twist angle (of about 10–15°) between the CNT alignment direction and the yarn axis. Of course, the most striking outcome seen in Fig. 14.4 is that all of these measured strength values are low (<~1 GPa) – considerably lower than conventional carbon fibres. This is unsurprising. These yarns actually have a structural resemblance to carbon fibres, except that they are not fully dense, they contain higher levels of what might be regarded as defects and they have their graphene planes

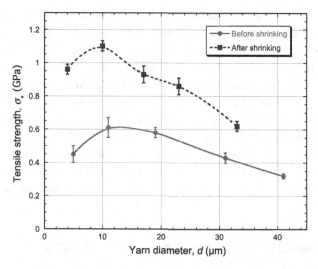

Fig. 14.4 Experimental data [8] for the tensile strength of CNT yarns produced using the set-up shown in Fig. 14.3, as a function of yarn diameter, with and without a shrinking procedure that pulls the individual CNT strands together.

slightly less well aligned with the fibre axis. Since they are far more expensive to produce than carbon fibres, they do not appear to offer any attractions.

It might be worth mentioning, however, that they do have at least one interesting feature. Since the individual CNTs are not bonded to each other (at least in most variants), they really do resemble ropes and, as with all ropes, they have high flexibility – i.e. they can adopt very high local curvature without breaking – see Section 12.2.1. This is potentially attractive, although it is not really relevant once the yarn has been incorporated into a composite. It is possible that applications could be found for such yarns, although, if they are compared with other similar material (e.g. conventional nylon ropes), their strength is not so much higher, their environmental stability is lower and, of course, their cost is much greater.

14.3   Composites Containing Nano-Scale Reinforcement

14.3.1   Production of Nano-Composites

An immediate problem facing researchers attempting to explore the potential of composite materials containing CNTs has been that their manufacture presents some serious difficulties. They have a strong tendency to agglomerate, so their dispersion into (polymeric) matrices is far from easy, at least if the volume fraction is to be higher than a few per cent. A number of papers [9,10] have summarised these difficulties, often as part of an overall review of the prospects for composites of this type. In fact, the majority of publications on CNT-containing composites do relate to material in which the CNT content is small (<~10%, and often <~5%).

Unsurprisingly, the property enhancements obtained have tended to be very small. In fact, even when material has been produced with reasonably high CNT contents (~10%) – usually the outcome of strenuous efforts to stimulate good wetting, bonding and dispersion (often involving functionalisation of the CNTs) – the mechanical properties have been disappointing (see Section 14.3.2). Nevertheless, the level of interest in CNT composites has remained relatively high, sometimes with the focus moving more towards a wide range of relatively esoteric applications and types of 'composite'. In fact, use of other types of matrix (metallic and ceramic) also presents fabrication difficulties, which apply to a range of fine-scale reinforcement types, but are particularly severe for carbon (mainly as a consequence of its relatively high reactivity). Some results are presented in Section 14.3.4 for ultra-fine (non-carbon) metal matrix composites (MMCs).

14.3.2   Mechanical Properties of Carbon Nanotube–Polymer Composites

There has been extensive experimental work in this area, summarised in numerous review papers [11–14]. Many publications emerged during a period when there was considerable optimism about the potential of this type of composite. However, it must be recognised that, while many results have been framed in a positive way – for example, tensile strengths being quoted as a fractional increase relative to that of the unreinforced polymer – they have in general been very disappointing. Typical increases have ranged up to a few hundred per cent at most, whereas a corresponding figure for a conventional carbon fibre composite would be a factor of at least several thousand per cent – i.e. a tensile strength of at least around 1 GPa, compared with a few tens of MPa for an epoxy resin. Of course, there is the issue of fibre (nanotube) content and, as outlined in Section 14.3.1, it is difficult to make composites with high volume fractions of nanotubes. Nevertheless, it is of interest to at least try to compare actual measured strengths with values that might have been expected.

It is difficult to select representative experimental data from the large number of studies that have been undertaken in this area, many of which probably involved use of composite material that was rather inhomogeneous and defective. Nevertheless, some results are presented in Fig. 14.5, from a study [15] in which the composites concerned contained reasonably high levels (up to ~10%) of single-walled nanotubes, which appeared to be at least fairly well aligned along the tensile axis, with the samples (cylinders of diameter 130 μm) also apparently being homogeneous and free of large-scale defects. Also, attempts were made to disperse nanotube bundles and to improve bonding with the (polypropylene) matrix (by fluorination of individual nanotubes). An indication of the coarse- and fine-scale microstructure is given by the fracture surface micrographs in Fig. 14.6. Despite these measures, it is clear from the data in Fig. 14.5 that the strength (and stiffness) values obtained are very disappointing. Even the best properties (for the fluorinated nanotubes at the level of 10%) are below 1 GPa in stiffness and below 100 MPa in tensile strength.

Since a rule of mixtures is expected to be at least broadly applicable for both of these properties (see Sections 3.1 and 8.1.1), this suggests that the *in situ* stiffness of these

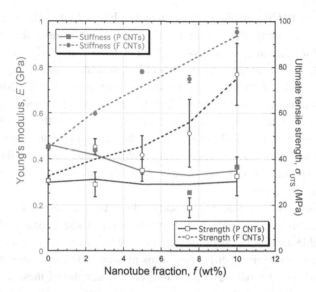

Fig. 14.5 Experimental data [15] for the Young's modulus and tensile strength of composites produced by dispersing single-walled CNTs, either pure (P) or after fluorination (F) of tube walls, into a polypropylene matrix, and extruding into cylindrical samples of 130 μm diameter.

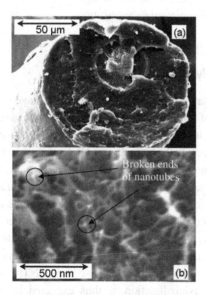

Fig. 14.6 SEM micrographs [15] at (a) low and (b) high magnification, of the fracture surface of a composite sample containing 10 wt% of fluorinated single-walled CNTs in a polypropylene matrix.

Reprinted with permission from McIntosh, D. et al., Nanocomposite Fiber Systems Processed from Fluorinated Single-Walled Carbon Nanotubes and a Polypropylene Matrix, Chemistry of Materials. Copyright 2006 American Chemical Society.

nanotubes is below 10 GPa and their *in situ* strength is below 1 GPa. These are certainly disappointing values, particularly since there appear to be no reliable experimental findings in the literature suggesting that composites have been made for which they have been significantly exceeded. It does seem likely that the nanotube alignment was not in fact very good, which would have quite a strong effect on the stiffness – see Section 4.3.3 – but the strength should be rather less affected and it seems clear that the effective strength of nanotubes in these composites is much lower than claimed values (which, as outlined in Section 14.2.1, have been in the range of tens or even hundreds of GPa). Even allowing for various doubts surrounding claims about single nanotube strengths, it seems unlikely that they would in fact be below those of conventional carbon fibres (i.e. ~ 3–5 GPa).

The explanation for this probably concerns the toughness of these composites, following the arguments put forward in Section 14.1.1. There appear to have been virtually no reliable measurements of the toughness of this type of composite (which is not so surprising in view of the difficulties of its measurement for small samples, and the practical challenges of making large, homogeneous samples of these composites). Nevertheless, it is almost certainly quite low. The fracture surfaces of Fig. 14.6 do provide some evidence of this, since it appears that pull-out aspect ratios are relatively low – i.e. actual pull-out lengths are extremely short. In general, these surfaces show little or no evidence of any energy-absorbing processes having accompanied fracture. Such brittleness is, of course, very undesirable. Not only is it likely to result in premature failure, and hence a low apparent tensile strength, but it is also a major drawback in its own right, for general engineering applications.

14.3.3   Functional Properties of Carbon–Polymer Nano-Composites

Perhaps partly in view of the very disappointing strength values that have been almost universally obtained for composites of this type, there has been a lot of interest in their functional properties. Various properties have been studied, but much of the attention has been focused on (thermal and electrical) conductivity. Of course, it is well established that carbon fibres can exhibit interesting properties of this type. Some aspects of this were mentioned in Section 2.1.1 and Fig. 10.20 shows that carbon fibres (of high perfection) can exhibit very high thermal conductivity. In fact, these values can be above those of all metals, although corresponding electrical conductivities are much lower. As explained in Section 2.1.1, this arises because, at least when the level of structural perfection is high, phonon transmission in the plane of graphene sheets is very efficient. This is certainly expected to be the case for CNTs. Very high thermal conductivity, at least in the axial direction, is thus expected in composites containing uniaxially aligned CNTs. This has formed the basis for a large number [9,10,16–18] of suggestions and developments, although it should be appreciated that it is the high structural perfection of CNTs, rather than their small size *per se*, that gives rise to these expectations. Many other functional properties have also been investigated.

14.3.4   Mechanical Properties of Nanoparticle-Reinforced MMCs

There has been interest [19] in the use of very fine-scale reinforcement for MMCs, although to a lesser extent than for polymer matrices. This is partly due to many metals being strong carbide-formers, making them totally unsuitable for fine carbon reinforcement, and partly a consequence of the manufacturing difficulties, which tend to be even worse with metals than for polymers. Nevertheless, fine-scale MMCs have certainly been produced, commonly with SiC or Al_2O_3 nanoparticles. There are differences in what might be expected with a metal matrix, as opposed to a polymeric one. For example, fine-scale reinforcement is likely to influence the grain structure of a metal, which is expected to affect the mechanical properties. Furthermore, if the particles are sufficiently small, and closely spaced, then they might directly affect the mobility of dislocations (forcing them to adopt high curvature to avoid these obstacles, and hence raising the stress needed to make them move). These effects have no parallels for polymeric matrices.

Such mechanisms are expected to operate even with relatively low reinforcement contents and such materials are more properly regarded as 'dispersion-strengthened' rather than being genuine composites in which property enhancement has arisen from load transfer between matrix and reinforcement – see Section 1.3. Effects of these types are apparent in Fig. 14.7, which shows data [20] from tensile testing of Al containing up to 7 vol% of Al_2O_3 nanoparticles, with some corresponding values for more conventional MMCs containing much coarser SiC particles. It can be seen that the finer particles do lead to greater rises in yield stress and tensile strength than coarser ones, although at a cost to ductility (and toughness).

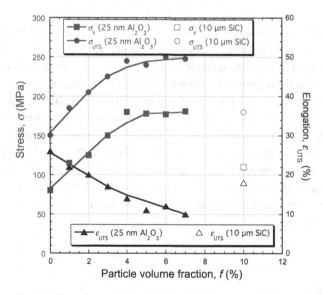

Fig. 14.7 Experimental data [20] for yield stress, tensile strength and ductility, for aluminium reinforced with either nanoparticles of Al_2O_3 or coarser particles of SiC, all having been produced via a similar (powder-based) manufacturing route.

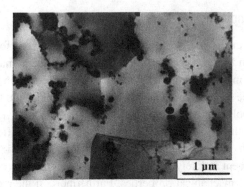

Fig. 14.8 TEM micrograph [20] of an Al-based MMC containing 4 vol% of Al_2O_3 nanoparticles. Reprinted from Materials Chemistry and Physics, 85, Yung-Chang Kang and Sammy Lap-Ip Chan, Tensile properties of nanometric Al_2O_3 particulate-reinforced aluminum matrix composites, 438–443, © 2004, with permission from Elsevier.

Insights into the origin of these effects can be obtained from the microstructure, such as that shown [20] in the TEM micrograph in Fig. 14.8. This suggests that the grain size was reduced by the presence of the nanoparticles and indeed it was confirmed [20] that a progressive reduction in grain size occurred as the nanoparticle content was raised. It can also be seen that particle separations in the sub-micron range are common. It is in this range that the shear stress, τ, needed for 'Orowan bowing' between obstacles starts to become significant

$$\tau \approx \frac{Gb}{L} \tag{14.1}$$

where G is the shear modulus, b is the Burgers' vector and L is the obstacle separation. (The G and b values of Al are around 25 GPa and 0.3 nm, so obstacles with a spacing of 200 nm are predicted to cause an increase in the yield stress of about 40 MPa.)

In general, while these property enhancements do look encouraging, and indeed dispersion strengthening of Al by fine Al_2O_3 particles is well-established, these cannot really be regarded as genuine composites (as opposed to dispersion-strengthened metals). In fact, as outlined in Section 9.2.5 (Figs 9.15 and 9.16), there are clear indications that low toughness becomes an issue as the particle contents of MMCs are raised, and this is often worse for finer particles. (This is again due to the reduced volume of the 'process zone' in which energy is absorbed during crack propagation.) Very fine-scale reinforcement is thus unlikely to be beneficial for MMCs, any more than for polymer composites.

References

1. Young, RJ, IA Kinloch, L Gong and KS Novoselov, The mechanics of graphene nanocomposites: a review. *Composites Science and Technology* 2012; **72**(12): 1459–1476.

2. Akinwande, D, CJ Brennan, JS Bunch, P Egberts, JR Felts, HJ Gao, R Huang, JS Kim, T Li, Y Li, KM Liechti, NS Lu, HS Park, EJ Reed, P Wang, BI Yakobson, T Zhang, YW Zhang, Y Zhou and Y Zhu, A review on mechanics and mechanical properties of 2D materials: graphene and beyond. *Extreme Mechanics Letters* 2017; **13**: 42–77.

3. Mensah, B, KC Gupta, H Kim, W Wang, KU Jeong and C Nah, Graphene-reinforced elastomeric nanocomposites: a review. *Polymer Testing* 2018; **68**: 160–184.

4. Lee, C, XD Wei, JW Kysar and J Hone, Measurement of the elastic properties and intrinsic strength of monolayer graphene. *Science* 2008; **321**(5887): 385–388.

5. Li, F, HM Cheng, S Bai, G Su and MS Dresselhaus, Tensile strength of single-walled carbon nanotubes directly measured from their macroscopic ropes. *Applied Physics Letters* 2000; **77** (20): 3161–3163.

6. Yu, MF, O Lourie, MJ Dyer, K Moloni, TF Kelly and RS Ruoff, Strength and breaking mechanism of multiwalled carbon nanotubes under tensile load. *Science* 2000; **287**(5453): 637–640.

7. Demczyk, BG, YM Wang, J Cumings, M Hetman, W Han, A Zettl and RO Ritchie, Direct mechanical measurement of the tensile strength and elastic modulus of multiwalled carbon nanotubes. *Materials Science and Engineering A: Structural Materials, Properties, Microstructure and Processing* 2002; **334**(1–2): 173–178.

8. Liu, K, YH Sun, RF Zhou, HY Zhu, JP Wang, L Liu, SS Fan and KL Jiang, Carbon nanotube yarns with high tensile strength made by a twisting and shrinking method. *Nanotechnology* 2010; 21(4): 045708.

9. Khan, ZU, A Kausar, H Ullah, A Badshah and WU Khan, A review of graphene oxide, graphene buckypaper, and polymer/graphene composites: properties and fabrication techniques. *Journal of Plastic Film & Sheeting* 2016; **32**(4): 336–379.

10. Mohan, VB, KT Lau, D Hui and D Bhattacharyya, Graphene-based materials and their composites: a review on production, applications and product limitations. *Composites Part B: Engineering* 2018; **142**: 200–220.

11. Coleman, JN, U Khan, WJ Blau and YK Gun'ko, Small but strong: a review of the mechanical properties of carbon nanotube–polymer composites. *Carbon* 2006; **44**(9): 1624–1652.

12. Shokrieh, MM and R Rafiee, A review of the mechanical properties of isolated carbon nanotubes and carbon nanotube composites. *Mechanics of Composite Materials* 2010; **46** (2): 155–172.

13. Spitalsky, Z, D Tasis, K Papagelis and C Galiotis, Carbon nanotube–polymer composites: chemistry, processing, mechanical and electrical properties. *Progress in Polymer Science* 2010; **35**(3): 357–401.

14. Al-Saleh, MH and U Sundararaj, Review of the mechanical properties of carbon nanofiber/ polymer composites. *Composites Part A: Applied Science and Manufacturing* 2011; **42**(12): 2126–2142.

15. McIntosh, D, VN Khabashesku and EV Barrera, Nanocomposite fiber systems processed from fluorinated single-walled carbon nanotubes and a polypropylene matrix. *Chemistry of Materials* 2006; **18**(19): 4561–4569.

16. Moisala, A, Q Li, IA Kinloch and AH Windle, Thermal and electrical conductivity of single- and multi-walled carbon nanotube-epoxy composites. *Composites Science and Technology* 2006; **66**(10): 1285–1288.

17. Yu, AP, P Ramesh, XB Sun, E Bekyarova, ME Itkis and RC Haddon, Enhanced thermal conductivity in a hybrid graphite nanoplatelet: carbon nanotube filler for epoxy composites. *Advanced Materials* 2008; **20**(24): 4740–4744.

18. Cui, W, FP Du, JC Zhao, W Zhang, YK Yang, XL Xie and YW Mai, Improving thermal conductivity while retaining high electrical resistivity of epoxy composites by incorporating silica-coated multi-walled carbon nanotubes. *Carbon* 2011; **49**(2): 495–500.

19. Tjong, SC, Novel nanoparticle-reinforced metal matrix composites with enhanced mechanical properties. *Advanced Engineering Materials* 2007; **9**(8): 639–652.

20. Kang, YC and SLI Chan, Tensile properties of nanometric Al_2O_3 particulate-reinforced aluminum matrix composites. *Materials Chemistry and Physics* 2004; **85**(2–3): 438–443.

15 Fabrication of Composites

An important aspect of composite materials concerns the technology by which they are produced. Depending on the nature of matrix and fibre, and the required architecture of fibre distribution, production at reasonable cost and with suitable microstructural quality can present a challenge. In most cases, manufacture of the final component and production of the composite material are carried out at the same time. This gives scope for optimal fibre placement and distribution of orientations, but also requires that the mechanical requirements of the application be well understood and that the processing route be tailored accordingly. Fabrication procedures for most commercially important (fibre-reinforced polymer) composites are technically mature, but there are some types of composite for which processing routes are still under development.

15.1 Polymer Matrix Composites

There are many commercial processes for the manufacture of polymer matrix composite (PMC) components. Reviews are available [1–6] covering different aspects of the technological issues involved. These processes may be subdivided in a variety of ways, but broadly speaking there are three main approaches to the manufacture of fibre-reinforced thermosetting resins and two distinct production methods for thermoplastic composites. These are briefly covered below under separate headings. In all cases, the main microstructural objectives are to ensure that the fibres are well wetted, uniformly distributed and correctly aligned. Practical considerations relating to capital cost, speed of production and component size and shape are often of paramount importance.

It may be noted at this point that the fibres are most commonly glass or carbon, usually with diameters in the range 7–12 μm. Most of the techniques can in principle be employed with much finer fibres, such as carbon nanotubes or nano-fibrils. In practice, however, there are major handling and processing obstacles, as outlined in Section 14.3.1. The consequence is severe difficulties in producing such composites with fibre contents above about 5–10%. Combined with some adverse expectations about achievable (mechanical) properties (Section 14.3.2), the outcome is that they are not in significant commercial use.

15.1.1   Liquid Resin Impregnation Routes

In the basic route, low-viscosity resin is impregnated into arrays of dry fibre. This can be achieved in several ways. The simplest is a wet lay-up procedure in which the fibres, usually in the form of a mat on a polished former or mould, are impregnated with resin by rolling or spraying. The resin and curing agent are mixed immediately prior to application. Curing usually takes place at ambient temperature. Variants include co-spraying of liquids so that mixing occurs as the liquids are introduced to the fibres. The main advantages of the process lie in its versatility. Virtually any shape can be produced and capital cost is limited to that of the mould. The technique has traditionally been used in a variety of small-scale operations, such as car body repair, commonly with glass fibres in the form of woven roving or chopped strand mat (see Figs 2.7 and 2.8). Larger-scale operation is, however, increasingly common. For example, the technique is widely used in the boat-building industry for production of relatively large craft such as minesweepers of up to about 50 m in length. The main disadvantages lie in the difficulty of ensuring complete impregnation and the labour-intensive nature of much of the work.

A process better suited to automation, although limited to certain component shapes, is *filament winding*. Fibre tows (bundles), are drawn through a bath of resin, before being wound onto a mandrel or former of the required shape. The equipment comprises (a) a creel stand, from which the fibre tows are fed under the required tension from a set of reels; (b) a bath of resin, through which the fibre tows pass via a set of guides; (c) a delivery eye, through which the fibres emerge, the position of which is controlled by a mechanical system; and (d) a rotating mandrel, onto which the fibre tows are drawn. The key parameters are the fibre tension, the resin take-up efficiency and the winding geometry. There are several possible designs for the movement of the delivery eye. A typical five-spindle machine is shown in Fig. 15.1(a) and a large filament-wound vessel can be seen in Fig. 15.1(b).

In most cases, the eye motion and mandrel rotation systems are computer-controlled. Component shapes can be fairly complex, although they usually exhibit a fairly high degree of symmetry. There are also some limitations on the paths that the fibres take over the surface of the component. On any curved surface, there will be a tendency for the fibres to follow a *geodesic* path – i.e. the shortest one. This can cause problems with some shapes, since it may be difficult to ensure that fibres cover some parts of the surface, or lie in certain orientations. It is, however, possible to ensure that fibres follow certain non-geodesic paths, provided the delivery eye is moved appropriately and there is sufficient friction between the tow and the underlying surface. Filament winding is often used to produce high-performance components and is obviously well suited to simple shapes such as tubes.

Pultrusion is a process that is similar in some respects to filament winding. This also involves fibre tows being passed through a resin bath. In this case, the impregnated tows are then fed into a heated, tubular die, in which they become consolidated and the resin is cured. The die may have a relatively complex sectional shape. The composite is pulled from the die by a frictional extraction system. The process generates stock material rather than finished components, and the product is similar in that respect to plastic and metallic material produced by conventional extrusion.

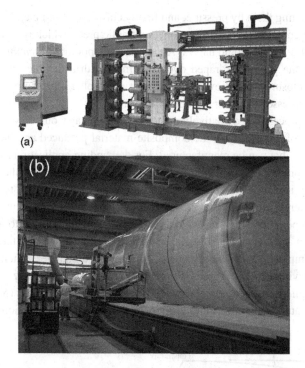

Fig. 15.1 (a) A five-spindle/four-axes filament winding machine (Mikrosam type MAW 20 MS4/5 machine – image from www.mikrosam.com/new/article/en/maw-20-ms45, courtesy of Mikrosam); and (b) a large pressure vessel being produced by filament winding (courtesy of Selip SpA).

Another impregnation route involves injection of resin into a mould in which the fibres are placed in position. The resin is fed in under gravity or external pressure. There are several different types of machine. A commonly used process is that termed *resin transfer moulding* (RTM), in which the fibres are enclosed in a die and pre-catalysed resin, having a low viscosity, is injected into the mould at relatively low pressure. Cure occurs within the mould, often assisted by heating. The mould is usually of metal, which gives good heat transfer and lasts for many moulding operations. To encourage good infiltration, the mould is sometimes evacuated prior to injection of the resin. Relatively large mouldings, including many body parts for the automobile industry, are made in this way.

15.1.2 Pressurised Consolidation of Resin Pre-Pregs

This approach involves production of a *pre-preg*, which is a tape or sheet of fibres impregnated with resin. Pre-preg is manufactured by laying the fibres and resin between sheets of siliconised paper or plastic film, which are pressed or rolled to ensure consolidation and wetting out of the fibres, and then partially cured to produce a flexible aggregate. The process allows excellent alignment of the fibres in unidirectional layers. A component is formed by stacking up the layers of pre-preg in predetermined directions, consolidating them by pressure, and finally curing by heating under pressure. The simplest arrangement involves a directional hydraulic press, which is used to apply pressure to a pair of matched mould halves. An alternative approach, *vacuum moulding*, involves enclosing the pre-preg (draped over a former) with a flexible, impermeable membrane. The enclosure is then evacuated so that atmospheric pressure acts on the membrane and compresses the pre-preg onto the former.

While the above processes are relatively quick and easy to carry out, they suffer from limitations in terms of the quality of the composite material produced and the size and shape of components that can be manufactured. For demanding applications, *autoclave moulding* is often preferred. This is similar to vacuum moulding, but the enclosed composite assembly is placed into a large chamber that can be pressurised, typically up to ~10–20 atmospheres. Heating is applied during pressurisation to cure the resin. In many cases, the temperature is first raised to an intermediate level in order to reduce the viscosity of the resin and ensure that all voids are removed. Further heating then ensures that the cure is complete. A schematic representation of the changes in temperature, pressure and resin viscosity during autoclave moulding is shown in Fig. 15.2. The process can be applied to large components. This is illustrated by Fig. 15.3, which shows a very large autoclave, with some technical information in the caption.

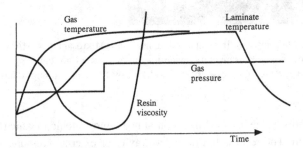

Fig. 15.2 Schematic of an autoclave cure cycle.

Fig. 15.3 Photograph of a large autoclave, with inside working diameter over 9 m and length over 23 m. The maximum temperature is ~230°C and the maximum pressure ~10 bar. It is used for production of large composite components, particularly in the aerospace industry. (Courtesy of ASC Process Systems.)

15.1.3   Consolidation of Resin Moulding Compounds

The third approach for thermosetting resins is a variation of the pre-preg approach. The intermediate products are *sheet moulding compound* (SMC) and *dough moulding compound* (DMC), which are usually based on polyester resins. To make SMC, resin containing thickening agents and particulate fillers such as calcium carbonate (chalk), are mixed with chopped fibres to form a slurry. This is then fed between two thermoplastic films and over a series of rolls, so that impregnation takes place and a consolidated sheet is produced. The fibres lie mainly in the plane of the sheet and the fibre volume fraction is usually in the range 15–40%. Sheets are typically 3–10 mm thick. DMCs are made by mixing together similar constituents to those in SMCs, but the mixing is carried out differently. Usually, a blender generating high shear rates is used for DMCs. This tends to break the fibres up into shorter lengths and also results in a more random orientation distribution.

Both SMC and DMC are processed in closed moulds within a hot press, where they undergo a final cure. SMCs are commonly converted directly into components by hot press or compression moulding. This is done by removing the thermoplastic films, cutting the sheets to suitable shapes and placing them in a heated mould. Similar operations are carried out with DMCs, although in this case the mixture is sometimes formed into more complex 3D shapes. A related process is *reaction injection moulding*, which can be done with or without fibre reinforcement. This also bears some similarity to RTM, but instead of using a pre-catalysed resin that cures slowly, as with RTM, two components that react quickly are mixed, together with some short fibres, and immediately injected into a mould. In all such processes, the presence of fibres raises the viscosity of the mixture dramatically and strongly inhibits its flow, even when they are relatively short, so that the moulding operation often becomes impractical unless the fibre content is low.

15.1.4   Injection Moulding of Thermoplastics

Short fibre-reinforced thermoplastic materials are used to make components by *injection moulding*, a process that is widely applied to unreinforced thermoplastics. Three polymers for which the process is common are polypropylene, nylon and polycarbonate. The first two are semi-crystalline with about 25–50% crystallinity and the latter is amorphous. Polymer pellets are fed into the heated barrel of a chamber containing an Archimedes screw for transporting the charge. The shear motion helps to homogenise and melt the polymer, which is then periodically injected into a mould. The same process is carried out using pellets containing short fibres, typically 1–5 mm long, intimately mixed and dispersed in the matrix. However, the comments in Section 15.1.3 about flow of a mixture becoming difficult when fibres are present also apply to these operations, and the volume fraction of fibres is usually no greater than about 10–20%.

Of interest for injection-moulded components is the fibre orientation distribution. For example, it may be possible to arrange for the fibres to be oriented parallel to the direction of stress in parts of the component that will be heavily loaded in service. Fibre

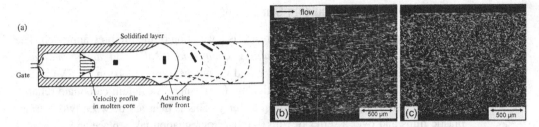

Fig. 15.4 Filling of the mould during injection moulding [7]: (a) a depiction of the process, showing the deformation of an initially square fluid element at successive positions of the advancing flow front; (b) a contact micro-radiograph of the longitudinal section of a polypropylene–15% glass fibre injection moulding; and (c) a transverse section from the same moulding.

Reprinted from Polymer, 21, M.J. Folkes and D.A.M. Russell, Orientation effects during the flow of short-fibre reinforced thermoplastics, 1252–1258, © 1980, with permission from Elsevier.

orientation is controlled by the nature of the flow field during filling of the mould. In general, fibres tend to become aligned parallel to the direction in which the material is becoming elongated (provided the flow is not too turbulent). This effect is illustrated by the flow pattern depiction and corresponding micro-radiographs shown in Fig. 15.4. In this case, the fibres are fairly well aligned in the outer layers of the moulding, but more randomly oriented towards the core. By predicting the flow behaviour under different injection conditions for specific components, a degree of control over the final fibre orientation pattern is often possible.

15.1.5   Hot Press Moulding of Thermoplastics

Thermoplastics reinforced with long or continuous fibres are commonly used in *laminated structures*. The first stage in manufacture is production of a pre-preg by *melt impregnation* of the fibres, which may be in the form of continuous aligned sheets, woven cloth, etc. The pre-preg sheets are then stacked in the required orientations and *hot pressed* to form the final product. Pre-pregs are available in many materials, including polypropylene, polysulphone and polyether ether ketone (PEEK). Processing is carried out in a similar manner to that involved during pressurised consolidation of resin pre-pregs (Section 15.1.2). For thermoplastics, there is no requirement for the matrix to be cured and the temperature is usually set to the minimum necessary for the matrix to melt and flow sufficiently for consolidation to occur. However, even when fully molten, thermoplastic melts are much more viscous than uncured resins, since thermoplastics are already fully polymerized. Substantial heating and pressure are therefore necessary, and impregnation distances into fibre arrays are always kept as short as possible.

15.2   Metal Matrix Composites

While there is significant commercial production of metal matrix composites (MMCs), it is on a far smaller scale than that of PMCs. To some extent, this depends on how an MMC is defined. For example, the cermet ('hardmetal') industry operates on a relatively large scale and these materials can certainly be classified as MMCs. A number of other types of MMC are also in fairly widespread use and there is a range of processing options. Most MMCs of commercial interest, including cermets, are based on particulate reinforcement and are manufactured by some kind of powder blending and sintering route. Other processes involve handling of bulk liquid metal. Common dangers include (excessive) porosity, interfacial reaction and inhomogeneous distribution of reinforcement.

15.2.1 Particulate MMCs

Excessive chemical reaction during processing can occur with some reinforcements, such as SiC in Al- and Ti-based MMCs. Reaction between SiC and liquid Al is well documented [8–10]. Titanium is likely to be processed into MMCs in the solid state (via powder routes), but SiC often reacts with Ti during such processing [11–13]. Since Ti is a strong carbide-former, this occurs to a greater extent than with C-free alternatives such as TiB_2 [14]. Coatings (such as B_4C [15] or TiB_2 [16]) are sometimes created on SiC for use in Ti, in order to control interfacial reactions.

Reactions also occur between SiC and ferrous matrices [17], sometimes leading to complete dissolution [18], although again TiB_2 has proved effective as a stable particulate reinforcement in ferrous MMCs. In fact, there is interest in 'high modulus steels' containing about 10–20 vol% of TiB_2 particulate [19–23]. The Young's modulus can thus be raised from 210 GPa to around 250–300 GPa. This system might turn out to be one of the more successful types of particulate MMC.

Alumina is less reactive than SiC in Al, but it does react strongly with Ti at high temperature. Magnesium is rather different from Al and Ti, in that it does not form a stable carbide, but it does have a very high affinity for oxygen. The greater stability of Al_2O_3, compared with SiC, in Al is therefore reversed for Mg matrices [24]. In general, while coatings or other surface treatments may be worth considering for fibres (particularly monofilaments), economic and practical considerations mean that particulate reinforcement is normally introduced into MMCs in the virgin state. This may, however, be such that a surface oxide layer is present and deliberate thickening of this layer, for example by heat treatment in air, has in some instances been found to have a beneficial effect on interfacial bonding or other characteristics.

Particulate MMCs are usually manufactured on a commercial basis either by melt incorporation and casting or by powder blending and consolidation. Typically, reinforcement particles are about 10–20 μm in diameter and constitute about 10–30% by volume of the material, although MMCs in which the values concerned lie outside of these ranges have been studied and are available commercially (particularly finer particles and higher particle contents). Of course, cermets constitute particulate MMCs with a very high loading (~90%) of reinforcement. The two micrographs in Fig. 15.5 allow comparison between structures typical of particulate MMCs and cermets. Fig. 15.5(a) shows an MMC produced by blending of (spheroidised) Al and Al_2O_3 particles, followed by extrusion and heat treatment, as part of a study on the effect of reinforcement in MMCs on recrystallisation [25].

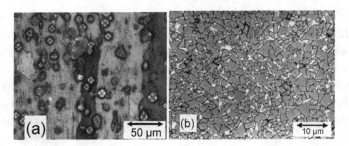

Fig. 15.5 Optical micrographs [26] of (a) a particle-reinforced MMC (Al–10% Al_2O_3) and (b) a cermet (Co–90% WC).
Reprinted from Elsevier Books, 21, T. William Clyne, Comprehensive Composite Materials II, 1–21, © 2018, with permission from Elsevier.

The degree of inhomogeneity (clustering) apparent here is typical of both powder and melt route particulate MMCs. As with the material shown, individual grains often contain a (large) number of particles. Also, such material is usually more or less free of porosity. The cermet shown in Fig. 15.5(b), on the other hand, was produced by blending of WC and Co powders, followed by moulding to a green compact and heating so as to melt the Co matrix. The volume fraction of ceramic particles is so high that they are mostly in contact with each other and there is little inhomogeneity. On the other hand, the solidification shrinkage of the Co almost inevitably leaves some residual porosity, which can be seen in this micrograph. Also, while it is not entirely clear in this micrograph, individual matrix grains rarely envelop even a single ceramic particle.

In fact, one of the keys to the commercial success of cermets is their ease of processing. They are commonly produced by blending of ceramic and metallic powders, followed by liquid-phase sintering, although there are many processing variants, including spark plasma sintering, application of pressure, etc. Typically, the ceramic particles used in cermet production are 1–10 μm in diameter. Blending involves a milling operation that tends to coat the ceramic particles with metal. This is usually followed by cold isostatic pressing, or injection moulding, to give the required shape, and then holding at a suitable temperature under vacuum, inert gas or hydrogen. During liquid-phase sintering, particle rearrangement occurs [27], driven by capillarity forces. This may or may not lead to effective densification, depending on the degree of wetting in the system. In some cases, notably for oxide-based cermets, it is often necessary to impose unixial or hydrostatic pressure in order to eliminate porosity.

15.2.2 Reactively Processed MMCs

A concept that has attracted interest is that of MMCs produced by passage of a molten metal infiltration front through a packed ceramic bed of some sort, with or without associated chemical reaction. Such materials sometimes bear a marked resemblance to the cermets described in the preceding section, but a distinction can be drawn in terms of certain differences in processing conditions and in the alloy and ceramic systems that are commonly involved.

Much of the work done in this area has its origins in attempts to facilitate the melt infiltration process, particularly with Al-based melts. When liquid metal is injected into a relatively fine array of particles or fibres, the applied pressure needed to generate the necessary meniscus curvature at the infiltration front can be very large [28–30] – typically at least several MPa. If the scale of the reinforcement is sub-micron, then even higher pressures are needed. Efforts to reduce this requirement have focused on 'wetting' and/or chemical reaction at the infiltration front, to promote spontaneous infiltration without the need for external pressure. While it has proved difficult to promote rapid infiltration under these conditions, MMC products can be made in this way with good near-net shape characteristics, particularly when the ceramic content is high. In Al melts, wetting can be promoted by chemical reactions arising from the introduction of Mg into the melt and nitrogen into the surrounding atmosphere (preferably in the absence of oxygen). Such conditions are experimentally fairly easy to arrange. Compounds such as AlN and $MgAl_2O_4$ are then formed [30–33].

While the details of the thermodynamic and kinetic characteristics involved are complex, the important point is that the reactions involved take place at locations and at rates such as to promote spontaneous penetration of an Al-based melt into an array of ceramic particles or fibres. Commonly, the melt employed is a binary or multi-component Al–Mg alloy and the process is carried out under a nitrogen atmosphere. This type of processing, for which trademarks such as 'DIMOX' and 'PRIMEX' have been used, are described in some detail [34–40]. There have also been claims [41] that arrays of carbon nanotubes can be infiltrated (by metallic melts) without the application of pressure, although an obvious problem with ultra-fine reinforcement is that almost any reaction (needed for infiltration) will entirely consume it. This is particularly problematic for carbon in Al, which rapidly forms a carbide (Al_4C_3) that is hygroscopic.

Unfortunately, the reaction kinetics and local melt flow characteristics are usually such that, even when infiltration does occur spontaneously, it tends to be relatively slow. Typically it might take many minutes, or even hours, for a preform to become fully infiltrated. This problem, together with the fact that the presence of relatively high reaction product contents may impair mechanical properties, has meant that reactive processing is not extensively used for production of 'conventional' low-cost MMCs, for applications such as pistons or brake disks. However, the process has clear attractions for more specialised applications, particularly those requiring high ceramic contents – which are rather difficult to produce by standard infiltration or powder processing methods. Again, there is clear overlap with cermet technology here. For example, the Al–70 vol% SiC particulate MMC shown in Fig. 15.6(a) was produced using the PRIMEX process, in the form of the components shown in Fig. 15.6(b). These are part of a specialised chuck used in the electronics industry for wafer handling, which has demanding stiffness, conductivity and thermal expansivity requirements [42]. The bimodal size distribution of the SiC particles may be noted: this facilitates the generation of high ceramic contents in the initial powder compact.

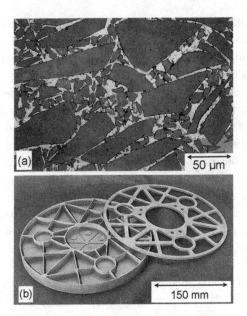

Fig. 15.6 Use of the PRIMEX process [42] to produce an Al–70 vol% SiC MMC, showing (a) an optical micrograph and (b) components of a chuck for securing Si wafers during processing of electronic devices.
Reprinted from Elsevier Books, 21, T. William Clyne, Comprehensive Composite Materials II, 1–21, © 2018, with permission from Elsevier.

15.2.3 Short Fibre- and Monofilament-Reinforced MMCs

While MMCs of these types have not yet found extensive commercial application, there is ongoing interest in their potential. For short fibres, the injection of liquid metal into a fibre 'preform' has been extensively explored [30], often termed *squeeze infiltration*. Commonly, the preform is designed with a specific shape to form an integral part of a finished product in the as-cast form. Preforms are usually fabricated by sedimentation of short fibres from liquid suspension, often using short alumina fibres, such as 'Saffil'^{TM}. An SEM micrograph of some (milled) Saffil fibres is shown in Fig. 15.7.

Various types of long (continuous) fibre with relatively small diameters (<~20 μm), mainly based on SiC or Al_2O_3, have been extensively explored for use in MMCs, but in general they have not been commercially successful. This is partly because of difficulties, using either liquid- or solid-state routes, in obtaining well-infiltrated arrays of such fibres, without excessive interfacial reaction or other fibre damage. Monofilaments (~100–150 μm diameter) are less susceptible to interfacial reaction, since it affects a lower proportion of the fibre volume. There is also greater scope for tailoring the surface chemistry and introducing surface coatings (as part of the fibre production process). Most are produced by chemical vapour deposition (CVD) of either SiC or boron onto a core of carbon fibre or tungsten wire. A consequence of the large diameter is that they have limited flexibility and so need to be handled as single

Fig. 15.7 SEM micrograph of 'Saffil' short fibres of Al_2O_3.
(Courtesy of Dr A. J. Houston.)

fibres rather than bundles. Virtually all fibre materials react with Ti at elevated temperature, so that use of coated monofilaments is one of the few approaches offering scope for control of this problem. Thick graphitic coatings (which are progressively consumed, but prevent defects from forming on the fibre itself) have been popular and there has also been work [43–45] on various duplex layers, such as TiB_2/C and Y/Y_2O_3, which are designed to slow interfacial reaction rates down to very low levels.

Composite production can be carried out by placing arrays of fibres between thin Ti (alloy) foils, often involving a filament winding operation, followed by hot pressing. An alternative is to pass fibres through an evacuated chamber, where evaporated Ti condenses onto them so as to produce a relatively thick coating. The vapour is produced by directing a high-power electron beam so as to melt the end of a solid bar feedstock. Fabrication is completed by assembling the coated fibres into a bundle and consolidating it by hot pressing or hot isostatic pressing (HIP). A very uniform distribution of fibres is produced, with fibre contents of up to about 80%. However, production is slow and the process is very expensive. There has been limited recent activity in the area, partly due to the very high cost of this type of processing. Viable niche applications for such composites may nevertheless emerge in due course, since they can offer some attractive combinations of properties.

15.3 Ceramic Matrix Composites

Fabrication of ceramic matrix composites (CMCs) presents some major challenges. This is partly due to the brittle nature of ceramic matrices, which makes deformation processing difficult. Furthermore, the matrix is often unable to accommodate the volume changes associated with consolidation, without cracks being formed. This can be particularly troublesome when fibres are being incorporated, since these tend to resist the contraction of the matrix as voids are eliminated. In addition, many CMCs, particularly most of those composed of ceramic fibres in a ceramic matrix, have exhibited mechanical properties that were rather disappointing. The whole concept of a CMC is largely focused on toughness, since 'reinforcements' are unlikely to confer benefits in terms of stiffness or 'strength' (ultimate tensile stress). Of course, toughness enhancement is a key benefit of introducing ceramic fibres into a polymer matrix. This occurs mainly via fibre pull-out, which might be expected to operate with a ceramic matrix. However, with CMCs it is often difficult to create a suitable type of interface for this to occur. Many of the routes employed have either left the fibres completely unbonded or have stimulated interfacial reactions that create such strong bonding that pull-out does not occur.

However, there are at least two approaches that are currently in significant commercial use. One involves the use of metallic fibres, with the ceramic matrix infiltrated as a slurry (and left with a relatively high porosity). The toughness arises largely from the scope for the fibres to absorb energy during crack propagation via plastic deformation, as well as by pull-out. The other, which is specific to carbon–carbon composites, involves infiltration of the matrix into an array of fibres via evaporation and condensation, avoiding the above difficulties related to cracking, etc. However, there is ongoing research interest in CMCs produced using a variety of routes, many of which do involve sintering.

15.3.1   Powder Sintering Routes with Ceramic Fibres

Production of monolithic ceramic components via powder-based routes is well established in industrial practice. Green bodies are produced by cold compaction of fine powders, often with a binder of some type, followed by pressureless sintering or hot pressing, such that voids are eliminated by diffusive processes. In some cases, a liquid is present during the consolidation, in which case void elimination is assisted by capillary action and consolidation is much faster. However, problems arise when an attempt is made to introduce fibres into such operations. Various reviews [46–49] cover the issues involved. The fibres can be mixed with ceramic powder particles, for example by blending or by dragging a fibre tow through a suitable slurry. However, subsequent consolidation is then strongly inhibited as the fibres resist the volume contraction of the matrix as it consolidates. Severe matrix cracking often results. In some cases, these contraction stresses can be at least partly offset by the imposition of a large hydrostatic compressive stress, as in the HIP process. However, this adds substantially to the processing costs and may not eliminate cracking entirely.

A possible way to resolve this problem is to employ a matrix that is partly, or completely, liquid at the consolidation temperature. This constrains the choice of matrix to those that are likely to show relatively poor properties at elevated temperatures. Nevertheless, there has been interest in making CMCs in this way, using glass or glass–ceramic matrices. Boro-silicate glasses and cordierite (glass–ceramic) have attracted particular attention. However, even when the consolidation can be achieved without severe matrix cracking, problems often arise from differential thermal contraction. The misfit strain is often large for systems in which the matrix is fluid at high temperatures, since in such cases it is common for it to exhibit an appreciably higher thermal expansivity than the fibres.

15.3.2 Powder Slurry Routes with Metallic Fibres

This type of composite material is marketed commercially, an example being that produced by Fiberstone Products Ltd under the tradename 'Fiberstone™'. It is manufactured using a slightly different approach to that outlined above, with no high-temperature processing involved, and the matrix in the finished product having a relatively high level of porosity. An assembly of short metallic fibres is infiltrated (without pressure) with a ceramic slurry, which subsequently consolidates via chemical reactions. Some information about the microstructure and properties of this type of material is provided in Section 9.2.7. The fibres confer high levels of toughness, mainly because they undergo plastic deformation during fracture of the composite. The material can be used at high temperatures (~1000°C).

The fibres are usually stainless steel, commonly AISI304, although other grades may be preferred if good resistance to degradation (oxidation) and creep at very high temperatures is required. A cost-effective way to produce such (short) fibres is the melt extraction process [50], which involves solidification of fibres onto surface protrusions on a water-cooled, rapidly rotating copper drum, which is in contact with a steel melt. This process is normally carried out on a relatively large scale, with a typical batch size for a single melt being ~1.5 tonnes. Fig. 15.8 shows a schematic depiction and a photograph of the process. There are, of course, more conventional routes, involving mechanical reduction in section (extrusion, drawing, swaging, etc.), but melt extraction requires a lower total energy input. The diameter and aspect ratio (length/diameter) of the fibres are variables, although in practice they are usually relatively coarse (hundreds of microns in diameter) and have aspect ratios of the order of 50–100. As outlined in Section 9.2.7, coarser fibres lead to composites with higher toughness.

Products are supplied in the form of finished ('cast') components, rather than as stock material. They are produced [51] by packing fibres into a mould and pouring a slurry

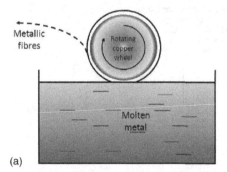
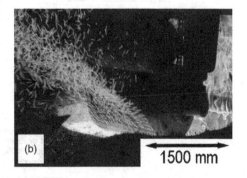

Fig. 15.8 Illustration [51] of the melt extraction process for the manufacture of short fibres (of stainless steel), showing (a) a schematic section and (b) a photograph of a unit in use. Reprinted from Elsevier Books, 18, T. William Clyne and Lee W. Marston, Comprehensive Composite Materials II, 464–481, © 2018, with permission from Elsevier.

into the cavity, using vibration to assist infiltration. 'Castings' are often subjected to (relatively mild) heat treatments – for example at ~300–600°C – to ensure that they contain no residual unbound water. The matrix is not fully dense and porosity levels are often ~10–15%. In most cases, the matrix is primarily either corrundum (α-alumina) or mullite, although there are also (cementitious) species of some sort present in the starting powder, which react with water to promote inter-particle bonding. The composite matrix therefore contains these hydration reaction products. The fibres usually occupy ~10–20 vol% and they have an approximately isotropic orientation distribution. An example application is described in Section 16.7.3.

15.3.3   Carbon–Carbon Composites

Carbon is an excellent high-temperature material, provided it is not exposed to oxidising environments. Composites made up of carbon fibres in a carbon matrix, with suitable interfacial properties, have found important applications, particularly aircraft brakes (Section 16.5.1). Two basic approaches [52–54] are used for their production. Both involve the infiltration of a carbon-bearing fluid into the interstices between an assembly of carbon fibres. In both cases, the main concern is with achieving complete infiltration in a reasonably short time. The two routes are shown schematically in Fig. 15.9. During liquid impregnation, a pitch or resin is injected and then heated so that it decomposes to leave a carbon deposit. Chemical vapour impregnation (CVI) involves injection of a suitable hydrocarbon gas, such as methane, together with hydrogen and nitrogen, which decomposes at the infiltration temperature to deposit carbon on the fibres. For both liquid and gaseous impregnation, several cycles of heating and cooling are necessary to complete the operation. Furthermore, it is common to set up a thermal gradient across

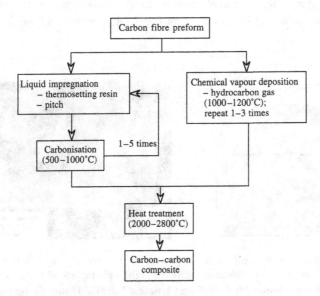

Fig. 15.9 Schematic overview of approaches to production of carbon–carbon composites.

the component in order to encourage complete infiltration before the supply of fluid becomes choked off by closing of channels near to the source of fluid. For these reasons, processing is time-consuming and costly.

References

1. Skartsis, L, JL Kardos and B Khomami, Resin flow through fiber beds during composite manufacturing processes: 1. review of newtonian flow through fiber beds. *Polymer Engineering and Science* 1992; **32**(4): 221–230.
2. Svensson, N, R Shishoo and M Gilchrist, Manufacturing of thermoplastic composites from commingled yarns: a review. *Journal of Thermoplastic Composite Materials* 1998; **11**(1): 22–56.
3. McCrary-Dennis, MCL and OI Okoli, A review of multiscale composite manufacturing and challenges. *Journal of Reinforced Plastics and Composites* 2012; **31**(24): 1687–1711.
4. Grunewald, J, P Parlevliet, and V Altstadt, Manufacturing of thermoplastic composite sandwich structures: a review of literature. *Journal of Thermoplastic Composite Materials* 2017; **30**(4): 437–464.
5. Parandoush, P and D Lin, A review on additive manufacturing of polymer–fiber composites. *Composite Structures* 2017; **182**: 36–53.
6. Bjornsson, A, M Jonsson and K Johansen, Automated material handling in composite manufacturing using pick-and-place systems: a review. *Robotics and Computer-Integrated Manufacturing* 2018; **51**: 222–229.
7. Folkes, MJ and DAM Russell, Orientation effects during the flow of short fiber reinforced thermoplastics. *Polymer* 1980; **21**(11): 1252–1258.
8. Skibo, M, PL Morris and DJ Lloyd. Structure and properties of liquid metal processed SiC reinforced aluminium, in *Cast Reinforced Metal Composites*. ASM, 1988.
9. Lloyd, DJ, H Lagace, A McLeod and PL Morris, Microstructural aspects of aluminium–silicon carbide particulate composites produced by a casting method. *Materials Science and Engineering A: Structural Materials, Properties, Microstructure, and Processing* 1989; **107**: 73–80.
10. Sritharan, T, LS Chan, LK Tan and NP Hung, A feature of the reaction between Al and SiC particles in an MMC. *Materials Characterization* 2001; **47**(1): 75–77.
11. Kooi, BJ, M Kabel, AB Kloosterman and JTM de Hosson, Reaction layers around SiC particles in Ti: an electron microscopy study. *Acta Materialia* 1999; **47**(10): 3105–3116.
12. Zou, Y, ZM Sun, H Hashimoto and L Cheng, Reaction mechanism in Ti-SiC-C powder mixture during pulse discharge sintering. *Ceramics International* 2010; **36**(3): 1027–1031.
13. Zhang, W, YQ Yang, GM Zhao, B Huang, ZQ Feng, X Luo, MH Li and JH Lou, Investigation of interfacial reaction in SiC fiber reinforced Ti-43Al-9V composites. *Intermetallics* 2013; **33**: 54–59.
14. Turner, SP, R Taylor, FH Gordon and TW Clyne, Thermal properties of Ti-SiC and Ti-TiB$_2$ reinforced composites. *International Journal of Thermophysics* 1996; **17**: 239–251.
15. Zhang, W, YQ Yang, GM Zhao, ZQ Feng, B Huang, X Luo, MH Li and YX Chen, Interfacial reaction studies of B4C-coated and C-coated SiC fiber reinforced Ti-43Al-9V composites. *Intermetallics* 2014; **50**: 14–19.

16. Ferraris, M, C Badini, F Marino, F Marchetti and S Girardi, Interfacial reactions in a Ti-6Al-4V based composite: role of the TiB$_2$ coating. *Journal of Materials Science* 1993; **28**(7): 1983–1987.

17. Pelleg, J, Reactions in the matrix and interface of the Fe–SiC metal matrix composite system. *Materials Science and Engineering A: Structural Materials, Properties, Microstructure, and Processing* 1999; **269**: 225–241.

18. Buytoz, S and M Ulutan, In situ synthesis of SiC reinforced MMC surface on AISI 304 stainless steel by TIG surface alloying. *Surface & Coatings Technology* 2006; **200**(12–13): 3698–3704.

19. Tanaka, K and T Saito, Phase equilibria in TiB$_2$-reinforced high modulus steel. *Journal of Phase Equilibria* 1999; **20**(3): 207–214.

20. Aparicio-Fernandez, R, H Springer, A Szczepaniak, H Zhang and D Raabe, In-situ metal matrix composite steels: effect of alloying and annealing on morphology, structure and mechanical properties of TiB$_2$ particle containing high modulus steels. *Acta Materialia* 2016; **107**: 38–48.

21. Springer, H, RA Fernandez, MJ Duarte, A Kostka and D Raabe, Microstructure refinement for high modulus in-situ metal matrix composite steels via controlled solidification of the system Fe–TiB$_2$. *Acta Materialia* 2015; **96**: 47–56.

22. Zhang, H, H Springer, R Aparicio-Fernandez and D Raabe, Improving the mechanical properties of Fe–TiB$_2$ high modulus steels through controlled solidification processes. *Acta Materialia* 2016; **118**: 187–195.

23. Baron, C, H Springer and D Raabe, Efficient liquid metallurgy synthesis of Fe–TiB$_2$ high modulus steels via in-situ reduction of titanium oxides. *Materials & Design* 2016; **97**: 357–363.

24. Hallstedt, B, ZK Liu and J Agren, Reactions in Al$_2$O$_3$-Mg metal-matrix composites during prolonged heat-treatment at 400-degrees-C, 550-degrees-C and 600-degrees-C. *Materials Science and Engineering A: Structural Materials, Properties, Microstructure, and Processing* 1993; **169**(1–2): 149–157.

25. Shahani, RA and TW Clyne, Recrystallization in fibrous and particulate metal matrix composites. *Materials Science and Engineering A: Structural Materials, Properties, Microstructure, and Processing* 1991; **135**: 281–285.

26. Clyne, TW, An introductory overview of metal matrix composites systems, types and developments, in *Comprehensive Composite Materials II*, Clyne, TW, editor. Elsevier, 2018, pp. 1–21.

27. German, RN, *Sintering Theory and Practice*. John Wiley & Sons, 1996.

28. Clyne, TW and JF Mason, The squeeze infiltration process for fabrication of metal matrix composites. *Metallurgical Transactions A* 1987; **18**: 1519–1530.

29. Mortensen, A and T Wong, Infiltration of fibrous preforms by a pure metal: part III. Capillary phenomena. *Metallurgical Transactions A*, 1990; **21**: 2257–2263.

30. Manu, KMS, LA Raag, TPD Rajan, M Gupta and BC Pai, Liquid metal infiltration processing of metallic composites: a critical review. *Metallurgical and Materials Transactions B: Process Metallurgy and Materials Processing Science* 2016; **47**(5): 2799–2819.

31. Aghajanian, MK, MA Rocazella, JT Burke and SD Keck, The fabrication of metal matrix composites by a pressureless infiltration technique. *Journal of Materials Science* 1991; **26**: 447–454.

32. Scholz, H and P Greil, Nitridation reactions of molten Al–(Mg, Si) alloys. *Journal of Materials Science* 1991; **26**: 669–677.

33. Scholz, H, R Gunther, J Rodel and P. Greil, Formation of Al_2O_3 fibre reinforced AlN/Al matrix composites by Al(Mg) melt nitridation. *Journal of Materials Science Letters* 1993; **12**: 939–942.

34. Schiroky, GH, DV Miller, MK Aghajanian and AS Fareed, Fabrication of CMCs and MMCs using novel processes, in *CMMC 96: Proceedings of the First International Conference on Ceramic and Metal Matrix Composites, Pts 1 and 2*, Fuentes, M, Martinez Esnaola, J M and Daniel, AM, editors. Trans Tech Publications, 1997, pp. 141–152.

35. Cui, CX, YT Shen, FB Meng and SB Kang, Review on fabrication methods of in situ metal matrix composites. *Journal of Materials Science & Technology* 2000; **16**(6): 619–626.

36. Rao, BS and V Jayaram, Pressureless infiltration of Al–Mg based alloys into Al_2O_3 preforms: mechanisms and phenomenology. *Acta Materialia* 2001; **49**(13): 2373–2385.

37. Ren, SB, XB He, XH Qu, IS Humail and Y Li, Effect of Mg and Si in the aluminum on the thermo-mechanical properties of pressureless infiltrated SiCp/Al composites. *Composites Science and Technology* 2007; **67**(10): 2103–2113.

38. Wittig, D, A Glauche, CG Aneziris, T Minghetti, C Schelle, T Graule and J Kuebler, Activated pressureless melt infiltration of zirconia-based metal matrix composites. *Materials Science and Engineering A: Structural Materials, Properties, Microstructure and Processing* 2008; **488**(1–2): 580–585.

39. Chen, JC, CY Hao and JS Zhang, Fabrication of 3D-SiC network reinforced aluminum-matrix composites by pressureless infiltration. *Materials Letters* 2006; **60**(20): 2489–2492.

40. Nakae, H and Y Hiramoto, Spontaneous infiltration of Al melts into SiC preform. *International Journal of Metalcasting* 2011; **5**(2): 23–28.

41. Zhou, SM, XB Zhang, ZP Ding, CY Min, GL Xu and WM Zhu, Fabrication and tribological properties of carbon nanotubes reinforced Al composites prepared by pressureless infiltration technique. *Composites Part A: Applied Science and Manufacturing* 2007; **38**(2): 301–306.

42. Singh, JR, The role of composites as an enabling materials technology for transition to 300 mm precision systems. *Future Fab* 1999; **2**(5). Available at: www.future-fab.com.

43. Kieschke, RR and TW Clyne, Development of a diffusion barrier for SiC monofilaments in titanium. *Materials Science and Engineering A: Structural Materials, Properties, Microstructure and Processing* 1991; **135A**: 145–149.

44. Haque, S and KL Choy, Push-out testing of SiC monofilaments with a TiC based functionally graded coating. *Journal of Materials Science* 2000; **35**(17): 4225–4229.

45. Shatwell, RA, Adhesion of SM1240+ coatings to silicon carbide substrate in sigma mono-filament. *Materials Science and Technology* 1995; **10**: 552–557.

46. Boccaccini, AR, Glass and glass–ceramic matrix composite materials. *Journal of the Ceramic Society of Japan* 2001; **109**(7): S99–S109.

47. Zok, FW, Developments in oxide fiber composites. *Journal of the American Ceramic Society* 2006; **89**(11): 3309–3324.

48. Nozawa, T, T Hinoki, A Hasegawa, A Kohyama, Y Katoh, LL Snead, CH Henager and JBJ Hegeman, Recent advances and issues in development of silicon carbide composites for fusion applications. *Journal of Nuclear Materials* 2009; **386–388**: 622–627.

49. Garshin, AP, VI Kulik, SA Matveev and AS Nilov, Contemporary technology for preparing fiber-reinforced composite materials with a ceramic refractory matrix (review). *Refractories and Industrial Ceramics* 2017; **58**(2): 148–161.

50. Maringer, RE and CE Mobley, Casting of metallic filament and fiber. *Journal of Vacuum Science and Technology* 1974; **11**(6): 1067–1071.

51. Clyne, TW and LW Marston, Metal fibre-reinforced ceramic composites and their industrial usage, in *Comprehensive Composite Materials II*, Clyne, TW, editor. Elsevier, 2018, pp. 464–481.

52. Windhorst, T and G Blount, Carbon–carbon composites: a summary of recent developments and applications. *Materials & Design* 1997; **18**(1): 11–15.

53. Blanco, C, J Bermejo, H Marsh and R Menendez, Chemical and physical properties of carbon as related to brake performance. *Wear* 1997; **213**(1–2): 1–12.

54. Byrne, C, Modern carbon composited brake materials. *Journal of Composite Materials* 2004; **38**(21): 1837–1850.

16 Applications of Composites

Usage of composite materials is ubiquitous in the modern world. While global tonnages are still well below those of steel, they now find a wider range of applications and their value is starting to become comparable to that of steel products. As low weight and energy efficiency become increasingly important, this trend is likely to accelerate. In this chapter, the objective is to identify some of the issues involved in commercial exploitation of composites. This is done by means of case studies drawn from various industrial sectors. The examples cover a range of composite type, engineering complexity, manufacturing route, market size and competitive position relative to more traditional materials.

16.1 Overview of Composite Usage

Significant commercial usage of composites began, mostly for military purposes, in the late 1940s and early 1950s. Since then, global use of composite materials has grown rapidly. Accurate figures are difficult to quote, for various reasons, but annual production by weight has risen from less than 0.2 Mtonnes in 1960 to over 5 Mtonnes in the early 2000s, reaching several tens of Mtonnes in the 2020s (with a value of well over US$100 billion). For comparison, with a reference date of 2020, annual world production of steel is around 1000 Mtonnes, worth about US$500 billion, while the market for timber is similar to that of steel in weight, with a value of around US$200 billion. Needless to say, such figures provide no more than rough guidelines, but there is an unmistakable message that composites have grown in half a century from a small niche product to one of the most important types of material in the world. Furthermore, while steel and timber can be regarded as commodities, with much of their consumption occurring along a relatively small number of well-defined pathways, composite materials are very diverse and are used in a huge variety of applications and industrial sectors.

While the reasons for selecting composite materials can vary somewhat between sectors, the attractions of low weight, high specific stiffness, good toughness, excellent corrosion resistance and high mouldability are relevant to many applications. They are thus the automatic materials of choice for many components in industries as diverse as aerospace, marine, automotive, sports goods, wind energy, construction, fluid storage and transmission, and more. Furthermore, usage is projected to expand in virtually all of these areas. The situation regarding metal matrix composites (MMCs) and ceramic matrix composites (CMCs) is a little different, since the markets in which their usage has become established are much more limited and the situation is in general more fluid. Nevertheless, the applications described below include several (in the section on high-temperature applications) that concern MMC and CMC usage and a number of other possible areas of exploitation are being actively investigated. While there is no prospect of such composites being used on a scale comparable to that of polymer matrix composites (PMCs), they are of commercial significance and their efficient usage requires an understanding of their characteristics in the context of composite theory.

16.2   Aerospace and Automotive Applications

There are many aerospace and automotive applications for which combinations of lightness, stiffness and strength are highly attractive. A graphic illustration of this is provided by Fig. 16.1, which shows a human-powered aircraft in flight. Virtually all of the components involved are composites of various types (mostly based on carbon fibre). A measure of how rapidly these attractions have transformed the aerospace industry is provided [1] by Fig. 16.2, which shows the increasing composite content of aircraft over recent decades. Much of the airframe (wings, fuselage, tailplane, etc.) of many modern aircraft, covering the complete size range, is routinely made of composite material.

16.2.1   Airframes

Aircraft wings are now routinely made of composite material, most commonly carbon fibre-reinforced epoxy. Both the wing spar framework and the panels covering them are often made of composite, although there may be some metallic components within the

Fig. 16.1 The Michelob Light Eagle in flight over Rogers Dry Lake at the NASA Dryden Flight Research Center, Edwards, California, in 1987.
Credit: Nasa

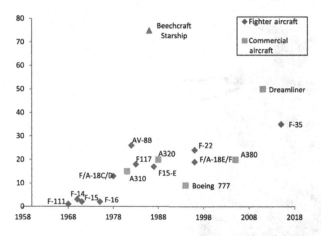

Fig. 16.2 Content in weight per cent of composite material in a selection of military and civilian aircraft, plotted against year of introduction [1], over the 50-year period up to 2015. Figure from Slayton, R. and Spinardi, G. 2016.

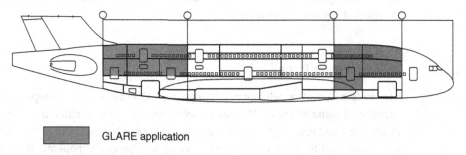

GLARE application

Fig. 16.3 Depiction [6] of the fuselage panels to be made of GLARE in the original Airbus A380 construction.
Reprinted by permission from Springer Nature: Springer Nature, Applied Composite Materials, Maintenance of Glare Structures and Glare as Riveted or Bonded Repair Material, H. J. M. Woerden, J. Sinke and P. A. Hooijmeijer, © 2003.

structure of the wing and metallic wire networks may be incorporated (to reduce local damage in the event of lightning strike). There is a strong move towards automated layup procedures for such structures, cutting production times and improving quality control. Stacking sequences and overall geometry may be quite complex. It may, however, be noted that there are some situations in which multilayer hybrid combinations of composite and aluminium [2–5] offer attractive combinations of properties, particularly in terms of fatigue performance (Section 9.3.1), and also other properties such as fire resistance. One such type of laminated construction is known as GLARE, which comprises alternate thin layers of Al alloy sheet and glass fibre composite laminae. Large fuselage panels have been routinely made of such materials for some time. Fig. 16.3 shows [6] the parts of the original Airbus A380 construction made of GLARE.

Fig. 16.4 Photograph of a helicopter, illustrating the shape of a typical main rotor blade. (Courtesy of Leonardo Helicopters.)

16.2.2   Helicopter Rotor Blade

The rotor blade of a helicopter provides a good example of a component requiring excellent specific stiffness. The blades act as aerofoils that generate lift. A typical rotor blade shape and rotor hub assembly configuration can be seen in Fig. 16.4. Composites have been used for rotor blades, and for other helicopter components, since the 1960s. Initial attractions of using composites included good fatigue resistance, as well as specific stiffness. Full use has also been made of the scope for tailoring the elastic properties via control of the fibre architecture and improved aerodynamic blade designs have emerged by stress and fluid dynamics modelling, utilising the anisotropic properties of the material [7,8].

A particular problem arises with helicopter blades from the combination of forward and rotational motion. Since the forward velocity of the aircraft may be up to about 100 m s^{-1} and the linear speed of the rotating blade, even at its tip, is often little more than 200 m s^{-1}, the airspeed of the blade during the advancing part of the rotational cycle is often substantially greater than that during the retreating phase. If the pitch angle of the blade were the same during each part of the cycle, then the uplift would vary substantially on the two sides of the aircraft and it would be tipped over. Compensation for this effect is achieved by altering the pitch angle of each blade during every rotation. Further changes in pitch angle are used to alter direction during manoeuvring. It is therefore very important that the blades have adequate torsional stiffness, since they must respond quickly and faithfully to pitch angle changes imposed at the rotor hub. The beam stiffness of the blade must also be high, to ensure that the tip does not lag behind during rotation or flap excessively under its own weight. The construction of a

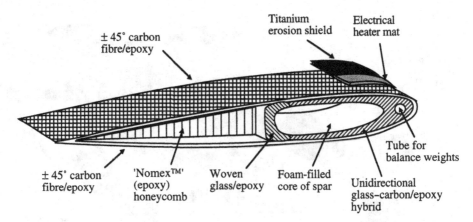

Fig. 16.5 Schematic section through a typical composite construction for a helicopter rotor blade.

typical blade section is shown in Fig. 16.5. The necessary torsional stiffness is provided by the carbon fibres at $\sim\pm45°$ to the blade axis. The carbon and glass fibres aligned parallel to the blade axis provide the beam stiffness necessary to minimise lag and flap. This construction also confers excellent fatigue resistance.

16.2.3 Car Body Panel

While the penetration of composites into the automotive market has been less rapid and deep than for aerospace, there is still extensive, and increasing, usage. The move towards electric vehicles [9], placing a further premium on low weight, is now accelerating this trend. The fact that Formula 1 racing cars are almost entirely made of composite materials provides confirmation that, in terms of pure performance, they are superior to metals. However, there are key issues related to the cost and ease of manufacture (and repair), in addition to factors such as visual appearance and crashworthiness. In general, steel is well suited to automated assembly of cars and other automotive products, with robotic welding being extensively used. However, partly due to technology transfer from the aerospace industry, similar levels of automation are now being applied to composite construction for automotive products. The technical issues involved are addressed in a number of publications [10–12].

In fact, there is a general trend towards hybrid construction, with a steel chassis but lightweight (composite or aluminium) body panels. A schematic concept design is shown in Fig. 16.6. The composite panels in this case are of (glass fibre-reinforced polyester) sheet moulding compound (SMC) – see Section 15.1.3. The floor panels are of carbon fibre-reinforced epoxy. It is, of course, difficult to predict exactly how the situation will evolve in an industry with very strong social, political and environmental drivers. An issue with composites, particularly for high-profile products such as cars, is that they are perceived as being difficult to recycle, and indeed there are some challenges in this area – see Section 13.4.2.

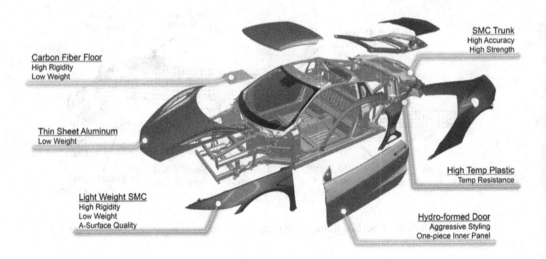

Carbon Fiber Floor
High Rigidity
Low Weight

SMC Trunk
High Accuracy
High Strength

Thin Sheet Aluminum
Low Weight

High Temp Plastic
Temp Resistance

Light Weight SMC
High Rigidity
Low Weight
A-Surface Quality

Hydro-formed Door
Aggressive Styling
One-piece Inner Panel

Fig. 16.6 Concept design for use of lightweight body panels in a car.
(Courtesy of Acura Division, American Honda Motor Co., Inc.)

16.3   Marine and Wind Energy Applications

16.3.1   Yacht Hull

While composites are very widely used for small marine craft, there is a tendency for larger vessels to be made with steel hulls. The cut-off is usually somewhere around a length of about 40–50 m, although, if there is a specific reason for avoiding metal (such as with minesweepers), then slightly larger vessels are sometimes made of composite. In most cases, glass fibre-reinforced polyester or polyvinyl is used. In general, the production methods used for composites in the marine industry are not well suited to very large hull sizes. There are also some concerns about fire hazards and recycling. In addition, the motivation for lighter materials is less strong in the marine industry than for aerospace and automotive. On the other hand, composites provide better thermal insulation and greater flexibility in terms of shaping, repair, etc. In any event, a 40 m yacht is a substantial construction: such a craft is shown in Fig. 16.7.

16.3.2   Wind Turbine Aerofoil

Wind turbines for energy production constitute one of the fastest growing industrial sectors, with recent growth rates of the order of 10% and an annual market size of around US$10 billion in the early 2020s. The aerofoils are almost exclusively made of composite materials. In fact, evolution in the details of design and materials is ongoing [13–15]. Of course, the technology is at least partially imported from the aerospace industry, since these aerofoils have a lot in common with aircraft wings and helicopter rotors. However, in view of the large size and quantity of these products, the cost is highly significant. Moreover, while low weight is an advantage, the premium on it is not as high as for

Fig. 16.7 Photo of a 40 m craft ('131 yacht'), which has a hull made of glass fibre, with a polyester resin matrix in internal regions and a vinyl ester resin matrix in the exterior regions. (Courtesy of Sunseeker International.)

Fig. 16.8 Photo [13] of a 5 MW wind turbine under construction in Germany. Reprinted from Reinforced Plastics, 49, George Marsh, Composites help improve wind turbine breed, 18–22, © 2005, with permission from Elsevier.

aircraft. Most turbines have therefore been made using glass fibres rather than the more expensive (and lighter) carbon fibres, although, as performance (power output) becomes more critical, this may change somewhat. A typical unit is shown [13] in Fig. 16.8.

16.4   Sports Goods

Many types of sports product, including skis, racquets, golf clubs, fishing rods, wind-surfing boards, high-performance cycles, etc., are routinely made of composite material.

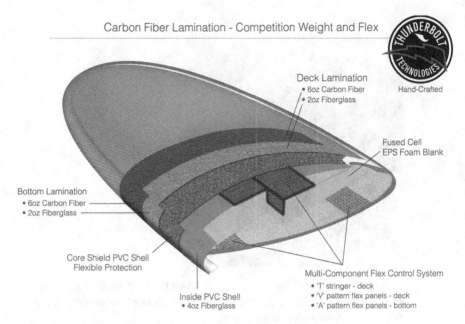

Fig. 16.9 Schematic representation of the construction of a surfboard.
Courtesy of Carve Sports.

The usual attractions of light weight, high stiffness and strength, corrosion resistance, mouldability, etc. are all relevant to this market. Some of the issues involved are highlighted here via a couple of examples, with slightly different constructions and requirements.

16.4.1   Surfboard

Surfboards, or at least those designed to support the full weight of a surfer, need quite a high level of buoyancy. They are also subjected to a variety of mechanical loads during use, and so need to be strong and tough. These requirements are often best satisfied by a relatively complex, hybrid construction that incorporates a foam core. Details are available in the literature [16]. An example structure is shown in Fig. 16.9. In this case, both glass fibre- and carbon fibre-reinforced materials are used. Such products are usually made using hand lay-up procedures.

16.4.2   Vaulting Pole

The requirements of a vaulting pole are simple, but mechanically demanding. They are just beams, designed for use in pure bending. They are also designed to sustain high curvatures, so that a lot of the incident kinetic energy of the athlete is converted into stored elastic strain energy, which can then be converted into potential energy in the form of increased height, as the pole straightens. There have been detailed studies [17] of the mechanics of this process. The switch (around 1960) from earlier pole materials

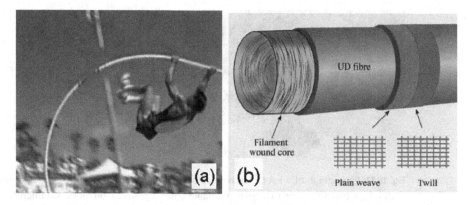

Fig. 16.10 Vaulting poles: (a) a vault at the point of maximum curvature and (b) typical construction [18] of a pole.
Reprinted from Davis, CL and SN Kukureka, Physics Education, 1996 © IOP Publishing.
Reproduced with permission. All rights reserved

(bamboo and Al) to composites (glass fibre-reinforced epoxy) promoted a step change in achievable heights [18].

Pole dimensions do vary – they are not regulated, but a typical radius, r, might be around 25 mm. During use (Fig. 16.10(a)), it might be bent to a radius of curvature, R, of about 1 m. The peak axial strain at the outer surface is then simply given by r/R (~2.5%), which is clearly a high level (for an elastic strain). The construction of poles (Fig. 16.10(b)) also varies, but the main element is uniaxial reinforcement, most commonly with glass fibres, along the length of the pole. Assuming a fibre content of 50%, and a fibre stiffness of 75 GPa, the (axial) composite stiffness is about 40 GPa, so the peak stress level is around 1 GPa (or perhaps a little less, since the axial reinforcement is not actually at the free surface – see Fig. 16.10(b)). This is a high stress level, but approximately within the strength range expected for a composite of this type. In practice, it is more relevant to consider the situation from a toughness point of view, and of course these composites tend to be relatively tough (Section 9.2.4). Nevertheless, the conditions are demanding and it is not unknown for vaulting poles to fracture during use.

16.5   High-Temperature Applications

Composite materials can offer various property combinations that are attractive under aggressive and demanding conditions, including high temperature. Of course, assuming that this means temperatures above about 200°C, PMCs are not suitable for such usage. The examples in this section are thus all MMCs or CMCs.

16.5.1   Aircraft Brakes

Aircraft brakes require a particularly demanding set of properties. During an emergency landing or aborted takeoff, large quantities of energy must be quickly absorbed, without

Fig. 16.11 Photograph of a set of (carbon–carbon) aircraft brakes, as used on the Airbus A350. Photograph taken by Julian Herzog. Available under Creative Commons 4.0 International License.

disintegration or seizure of the brakes. A typical construction is based on multiple rotating and stationary disks – see Fig. 16.11. The temperatures reached during operation naturally depend on a number of factors [19], but can approach 1000°C, and possibly higher during very severe usage. The disk material must therefore have excellent thermal shock resistance and good high-temperature strength. High thermal conductivity is also advantageous, to avoid overheating of the disk surfaces. Carbon can satisfy these requirements. Solid graphite, however, has inadequate strength and toughness. Brake disks are actually made by vapour phase infiltration of carbon into carbon fibre preforms, creating composite material with a high toughness. This is a slow and expensive process, often requiring high thermal gradients and several stages. Issues relating to control over the microstructure are covered in the literature [20,21].

These disks are manufactured via the infiltration techniques described in Section 15.3.3. They are used for aircraft partly because of the significance of the weight saving compared with alternatives, which are mainly steel-based. Also, aircraft are slightly unusual, in that brake usage is relatively moderate and infrequent (except in emergencies, after which the brakes are replaced). In contrast, usage of brakes in automotives and trains can routinely be prolonged and fairly severe. Carbon–carbon has the disadvantage that such usage can lead to excessive oxidation, which of course is progressive because the oxides are gases.

16.5.2   Diesel Engine Piston

This application stimulated a lot of interest in industrial use of MMCs, starting in the 1980s. It was demonstrated that components such as pistons could successfully be manufactured via squeeze infiltration of aluminium into permanent moulds containing alumina fibre preforms. It also became clear that such selective reinforcement – for example, in the region around the ring groove [22], as shown [23] in Fig. 16.12, could prolong component lifetimes, particularly in diesel engines, where combustion conditions are severe. Detailed information is available in the literature [24,25] concerning such applications. Of course, the introduction of such reinforcement does add to costs

Fig. 16.12 Photograph [23] of a diesel engine piston, showing the presence of a region of reinforcement (with short alumina fibres) around the upper ring groove.

and usage has not become very widespread. Nevertheless, it has become standard over an extended period in several high-volume models, particularly those of Japanese producers such as Toyota and Honda.

16.5.3 Top Ring of Ladle for Continuous Casting of Steel

Ladles are used to transfer molten steel from a blast furnace, where it is produced, to a tundish, where a constant metallostatic head is maintained during the continuous casting process. It is common for controlled (deoxidation) reactions to be carried out in the ladle, raising the temperature and creating energetic liquid flows. There are severe thermo-mechanical demands on the top ring of the ladle, where flow of the liquid steel into the tundish creates mechanical loads on the spout region. The steel temperature during these processes [26] is ~1600°C. While the bulk of the ladle is made of a suitable refractory (requiring little mechanical strength), with a metallic load-bearing enclosure that stays relatively cool, the top ring requires both strength and resistance to very high temperatures. They are commonly made of a low alloy steel, but can be damaged by the molten steel and/or the slag and tend to have short lifetimes. A photograph [27] of a ladle top ring made of CMC (metal fibre-reinforced ceramic) is shown in Fig. 16.13. It comprises four separate 'castings' bolted together. These are made using the procedure described in Section 15.3.2. One attraction of this manufacturing procedure is that bolts, and also threaded bolt holes, can be cast into components. This ring has a considerably greater lifetime than that of conventional steel rings.

16.5.4 Atmosphere Re-entry Heat Shield

Thermal protection of spacecraft during atmosphere re-entry is an important and demanding requirement. Exposure times tend to be relatively short (~<30 minutes), but atmospheric friction creates very high temperatures (~1500°C) on outer surfaces. The material shown in Fig. 12.4 was used for space shuttle tiles, although it still

Fig. 16.13 Photograph [27] of the top ring of a ladle for continuous casting of steel, made of Fiberstone™ comprising 10 vol% of stainless steel fibres in an alumina-rich matrix. Note the threaded holes for fixing bolts.
Reprinted from Elsevier Books, T. William Clyne and Lee W. Marston, Comprehensive Composite Materials II, 464–481, © 2018, with permission from Elsevier.

represents state-of-the-art design for this type of requirement. It is not a conventional composite, but rather a fibre network material of the type described in Section 12.1.3. Detailed design requirements and performance data are available in the literature [28,29]. Space shuttle tiles are ~70–80 mm thick, comprising an assembly of amorphous silica fibres, held together with a colloidal silica binder. Fibre diameter is ~10 μm and the porosity level is ~90%. Since the air gaps between fibres are relatively large, and the gas pressure is relatively low, conduction through the air can be neglected (even for relatively insulating fibres). Convective heat transfer is also negligible, since the tile surfaces are sealed (see Fig. 12.4(a)), so that there is little gas flow through the fibre assembly. Radiative transfer, on the other hand, is significant, in view of the high temperatures and low material density [30–35].

References

1. Slayton, R and G Spinardi, Radical innovation in scaling up: Boeing's Dreamliner and the challenge of socio-technical transitions. *Technovation* 2016; **47**: 47–58.
2. Schijve, J. Development of fiber-metal laminates (ARALL and GLARE) – new fatigue-resistant materials, in *Fracture '93*. Engineering Materials Advisory Service, 1993.
3. Fatt, MSH, C Lin, DM Revilock and DA Hopkins, Ballistic impact of GLARE fiber-metal laminates. *Composite Structures* 2003; **61**: 73–88.
4. Moriniere, FD, RC Alderliesten and R Benedictus, Low-velocity impact energy partition in GLARE. *Mechanics of Materials* 2013; **66**: 59–68.
5. Kotik, HG and JEP Ipina, Short-beam shear fatigue behavior of fiber metal laminate (Glare). *International Journal of Fatigue* 2017; **95**: 236–242.
6. Woerden, HJM, J Sinke and PA Hooijmeijer, Maintenance of Glare structures and Glare as riveted or bonded repair material. *Applied Composite Materials* 2003; **10**(4–5): 307–329.
7. Jung, SN, VT Nagaraj and I Chopra, Assessment of composite rotor blade modeling techniques. *Journal of the American Helicopter Society* 1999; **44**(3): 188–205.
8. Rafiee, M, F Nitzsche and M Labrosse, Dynamics, vibration and control of rotating composite beams and blades: a critical review. *Thin-Walled Structures* 2017; **119**: 795–819.

9. Delogu, M, L Zanchi, CA Dattilo and M Pierini, Innovative composites and hybrid materials for electric vehicles lightweight design in a sustainability perspective. *Materials Today Communications* 2017; **13**: 192–209.

10. Mangino, E, J Carruthers and G Pitarresi, The future use of structural composite materials in the automotive industry. *International Journal of Vehicle Design* 2007; **44**(3–4): 211–232.

11. Koronis, G, A Silva and M Fontul, Green composites: a review of adequate materials for automotive applications. *Composites Part B: Engineering* 2013; **44**(1): 120–127.

12. Balakrishnan, VS and H Seidlitz, Potential repair techniques for automotive composites: a review. *Composites Part B: Engineering* 2018; **145**: 28–38.

13. Marsh, G, Composites help improve wind turbine breed. *Reinforced Plastics* 2005; **49**(4): 18–22.

14. Brondsted, P, H Lilholt and A Lystrup, Composite materials for wind power turbine blades, in *Annual Review of Materials Research*. Annual Reviews, 2005, pp. 505–538.

15. Mishnaevsky, L, K Branner, HN Petersen, J Beauson, M McGugan and BF Sorensen, Materials for wind turbine blades: an overview. *Materials* 2017; **10**(11). DOI: 10.3390/ma10111285.

16. Poodts, E, R Panciroli and G Minak, Design rules for composite sandwich wakeboards. *Composites Part B: Engineering* 2013; **44**(1): 628–638.

17. Ekevad, M and B Lundberg, Influence of pole length and stiffness on the energy conversion in pole-vaulting. *Journal of Biomechanics* 1997; **30**(3): 259–264.

18. Davis, CL and SN Kukureka, Effect of materials and manufacturing on the bending stiffness of vaulting poles. *Physics Education* 2012; **47**(5): 524–529.

19. Xu, HJ, BY Huang, MZ Yi, XA Xiong and BL Lei, Influence of matrix carbon texture on the temperature field of carbon/carbon composites during braking. *Tribology International* 2011; **44**(1): 18–24.

20. Su, JM, J Yang, ZC Xiao, SJ Zhou, ZG Peng, JG Xin, R Li, M Han, SL Zhao and LM Gu, Structure and properties of carbon/carbon composite materials for aircraft brake discs. *New Carbon Materials* 2006; **21**(1): 81–89.

21. Hao, MY, RY Luo, ZH Hou, W Yang, QA Xiang and CL Yang, Effect of fiber-types on the braking performances of carbon/carbon composites. *Wear* 2014; **319**(1–2): 145–149.

22. Feest, EA, Exploitation of the metal matrix composites concept. *Metals & Materials* 1988; **4**: 273–278.

23. Clyne, TW and PJ Withers, *An Introduction to Metal Matrix Composites*. Cambridge University Press, 1993.

24. Miracle, DB, Metal matrix composites: from science to technological significance. *Composites Science and Technology* 2005; **65**(15–16): 2526–2540.

25. Manu, KMS, LA Raag, TPD Rajan, M Gupta and BC Pai, Liquid metal infiltration processing of metallic composites: a critical review. *Metallurgical and Materials Transactions B: Process Metallurgy and Materials Processing Science* 2016; **47**(5): 2799–2819.

26. Zimmer, A, ANC Lima, RM Trommer, SR Braganca and CP Bergmann, Heat transfer in steelmaking ladle. *Journal of Iron and Steel Research International* 2008; **15**(3): 11–14, 60.

27. Clyne, TW and LW Marston, Metal fibre-reinforced ceramic composites and their industrial usage, in *Comprehensive Composite Materials II*, Clyne, TW, editor. Elsevier, 2018, pp. 464–481.

28. Daryabeigi, K, Thermal analysis and design optimization of multilayer insulation for reentry aerodynamic heating. *Journal of Spacecraft and Rockets* 2002; **39**(4): 509–514.

29. Nakamura, T and T Kai, Combined radiation-conduction analysis and experiment of ceramic insulation for reentry vehicles. *Journal Thermophysics and Heat Transfer* 2004; **18**(1): 24–29.

30. Tong, TW and CL Tien, Radiative heat transfer in fibrous insulations: part 1. Analytical study. *Journal of Heat Transfer: Transactions of the ASME* 1983; **105**(1): 70–75.

31. Tong, TW, QS Yang and CL Tien, Radiative heat transfer in fibrous insulations: part 2. Experimental study. *Journal of Heat Transfer: Transactions of the ASME* 1983; **105**(1): 76–81.

32. Petrov, VA, Combined radiation and conduction heat transfer in high temperature fiber thermal insulation. *International Journal of Heat Mass Transfer* 1997; **40**(9): 2241–2247.

33. Lee, SC and GR Cunnington, Heat transfer in fibrous insulations: comparison of theory and experiment. *Journal of Thermophysics and Heat Transfer* 1998; **12**(3): 297–303.

34. Lee, SC and GR Cunnington, Conduction and radiation heat transfer in high porosity fiber thermal insulation. *Journal of Thermophysics and Heat Transfer* 2000; **14**(2): 121–136.

35. Daryabeigi, K, Heat transfer in high temperature fibrous insulation. *Journal of Thermophysics and Heat Transfer* 2003; **17**(1): 10–20.

Appendix: Questions

The 18 questions below are suitable for use in undergraduate and Masters level courses on composite materials. They have been collected into groups, mostly according to the chapter in this book that relates most closely to the coverage. However, some of them draw on topics from more than one chapter and in such cases they are included in the group for the chapter closest to the end of the book. This is consistent with the sequence of topics in the book being, very broadly, designed as a progression that might be followed in a course. Model answers for these questions are available from the publisher at cambridge.org/compositematerials.

Chapter 4 Tensor Analysis of Anisotropic Materials and the Elastic Deformation of Laminae

Q4.1 As shown in Section 4.3.3, the standard procedure for transforming second-rank tensors (Section 4.1.3) can be used to obtain Eqn (4.47), i.e. the three equations for converting a set of applied stresses (σ_x, σ_y and τ_{xy}) to those referred to the fibre axis (σ_1, σ_2 and τ_{12}). Show that the same set of equations can be obtained using the Mohr's circle construction (although it is rather more cumbersome).

Q4.2 Use Eqn (4.56) to show that the Young's modulus of a composite lamina (with the elastic constants, referred to the fibre axis, given below) falls by ~50% if it is loaded at 7° to the fibre axis, compared with the on-axis value. What is the minimum Young's modulus that the lamina can exhibit and at what loading angle does this occur ?

$$[E_1 = 200 \text{ GPa}, E_2 = 7 \text{ GPa}, G_{12} = 3 \text{ GPa}, \nu_{12} = 0.3]$$

Chapter 5 Elastic Deformation of Laminates

Q5.1 This question involves use of the DoITPoMS TLP 'Mechanics of Fibre Composites'. On the page 'Stiffness of Laminates', use the facility at the end to create an epoxy–50% glass composite (dragging the materials icons concerned to the matrix and reinforcement boxes) and to estimate the ratio of maximum to minimum Young's modulus it exhibits when loaded at different angles to the fibre axis. Now create a 0/90 (cross-ply) laminate of the same composite and repeat the operation. Find a sequence giving complete in-plane isotropy and confirm that the Young's modulus in this case is about 22 GPa for all in-plane directions.

Q5.2   Fig. 5.6 shows how the stresses within a cross-ply laminate depend on the angle between a reference direction (parallel to the fibres in one of the plies) and the direction of an applied tensile stress. Using information shown there, answer the following questions.

(a)   Why do stresses arise transverse to the applied stress, for loading angles of 0° and 90°?

(b)   For this composite material, the critical stresses for failure (parallel, transverse and in shear with respect to the fibre axis) are respectively 700, 20 and 50 MPa. Stating any assumptions, describe the sequence of (failure-related) events as an applied stress (parallel to one of the fibre directions) is progressively increased from zero

(c)   A cylindrical pressure vessel (having a radius/wall thickness ratio of 20) is made from this composite material, with one ply aligned along the cylinder axis and the other in the hoop direction. The pressure in the vessel is progressively raised. Use the maximum stress criterion to predict the type of failure that will occur first and the corresponding internal pressure.

(d)   The Tsai–Hill criterion (Eqn (8.16)) is usually regarded as more reliable than the maximum stress criterion. Use it to obtain a value for the critical pressure at the onset of failure in this case and comment on any difference between this value and that obtained in (c).

Chapter 8 Stress-Based Treatment of the Strength of Composites

Q8.1   Fig. 8.17(a) shows how the stresses within angle-ply laminates depend on the ply angle, when subjected to uniaxial tensile loading. Using information in this figure, answer the following questions.

(a)   Using the transform equation (Eqn (4.47)), and taking the example of a $\pm 50°$ laminate, show that the predicted stresses within a ply satisfy force balances with the external load in both x and y directions.

(b)   For this composite, the critical stresses for failure (parallel, transverse and in shear with respect to the fibre axis) are respectively 700, 30 and 40 MPa. For a $\pm 50°$ laminate, being loaded in tension as shown in the figure, use the maximum stress theory to predict the type of failure that will occur first as the applied stress is progressively increased, and the corresponding value of this stress. Without doing any further calculations, indicate why this stress level is likely to be an overestimate.

(c)   When a flat component made of this type of laminate is used at very low ambient temperatures, it is reported by the user to undergo significant out-of-plane distortion. Explain how this is likely to have arisen and suggest a simple alteration to its specification that is likely to remedy the problem.

Q8.2 Fig. 8.21(a) shows the stresses (parallel and transverse to the fibre direction) within one ply of an angle-ply laminate subjected to unequal biaxial tension ($\sigma_x = 2\sigma_y$). The stresses (ratios to σ_x) are shown as a function of the ply angle, ϕ (measured relative to the direction of σ_x). The critical stresses for failure of this composite axially, transversely and in shear, i.e. σ_{1*}, σ_{2*} and τ_{12*}, are respectively 700 MPa, 50 MPa and 30 MPa. Using Fig. 8.21(a), and the maximum stress criterion for failure, find the pressure at which an internally pressurised tube (radius = 50 mm, wall thickness = 2 mm) of this composite, wound with a ply angle, ϕ of $\pm40°$ (to the hoop direction) is predicted to fail.

Find the failure pressure according to the Tsai–Hill failure criterion (Eqn (8.16)). Explain any difference between this value and that obtained previously. Using the criterion you consider most reliable in this case, obtain an approximate estimate of the ply angle that would give the largest failure pressure.

Q8.3 An angle-ply ($\pm50°$) laminate of a polyester–50% glass composite is subjected to an increasing tensile stress in the σ_x ($\Phi=0°$) direction. Use the facility at the end of the section 'Failure of Laminates and the Tsai–Hill Criterion' within the 'Mechanics of Fibre Composites' TLP (www.doitpoms.ac.uk/tlplib/fibre_composites/index.php) to establish the applied stress at which the laminate will fail (according to the maximum stress criterion), given that $\sigma_{1*} = 700$ MPa, $\sigma_{2*} = 20$ MPa and $\tau_{12*} = 50$ MPa. Carry out the same calculation, using analytical equations, for one of the two plies (i.e. ignore the presence of the other) and compare this value with the first result. Account for any difference between the two.

Q8.4 Candidate materials for a gas pipeline are steel and a glass fibre-reinforced polymer composite. The (outer) diameter of the pipeline will be 1 m and the maximum gas pressure will be 100 bar (10 MPa). The composite would be filament-wound, at $\pm45°$ to the hoop direction. There are no concerns about stiffness, so the key design criterion is to avoid phenomena that could lead to failure (which would be likely to be plasticity in the case of the steel and some type of microstructural damage in the case of the composite). The main design variable will be the wall thickness.

(a) Using the von Mises yield criterion (Eqn (8.13)) for steel and the Tsai–Hill failure criterion (Eqn (8.16)) for the composite, and ignoring the issue of safety factors, estimate the minimum wall thickness in each case and hence deduce which material would allow the lighter pipeline.

(b) Comment on the assumptions and sources of error in your calculation and on whether there might be a danger of any other types of failure. Without carrying out any further calculations, indicate whether and how you would recommend changing the fibre-winding angle of the composite in order to make it more effective for this application.

[Steel density = 7.8 Mg m^{-3}. Composite density = 1.8 Mg m^{-3}. Critical stresses parallel, transverse and in shear relative to the fibre axis, σ_{1*}, σ_{2*} and τ_{12*}, are measured respectively to be 900 MPa, 30 MPa and 40 MPa.]

Q8.5 This question involves use of the DoITPoMS TLP 'Mechanics of Fibre Composites'. On the page 'Failure of Laminates and the Tsai–Hill Criterion', use the facility at the end to create a polyester–50% glass angle-ply laminate ($\pm40°$). Taking this to be a filament-wound tube, with the plies at $\pm40°$ to the hoop direction, and a radius/wall thickness ratio of 20, subjected to internal pressure, P, estimate the value of P at which failure will occur, according to the Tsai–Hill criterion, given that $\sigma_{1*} = 700$ MPa, $\sigma_{2*} = 20$ MPa and $\tau_{12*} = 50$ MPa. Using analytical equations, carry out the same calculation for one of the two plies (ignoring the presence of the other). Account for the difference between this value and the one you obtained treating the laminate as a whole (using the numerical procedure in the TLP).

Q8.6 **(a)** For a small aircraft, a choice must be made between an Al alloy and a composite for the fuselage material. The fuselage will approximate to a cylinder of diameter of 2 m and will experience internal pressures up to 0.6 atm (0.06 MPa) above that of the surrounding atmosphere, axial bending moments of up to 500 kN m and torques of up to 600 kN m. The composite fuselage would be produced by filament winding at $\pm45°$ to the hoop direction. It may be assumed that this is a strength-critical application, with the airframe stiffness expected to be adequate in any event. Using the Tresca yield criterion (Al) and the Tsai–Hill failure criterion (composite), and ignoring the issue of safety factors, estimate the minimum wall thickness in each case and hence deduce which material would allow the lighter fuselage.
(b) Comment on the main sources of error in your calculation and also on whether there might be a danger of any other type of failure.

[For Al alloy, yield stress in uniaxial tension = 250 MPa. For composite, failure stresses for loading transverse and in shear relative to the fibre axis are both 50 MPa; failure by fracture of the fibres can be neglected. Densities: Al = 2.70 Mg m^{-3}, composite = 1.50 Mg m^{-3} The peak axial stress in a thin-walled cylinder subjected to a bending moment M is $R\,M/I$, where R is the radius and I is the moment of inertia, which is given by $\pi R^3\,t$, where t is the wall thickness.]

Chapter 10   Thermal Effects in Composites

Q10.1 Fig. 10.9 shows measured lattice strains within an Al–SiC composite during heating and cooling. Using information in this figure, and the caption, answer the following.
(a) Since this sample was unloaded, a null internal force balance should apply at all temperatures. In fact, this may not be accurately reliable during the thermal cycling, since this had to be done fairly slowly (to allow time to acquire the diffraction data at selected temperatures), leading to some dynamic relaxation of the stresses. However, it should hold more accurately at ambient temperature, both before heating and after cooling. Using the property data

provided below, carry out a simple calculation to check whether this is in fact the case and comment on the outcome.

(b) These fibres (SiC whiskers) have average aspect ratios of ~10. A simple calculation is suggested, to check whether any fibres might have fractured during the cycling. In order to break even the weakest of them, a tensile stress of at least ~1.5 GPa is needed. It can be seen from the above figure that this would be most likely at a temperature of ~300–400°C, at which the matrix shear yield stress, τ_{mY}, is known to be ~50 MPa (and work hardening is negligible). Use a shear lag treatment of the stress distribution along the fibres (with matrix yielding throughout) to estimate the maximum stress that could be created in the fibres at this temperature and hence decide whether any fibre fracture would be likely. Check whether your deduced fibre stress distribution is consistent with data in Fig. 10.9.

[Young's moduli: Al = 70 GPa, SiC whisker axis (<111>) = 550 GPa.]

Chapter 11 Surface Coatings as Composite Systems

Q11.1 **(a)** A 1 mm thick unidirectional ply of epoxy–25 vol% glass fibre composite is bonded at 120°C to a steel plate with the same dimensions, and curing goes to completion at this temperature. The bonded pair is then cooled (elastically) to room temperature (20°C). Describe the out-of-plane distortion that arises and calculate the associated curvature(s).

(b) When the bonded pair is loaded in compression parallel to the fibre axis of the ply, it is observed that the curvature(s) it exhibits starts to reduce. Account for this effect. Calculate the applied stress at which the specimen would become flat and comment on whether this is likely to be achievable.

[For glass fibres: $E = 76$ GPa, $\alpha = 5 \times 10^{-6}$ K^{-1}, $\nu = 0.22$; for epoxy resin: $E = 3.5$ GPa, $\alpha = 58 \times 10^{-6}$ K^{-1}, $\nu = 0.40$; for steel: $E = 210$ GPa, $\alpha = 11.4 \times 10^{-6}$ K^{-1}, $\nu = 0.26$.]

Q11.2 **(a)** A 'vibration-damped' sheet material is made by bonding a 1 mm thick rubber layer between two steel plates of thickness 1 mm. The sheet is pushed against the surface of a large cylindrical former, which has a radius of 0.5 m. Sketch the through-thickness distributions of strain and stress in the sheet, assuming that both the steel and the rubber remain elastic.

(b) This forming operation is actually designed to generate plastic deformation, creating a shaped component with a uniform curvature in one plane. Taking the steel to have a yield stress of 300 MPa (in compression or tension), and assuming that the rubber remains elastic, show that the above operation would in fact induce plastic deformation in outer layers of both metal sheets and calculate the thickness of the layers that would yield in this way and the plastic strain at the free surfaces.

(c)   Show that, if the width of the sheet (length along the axis of the cylinder) is 0.5 m, then the beam stiffness ($\Sigma = EI$) of the sheet is 216.7 N m^2 and the bending moment that would be needed in order to bring the sheet into contact with the cylindrical former would be 433 N m, assuming that the steel remained elastic. Calculate the required bending moment for the actual case, with the steel undergoing plastic deformation at a yield stress of 300 MPa (but neglecting any work hardening).

[Steel: Young's modulus, E = 200 GPa; Rubber: Young's modulus value more than four orders of magnitude lower.]

Q11.3   (a)   A glass sheet of thickness 3 mm has a 10 μm layer of Al evaporated onto one side, to form a mirror. Production generates negligible stress in the coating. The sheet is subsequently heated from room temperature (20°C) to 170°C. Calculate the curvature exhibited by the sheet after heating, assuming that the system remained elastic.

(b)   Decide, stating any assumptions, whether yielding is in fact likely to occur in the Al layer during heating, given that it has a uniaxial yield stress at 170°C of 100 MPa.

(c)   Hence decide whether any detectable distortion of the reflective characteristics of the mirror is likely to be present after it has cooled to room temperature.

[For the glass: E = 75 GPa, α = 8.5 × 10^{-6} K^{-1}, ν = 0.20; for the Al: E = 70 GPa, ν = 0.33, α = 24.0 × 10^{-6} K^{-1}.]

Q11.4   (a)   John Harrison, the famous clock-maker credited with developing a time-keeping system sufficiently reliable to establish longitude at sea, was reportedly the first to create a bi-metallic strip (for compensation of the effects of temperature change), which he did by casting a thin brass layer onto a thin steel sheet. Show that, if both layers have a thickness of 0.1 mm, and the strip is 100 mm long, then the temperature change required to generate a lateral deflection of 1 mm at its end is about 4.6 K, assuming that the system remains elastic.

(b)   Sketch the (approximate) through-thickness distributions of stress and strain within the above strip, after it had been heated by 100 K. Give your view as to whether such heating would be likely to cause plastic deformation within the strip, given that the yield stresses of both constituents are expected to be of the order of 100 MPa.

[The relationship between curvature, κ, end deflection, y, and length, x, of a bimaterial strip may be expressed as

$$\kappa = \frac{2 \sin\left[\tan^{-1}\left(y/x\right)\right]}{\sqrt{(x^2 + y^2)}}$$

For steel: Young's modulus, E = 200 GPa; thermal expansivity, α = 13 × 10^{-6} K^{-1}
For brass: Young's modulus, E = 100 GPa; thermal expansivity, α = 19 × 10^{-6} K^{-1}]

Q11.5 **(a)** A thick metal sheet was held at 1000°C in air for several hours, after which an oxide film had formed (on both sides), with a thickness of 100 μm. No significant stresses were created in metal or oxide during this process. During subsequent cooling, the oxide spalled off from the substrate at 300°C. Estimate the fracture energy of the interface between the metal and the oxide, stating your assumptions.

(b) The above thermal treatment was repeated on a different sheet of the same metal, in the form of a relatively narrow strip of a thinner sheet and in a configuration such that only one side of the strip was exposed to air. In this case, it was observed that spallation did not occur, even after cooling to ambient temperature (20°C), and that the strip exhibited noticeable curvature at this stage. Give a qualitative explanation of the fact that spallation occurred in the first experiment (part (a)), but not in the second.

(c) In the curved strip obtained after the above experiment (part (b)), would the oxidised side be convex or concave? The residual thickness of the metal was found to be 1 mm. What magnitude of curvature would be expected? Is this significantly different from the value that would be obtained if the Stoney approximation were used?

[Thermal expansivities: metal $= 15 \times 10^{-6}$ K^{-1}; oxide $= 7 \times 10^{-6}$ K^{-1}; Young's moduli: metal $= 100$ GPa; oxide $= 200$ GPa; Poisson ratios: metal $= 0.3$; oxide $= 0.2$.]

Q11.6 **(a)** Steel sheet of thickness 1 mm is given a thin protective layer of vitreous enamel. This coating is created by adding glassy powder to the surface and holding at around 700–800°C, causing the powder to fuse and form a layer of uniform thickness. The sheet is then furnace cooled, taking several hours to reach room temperature, such that the thermal misfit strain is completely relaxed by creep down to about 220°C, after which cooling is elastic. Assuming that the coating/substrate thickness ratio, h/H, is sufficiently small for the Stoney equation to be valid, estimate the elastic strain in the coating, stating your assumptions.

(b) The adhesion of the enamel to the steel is excellent, so the system is highly resistant to debonding, but it is found that, if the coated sheet is progressively bent in one plane (with the steel undergoing plastic deformation), then through-thickness cracks appear in the enamel layer (on the convex side) when the local radius of curvature reaches 60 mm. Assuming that such cracking starts when the tensile strain in the enamel reaches a certain level, use this information to estimate this critical strain.

(c) A fabrication procedure requires bending of the coated sheet to a radius of curvature of 50 mm. The suggestion is made that, instead of furnace cooling the sheet after formation of the coating, it should be removed from the furnace and cooled more quickly, such that elastic cooling occurs below about 420°C (and stress relaxation is complete until this point). Would you expect this measure to result in the elimination of through-thickness cracking during bending of the sheet to this curvature?

(d) For the latter case (i.e. the rapidly cooled sheet), what are the principal stresses within the coating, before and after the bending operation? (The deformation of the steel sheet can be taken as entirely plastic.)

[Steel: $\alpha = 14\ 10^{-6}\ \mathrm{K}^{-1}$, enamel: $\alpha = 5\ 10^{-6}\ \mathrm{K}^{-1}$; $E = 70$ GPa; $\nu = 0.2$.]

Chapter 12   Highly Porous Materials as Composite Systems

Q12.1   **(a)** Explain briefly (without equations) how the thermal expansivity of a composite material can be predicted, and use this approach to deduce the effect on the expansivity of introducing pores into a material.

(b) A ceramic material is being used for a component in the form of a square section beam, with fixed length. It is required to have a certain beam stiffness, Σ $(= E\,I$, where E is the Young's modulus and I is the second moment of area of the beam section). It is proposed that, by the introduction of a randomly distributed set of spherical pores into the material, it may be possible to reduce the mass of the beam. Using the expressions provided at the end of the question for the second moment of area of a square beam and the Young's modulus of such a porous material, show that a (slight) reduction in weight could be achieved in this way.

(c) While the proposed change is not expected to allow a large reduction in mass, a key property required for the component concerned is resistance to thermal shock. It is suggested that this will be improved by the introduction of porosity (as a result of the associated reduction in stiffness). A commonly used merit index for thermal shock resistance is given by the expression

$$M = \frac{G_{\mathrm{c}}(1 - 2\nu)k}{E\alpha}$$

where G_{c} is the fracture energy, ν is the Poisson ratio, k is the thermal conductivity, E is the Young's modulus and α is the thermal expansivity. Using information provided at the end of the question, check whether the introduction of pores is expected to raise this merit index and, if so, estimate the optimal pore content and the increase in the value of M achieved at this level.

[The second moment of area of a square section beam, of side b, is given by

$$I = \frac{b^4}{12}$$

The (Halpin–Tsai) expression for the Young's modulus of a porous material containing a random distribution of spherical pores, with a volume fraction of f, may be written

$$E_{\mathrm{p}} = E_{\mathrm{m}}\left(\frac{1-f}{1+f}\right)$$

where E_{m} is the Young's modulus of the (fully dense) matrix. The effect of pores on the Poisson ratio is negligible. The thermal conductivity of such a material is given approximately by

$$k_{\mathrm{p}} = k_{\mathrm{m}}(1-f)$$

and its fracture energy by

$$G_{c,\mathrm{p}} = G_{c,\mathrm{m}}(1-f^2)]$$

Index

Printed in the United States
by Baker & Taylor Publisher Services